Progress in Polyurethane
and Composites

Progress in Polyurethane and Composites

Editors

Chang-An Xu
Zhuohong Yang

Basel • Beijing • Wuhan • Barcelona • Belgrade • Novi Sad • Cluj • Manchester

Editors

Chang-An Xu
Institute of Chemical Engineering
Guangdong Academy of Sciences
Guangzhou
China

Zhuohong Yang
College of Materials and Energy
South China Agricultural University
Guangzhou
China

Editorial Office
MDPI AG
Grosspeteranlage 5
4052 Basel, Switzerland

This is a reprint of articles from the Special Issue published online in the open access journal *Polymers* (ISSN 2073-4360) (available at: www.mdpi.com/journal/polymers/special_issues/1XLW7TZ9C6).

For citation purposes, cite each article independently as indicated on the article page online and as indicated below:

Lastname, A.A.; Lastname, B.B. Article Title. *Journal Name* **Year**, *Volume Number*, Page Range.

ISBN 978-3-7258-1816-7 (Hbk)
ISBN 978-3-7258-1815-0 (PDF)
doi.org/10.3390/books978-3-7258-1815-0

© 2024 by the authors. Articles in this book are Open Access and distributed under the Creative Commons Attribution (CC BY) license. The book as a whole is distributed by MDPI under the terms and conditions of the Creative Commons Attribution-NonCommercial-NoDerivs (CC BY-NC-ND) license.

Contents

About the Editors vii

Chang-An Xu and Zhuohong Yang
Progress in Polyurethane and Composites
Reprinted from: *Polymers* 2024, *16*, 2031, doi:10.3390/polym16142031 1

Youmin Tuo, Xubiao Luo, Yahong Xiong, Chang-An Xu and Teng Yuan
A Novel Polyfunctional Polyurethane Acrylate Derived from Castor Oil-Based Polyols for Waterborne UV-Curable Coating Application
Reprinted from: *Polymers* 2024, *16*, 949, doi:10.3390/polym16070949 7

Víctor M. Serrano-Martínez, Carlota Hernández-Fernández, Henoc Pérez-Aguilar, María Pilar Carbonell-Blasco, Avelina García-García and Elena Orgilés-Calpena
Development and Application of a Lignin-Based Polyol for Sustainable Reactive Polyurethane Adhesives Synthesis
Reprinted from: *Polymers* 2024, *16*, 1928, doi:10.3390/polym16131928 20

Wejdan Al-otaibi, Naser M. Alandis, Yasser M. Al-Mohammad and Manawwer Alam
Advanced Anticorrosive Graphene Oxide-Doped Organic-Inorganic Hybrid Nanocomposite Coating Derived from *Leucaena leucocephala* Oil
Reprinted from: *Polymers* 2023, *15*, 4390, doi:10.3390/polym15224390 35

Ayşe Durmuş-Sayar, Murat Tansan, Tuğçe Çinko-Çoban, Dilay Serttan, Bekir Dizman, Mehmet Yildiz and Serkan Ünal
Incorporation of Graphene Nanoplatelets into Fiber-Reinforced Polymer Composites in the Presence of Highly Branched Waterborne Polyurethanes
Reprinted from: *Polymers* 2024, *16*, 828, doi:10.3390/polym16060828 56

Sadi Ibrahim Haruna, Yasser E. Ibrahim, Zhu Han and Abdulwarith Ibrahim Bibi Farouk
Flexural Response of Concrete Specimen Retrofitted with PU Grout Material: Experimental and Numerical Modeling
Reprinted from: *Polymers* 2023, *15*, 4114, doi:10.3390/polym15204114 76

Yong Rok Kwon, Seok Kyu Moon, Hae Chan Kim, Jung Soo Kim, Miyeon Kwon and Dong Hyun Kim
Effects of Grafting Degree on the Formation of Waterborne Polyurethane-Acrylate Film with Hard Core–Soft Shell Structure
Reprinted from: *Polymers* 2023, *15*, 3765, doi:10.3390/polym15183765 93

Ju-Hong Lee, Won-Bin Lim, Jin-Gyu Min, Jae-Ryong Lee, Ju-Won Kim, Ji-Hong Bae and Pil-Ho Huh
Synthesis of Room Temperature Curable Polymer Binder Mixed with Polymethyl Methacrylate and Urethane Acrylate for High-Strength and Improved Transparency
Reprinted from: *Polymers* 2024, *16*, 1418, doi:10.3390/polym16101418 107

Junhua Chen, Xiaoting Lu, Jinlian Chen, Shiting Li, He Zhang, Yinping Wu, et al.
Synthesis and Properties of Cationic Core-Shell Fluorinated Polyurethane Acrylate
Reprinted from: *Polymers* 2024, *16*, 86, doi:10.3390/polym16010086 120

Aysel Ersoy, Fatih Atalar and Alper Aydoğan
Investigation of Novel Solid Dielectric Material for Transformer Windings
Reprinted from: *Polymers* 2023, *15*, 4671, doi:10.3390/polym15244671 136

Junhua Chen, Zhihao Zeng, Can Liu, Xuan Wang, Shiting Li, Feihua Ye, et al.
Aqueous Cationic Fluorinated Polyurethane for Application in Novel UV-Curable Cathodic Electrodeposition Coatings
Reprinted from: *Polymers* **2023**, *15*, 3725, doi:10.3390/polym15183725 . . . . **150**

Wenzi Liang, Na Ni, Yuxin Huang and Changmin Lin
An Advanced Review: Polyurethane-Related Dressings for Skin Wound Repair
Reprinted from: *Polymers* **2023**, *15*, 4301, doi:10.3390/polym15214301 . . . . **165**

About the Editors

Chang-An Xu

Xu Chang'an graduated from the University of Chinese Academy of Sciences, majoring in polymer chemistry and physics. Currently, he is working at the Institute of Chemical Engineering, Guangdong Academy of Sciences, engaged in research related to polymer materials. Presently, he is a Guest Editor of the international journal *Polymers*, an Editorial Board Member of the *Journal of Liaocheng University* (Natural Science Edition), and the first Young Editorial Board Member of the Chinese core journal *Engineering Plastics Application*. He has published more than 40 research papers, including 22 as the first author/corresponding author. He won the "President's Award of the Chinese Academy of Sciences" in 2022, the "Excellent Doctoral Graduates" of Beijing Ordinary Colleges and Universities in 2022, and the "National Doctoral Scholarship" of the Ministry of Education in 2021.

Zhuohong Yang

Yang Zhuohong is a Professor and a Vice Dean. He is mainly engaged in the research of light curing materials, water-based coatings, and biomass materials. He has published more than 40 papers, including 30 papers indexed by SCI and EI, and applied for 12 patents at home and abroad. He has chaired and participated in more than 10 projects, including the National Natural Science Foundation, the Ministry of Personnel's Returning Overseas Students Fund, the Guangdong Natural Science Foundation, the Hong Kong–Guangdong Cooperation Project, and the Hong Kong Polytechnic University Cooperation Fund. He was selected to be involved in the New Century Talent Project of Hubei Province and the thousand and ten Talents Training plan of Guangdong Province, and he won the first prize of Science and Technology Progress of Hubei Province.

Editorial

Progress in Polyurethane and Composites

Chang-An Xu [1,*] and Zhuohong Yang [2]

[1] Institute of Chemical Engineering, Guangdong Academy of Sciences, Guangzhou 510665, China
[2] Key Laboratory for Biobased Materials and Energy of Ministry of Education, College of Materials and Energy, South China Agricultural University, Guangzhou 510642, China; yangzhuohong@scau.edu.cn
* Correspondence: xuca2020@163.com

1. Introduction

Polyurethane materials have received increasing attention as daily materials due to their unique structures and properties. Polyurethane polymers are a class of polymers containing repeated carbamate groups in their main chain, which are formed according to the polyaddition of isocyanates and polyols [1]. By changing the type and composition of the raw materials, the shape and properties of polyurethane products can be adjusted. Due to this adaptivity, along with their excellent wear resistance, impact resistance, biocompatibility, adhesion, and mechanical properties, they are widely used as adhesives, elastomers, foam, and medical materials. However, pure polyurethane materials are limited by their structure, with their performance unable to meet the application requirements in some specific environments. Therefore, this is encouraging researchers to continue to explore and innovate polyurethane materials through the modification of the polyurethane structure or by creating composites with other materials to improve its performance and expand its potential applications.

With the increasing depletion of fossil fuel resources, enhanced environmental awareness, and China's proposal of its "dual carbon" goals, sustainable and green polyurethane materials are becoming increasingly popular. Waterborne polyurethane is a type of polyurethane material based on water as the dispersing medium, making it a green, environmentally protective, and pollution-free product expected to replace traditional oily polyurethane. Waterborne polyurethane reduces the volatilization of organic solvents and reduces harm to the human body. Equally, as its use is in line with green chemistry and sustainable development strategies, its study is sure to become a future direction of research. However, waterborne polyurethane is not as efficient as solvent-based polyurethane in terms of its resistance to water and high temperatures. With expansion of the research, these shortcomings of waterborne polyurethane are expected to be resolved.

In view of analyzing polyurethane materials and their development trends, 10 research papers and 1 review article on the synthesis and modification of polyurethane materials were published in this Special Issue "Progress in Polyurethane and Composites". Herein, the properties of polyurethane were characterized using various research methods and discussed, which expanded the possibilities for its application in different fields and provided a valuable point of reference for polyurethane-related research. These articles are introduced as follows.

2. An Overview of the Published Articles

In the study by Serrano-Martinez et al., in order to alleviate the crisis of non-renewable fossil fuel usage and reduce the harm of traditional polyurethane adhesives to the environment [2], the authors used a high-temperature treatment to extract lignin from rice straw as a raw polyol material and synthesize a polyurethane adhesive, effectively substituting traditional polyol PPG with lignin. The performance of this environmentally friendly adhesive and its application in the footwear industry were evaluated through thermogravimetric analysis, rheological analysis, and T-peel testing. The results showed that when the lignin content was 7.5%, the thermal and mechanical properties of the adhesive were effectively maintained, and the adhesive showed excellent thermal stability. This work increased the added value of lignin and expanded its application in the field of adhesives, in line with green chemistry and sustainable development strategies.

In Lee et al.'s study, a polyurethane acrylate prepolymer (UA) was prepared using polytetrahydrofuran, polyethylene glycol, and polypropylene glycol as polyols and acrylic monomer as the end-sealing agent. This was then polymerized with polymethyl methacrylate (PMMA) to synthesize a room-temperature binder with high strength and high transparency [3]. When PMMA was blended with UA, the physical entanglement between the two polymers formed a three-dimensional network structure, which improved the mechanical properties of the adhesive. By using molecular design to change the type and content of polyols used, the polymer adhesives were prevented from yellowing during curing reactions. The results showed that when the UA value was 5~10%, the light transmittance, shear strength, and tensile strength of the prepared polymer adhesive were high, and its properties were customizable. The PMMA/UA binder prepared in this work is expected to be a promising candidate for future road marking polymer binders.

Because of their unique molecular structure and renewable properties, vegetable oils have gradually become a research focus. Using castor oil, maleic anhydride, and glyceryl methacrylate as their raw materials, Tuo et al. successfully prepared a new multifunctional castor-oil-based acrylate (MACOG) through two-step chemical modification. They then prepared a castor-oil-based waterborne polyurethane acrylate emulsion and finally prepared a series of coating materials under UV curing [4]. The results showed that with an increase in the MACOG content, the glass transition temperature increased from 20.3 °C to 46.6 °C, and the surface water contact angle increased from 73.85 °C to 90.57 °C. In addition, the thermal decomposition temperature, mechanical strength, and water resistance of the sample were also greatly improved, mainly on account of the introduction of MACOG, which improved the system's cross-linking density. This study not only provided new ideas for the preparation of waterborne polyurethane coatings with excellent comprehensive properties but also expanded the application of castor oil in coatings.

Enhancing the mechanical properties of polymer composites by using fiber materials to enhance the interfacial interaction has been deemed an efficient method. In the study by Ünal et al., non-functional graphene nanosheets (GNPs) were combined with a water-based, highly branched, multifunctional polyurethane dispersion (HBPUD) to effectively regulate the fiber–matrix interface in FRPCs [5]. By means of a unique ultrasonic spray deposition technique, the GNP/HBPUD aqueous-phase mixture was deposited on the surface of carbon fiber fabric to prepare epoxy prepreg sheets and corresponding FRPC laminates. The influence of the polyurethane (PU) and GNP contents and their ratio at the fiber–matrix interface on the tensile properties of the resulting high-performance composites was systematically investigated using stress–strain analysis of the produced FRPC plates and SEM analysis of their fractured surfaces. Synergistic stiffening and toughening effects were observed when as low as 20 to 30 mg of the GNPs was deposited per square meter on each side of the carbon fiber fabric in the presence of the multi-functional PU layer. This resulted in a significant improvement in the tensile strength from 908 to 1022 MPa while maintaining the initial Young's modulus or slightly improving it from approximately 63 to 66 MPa. This study underscored the importance of carefully tuning the GNP content and the PU:GNP ratio to tailor the tensile properties of high-performance CFRPCs.

In Chen et al.'s study, vinyl-capped cationic waterborne polyurethane (CWPU) was prepared using isophorone diisocyanate (IPDI), polycarbonate diol (PCDL), trimethylolpropane (TMP), and N-methyldiethanolamine (MDEA) as the raw materials and hydroxyethyl methacrylate (HEMA) as a capping agent [6]. Then, a crosslinked FPUA composite emulsion was prepared according to core–shell emulsion polymerization, with polyurethane (PU) as the shell, fluorinated acrylate (PA) as the core, and CWPU as the seed emulsion, together with dodecafluoroheptyl methacrylate (DFMA), diacetone acrylamide

(DAAM), and methyl methacrylate (MMA). The effects of the core–shell ratio of PA/PU on the surface properties, mechanical properties, and heat resistance of the FPUA emulsions and films were investigated. The results showed that when w(PA) = 30~50%, the highest-stability FPUA emulsion was generated, and under TEM, the particles displayed a core–shell structure with bright and dark intersections. When w(PA) = 30%, the tensile strength reached 23.35 ± 0.08 MPa. When w(PA) = 50%, the fluorine content on the surface of the coating film was 14.75%, and the contact angle was as high as 98.5°, indicative of good hydrophobicity. AFM was used to observe the surface flatness of the films. It was found that the tensile strength of the films increased and then decreased with an increase in the core–shell ratio, and the heat resistance of the FPUA films gradually increased. The FPUA film had excellent properties, such as good impact resistance, high flexibility, high adhesion, and corrosion resistance.

Enhancing the dielectric strength and minimizing the dielectric loss of insulation materials have piqued the interest of many researchers as enhancement techniques. It is worth noting that the electrical breakdown of insulation material is determined by its electrochemical and mechanical performance. Optimizing the mechanical, electrical, and chemical properties of new materials is considered during the generation process. Thermoplastic polyurethane (TPU) is often used as a high-voltage insulator due to its favorable mechanical properties, high insulation resistance, lightweight nature, recovery, large actuation strain, and cost-effectiveness. Its elastomer structure enables its application in a broad range of high-voltage (HV) insulation systems. In the study by Ersoy et al., the feasibility of using TPU as a solid insulator instead of a pressure plate for transformer windings was evaluated [7]. Their experimental investigation shed light on the potential of TPU to expand the range of insulating materials used in HV transformers. Transformers play a crucial role in HV systems; hence, the selection of suitable materials, like cellulose and polyurethane, is of the utmost importance. This study involved the preparation of an experimental laboratory setup. Breakdown tests were conducted by generating a non-uniform electric field using a needle–plane electrode configuration in a test chamber filled with mineral oil. Various voltages ranging from 14.4 kV to 25.2 kV were applied to induce electric field stress with a step rise of 3.6 kV. The partial discharges and peak numbers were measured based on predetermined threshold values. This study investigated and compared the behaviors of two solid insulating materials under differing non-electric field stress conditions. Harmonic component analysis was utilized to observe the differences between the two materials. Notably, at 21.6 kV and 25.2 kV, polyurethane demonstrated superior performance compared to the pressboard with regards to the threshold value for the leakage current.

Metal corrosion poses a substantial economic challenge in a technologically advanced world. In the study by Al-otaibi., novel, environmentally friendly, anticorrosive graphene oxide (GO)-doped organic–inorganic hybrid polyurethane (LFAOIH@GO-PU) nanocomposite coatings were developed using Leucaena leucocephala oil (LLO) [8]. The formulation was produced through the amidation of LLO to form diol fatty amide, followed by the reaction of tetraethoxysilane and a dispersion of GO_x (X = 0.25, 0.50, and 0.75 wt%), along with the reaction of isophorane diisocyanate (25–40 wt%) to form LFAOIH@GO_x-PU_{35} nanocomposites. A detailed examination of the LFAOIH@$GO_{0.5}$-PU_{35} morphology was conducted using X-ray diffraction, scanning electron microscopy, energy-dispersive X-ray spectroscopy, and transmission electron microscopy. These studies revealed their distinctive surface roughness features, along with a contact angle of around 88 G.U, preserving their structural integrity at temperatures of up to 235 °C with minimal loading of GO. Additionally, with the dispersion of GO, their mechanical properties were improved, including their scratch hardness (3 kg), pencil hardness (5H), impact resistance, bending, gloss value (79), crosshatch adhesion, and thickness. Electrochemical corrosion studies involving Nyquist, Bode, and Tafel plots provided clear evidence of the coatings' outstanding anticorrosion performance.

Polyurethane (PU) composites are increasingly used as repair materials for civil engineering infrastructure, including runways, road pavements, and buildings. Evaluation of polyurethane grouting (PUG) material is critical for its maintenance. In Haruna et al.'s study, the flexural behavior of normal concrete repaired with polyurethane grout (NC-PUG) was evaluated under three-point bending tests [9]. A finite element (FE) model was developed to simulate the flexural response of the NC-PUG specimens. The equivalent principle response of the NC-PUG was analyzed through a three-dimensional finite element model (3D FEM). The NC and PUG's properties were simulated using the stress–strain relationships determined in compressive and tensile tests. The overlaid PUG material was prepared by mixing PU and quartz sand and overlayed on either the top or bottom surface of a concrete beam. Two different overlaid thicknesses were applied, namely 5 mm and 10 mm. The composite NC-PUG specimens were formed by casting a PUG material using different overlaid thicknesses and configurations. The reference specimen showed the highest average ultimate flexural stress of 5.56 MPa $\pm$ 2.57% at a 95% confidence interval, with a corresponding midspan deflection of 0.49 mm $\pm$ 13.60%. However, due to the strengthening effect of the PUG layer, the deflection of the composite specimen was significantly improved. The concrete specimens with PUG retrofitted at the top surface demonstrated a typical linear pattern from the initial loading stage to complete failure. Moreover, the concrete specimens with PUG retrofitted at the bottom surface exhibited two deformation regions before complete failure. The FE analysis showed good agreement between the modeled numerical and experimental test results. The numerical model accurately predicted the flexural strength of the NC-PUG beam, slightly underestimating Ke by 4% and overestimating the ultimate flexural stress by 3%.

To improve the film-forming ability of hard-type acrylic latex, waterborne polyurethane-acrylate (WPUA) was grafted with polyurethane. To balance its film-forming ability and hardness, the WPUA latex was designed with a hard core (polyacrylate) and a soft shell (polyurethane) [10]. The grafting ratio was controlled by varying the content of 2-hydroxyethyl methacrylate (HEMA) used to cap the ends of the polyurethane prepolymer. The morphologies of the latex particles, the film surface, and the fracture surface of the film were characterized through transmission electron microscopy, atomic force microscopy, and scanning electron microscopy, respectively. An increase in the grafting ratio resulted in enhanced miscibility of the polyurethane and polyacrylate but reduced the adhesion between particles and increased the minimum temperature for film formation. In addition, grafting was essential to obtain transparent WPUA films. Excessive grafting induced defects such as micropores within the film, leading to decreased hardness and adhesive strength. The optimal HEMA content for the preparation of a WPUA coating with an excellent film-forming ability and high hardness in ambient conditions was 50%. The final WPUA film was prepared without coalescence agents that generate volatile organic compounds.

Aqueous polyurethane is an environmentally friendly, low-cost, high-performance resin with good abrasion resistance and strong adhesion. Cationic aqueous polyurethane has limited use in cathodic electrophoretic coatings due to its complicated preparation process and its poor stability and performance after emulsification and dispersion. The introduction of perfluoropolyether alcohol (PFPE-OH) and the application of light curing technology can effectively improve the stability of aqueous polyurethane emulsions and thus enhance the functionality of coating films. In Chen et al.'s study, a new UV-curable fluorinated polyurethane-based cathodic electrophoretic coating was prepared by using cationic polyurethane as a precursor, introducing PFPE-OH capping, and grafting hydroxyethyl methacrylate (HEMA) [11]. The results showed that the presence of perfluoropolyether alcohol in the structure affected the variation in the moisture content of the paint film after flash evaporation. Based on the emulsion particle size and morphology tests, the fluorinated cationic polyurethane emulsion was identified as a core–shell structure with its hydrophobic ends encapsulated in the polymer and its hydrophilic ends on the outer surface. After abrasion testing and baking, the fluorine atoms of the coating were found to increase from 8.89% to 27.34%. The static contact angle of the coating to water was 104.6° ± 3°, and water droplets rolled off it without traces, indicating that the coating was hydrophobic. The coating had excellent thermal stability and tensile properties and proved effective when its impact resistance, flexibility, adhesion, and resistance to chemical corrosion in extreme environments were tested. This study provided novel insights into the construction of a new and efficient cathodic electrophoretic coating system, as well as a greater scope for the promotion of cationic polyurethane in practical applications.

The inability of wounds to heal effectively through normal repair has become a burden that seriously affects socio-economic development and human health. Therapy for acute and chronic skin wounds still poses great clinical difficulty due to the lack of suitable functional wound dressings. While dressings made of polyurethane exhibit excellent diverse biological properties, they are not functional for clinical needs, and most dressings are unable to dynamically adapt to microenvironmental changes during the healing process of chronic wounds at different stages. Therefore, the development of multifunctional polyurethane composite materials has become a hot research topic. In light of this, the final review describes how the incorporation of different polymers and fillers into polyurethane dressings changes their physicochemical and biological properties and describes their applications in wound repair and regeneration. Liang et al. [12] cover several polymers, mainly natural-based polymers (e.g., collagen, chitosan, and hyaluronic acid) and synthetic-based polymers (e.g., polyethylene glycol, polyvinyl alcohol, and polyacrylamide), and some other active ingredients (e.g., LL37 peptide, platelet lysate, and exosomes). The design, conversion, use, and application of advanced functional polyurethane-related dressings are discussed, alongside future development directions, providing reference for novel designs and applications.

3. Conclusions

This Special Issue covers research on the structural design and performance characterization of polyurethane materials and their composites, as well as their applications in adhesives, coatings, and UV resins. Effective design and characterization are essential in creating performant polyurethane materials with wide applications. With continuous research on polyurethane materials, new multifunctional and high-performance polyurethane materials will continue to emerge. In addition, given national and international interest in green and environmentally friendly materials, the preparation of sustainable water-based materials and bio-based polyurethane materials will be the future development trend. In short, the research on polyurethane materials collated in this Special Issue constitutes a valuable point of reference for researchers working with related polyurethane materials.

Acknowledgments: As Guest Editors of the Special Issue "Progress in Polyurethane and Composites", we would like to express our deep appreciation to all the authors whose valuable work was published in this issue and who thus contributed to the success of the edition.

Conflicts of Interest: The authors declare no conflicts of interest.

References

1. Xu, C.-A.; Lu, M.; Wu, K.; Shi, J. Effects of polyether and polyester polyols on the hydrophobicity and surface properties of polyurethane/polysiloxane elastomers. *Macromol. Res.* **2020**, *28*, 1032–1039. [CrossRef]
2. Serrano-Martínez, V.M.; Hernández-Fernández, C.; Pérez-Aguilar, H.; Carbonell-Blasco, M.P.; García-García, A.; Orgilés-Calpena, E. Development and application of a lignin-based polyol for sustainable reactive polyurethane adhesives synthesis. *Polymers* **2024**, *16*, 1928. [CrossRef]
3. Lee, J.-H.; Lim, W.-B.; Min, J.-G.; Lee, J.-R.; Kim, J.-W.; Bae, J.-H.; Huh, P.-H. Synthesis of room temperature curable polymer binder mixed with polymethyl methacrylate and urethane acrylate for high-strength and improved transparency. *Polymers* **2024**, *16*, 418. [CrossRef] [PubMed]
4. Tuo, Y.; Luo, X.; Xiong, Y.; Xu, C.-A.; Yuan, T. A novel polyfunctional polyurethane acrylate derived from castor oil-based polyols for waterborne UV-curable coating application. *Polymers* **2024**, *16*, 949. [CrossRef]
5. Durmuş-Sayar, A.; Tansan, M.; Çinko-Çoban, T.; Serttan, D.; Dizman, B.; Yildiz, M.; Ünal, S. Incorporation of graphene nanoplatelets into fiber-reinforced polymer composites in the presence of highly branched waterborne polyurethanes. *Polymers* **2024**, *16*, 828. [CrossRef] [PubMed]

6. Chen, J.; Lu, X.; Chen, J.; Li, S.; Zhang, H.; Wu, Y.; Zhu, D.; Hao, X. Synthesis and properties of cationic core-shell fluorinated polyurethane acrylate. *Polymers* **2023**, *16*, 86. [CrossRef] [PubMed]
7. Ersoy, A.; Atalar, F.; Aydoğan, A. Investigation of novel solid dielectric material for transformer windings. *Polymers* **2023**, *15*, 4671. [CrossRef]
8. Al-otaibi, W.; Alandis, N.M.; Al-Mohammad, Y.M.; Alam, M. Advanced anticorrosive graphene oxide-doped organic-inorganic hybrid nanocomposite coating derived from Leucaena leucocephala oil. *Polymers* **2023**, *15*, 4390. [CrossRef]
9. Haruna, S.I.; Ibrahim, Y.E.; Han, Z.; Farouk, A.I.B. Flexural response of concrete specimen retrofitted with PU grout material: Experimental and numerical modeling. *Polymers* **2023**, *15*, 4114. [CrossRef]
10. Kwon, Y.R.; Moon, S.K.; Kim, H.C.; Kim, J.S.; Kwon, M.; Kim, D.H. Effects of grafting degree on the formation of waterborne polyurethane-acrylate film with hard core–soft shell structure. *Polymers* **2023**, *15*, 3765. [CrossRef] [PubMed]
11. Chen, J.; Zeng, Z.; Liu, C.; Wang, X.; Li, S.; Ye, F.; Li, C.; Guan, X. Aqueous cationic fluorinated polyurethane for application in novel UV-curable cathodic electrodeposition coatings. *Polymers* **2023**, *15*, 3725. [CrossRef] [PubMed]
12. Liang, W.; Ni, N.; Huang, Y.; Lin, C. An advanced review: Polyurethane-related dressings for skin wound repair. *Polymers* **2023**, *15*, 4301. [CrossRef] [PubMed]

Disclaimer/Publisher's Note: The statements, opinions and data contained in all publications are solely those of the individual author(s) and contributor(s) and not of MDPI and/or the editor(s). MDPI and/or the editor(s) disclaim responsibility for any injury to people or property resulting from any ideas, methods, instructions or products referred to in the content.

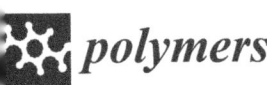

A Novel Polyfunctional Polyurethane Acrylate Derived from Castor Oil-Based Polyols for Waterborne UV-Curable Coating Application

Youmin Tuo, Xubiao Luo, Yahong Xiong, Chang-An Xu * and Teng Yuan *

Key Laboratory for Biobased Materials and Energy of Ministry of Education, College of Materials and Energy, South China Agricultural University, Guangzhou 510642, China
* Correspondence: xuchangan@scau.edu.cn (C.-A.X.); yuant@scau.edu.cn (T.Y.)

Abstract: Because of its unique molecular structure and renewable properties, vegetable oil has gradually become the focus of researchers. In this work, castor oil was first transformed into a castor oil-based triacrylate structure (MACOG) using two steps of chemical modification, then it was prepared into castor oil-based waterborne polyurethane acrylate emulsion, and finally, a series of coating materials were prepared under UV curing. The results showed that with the increase in MACOG content, the glass transition temperature of the sample was increased from 20.3 °C to 46.6 °C, and the water contact angle of its surface was increased from 73.85 °C to 90.57 °C. In addition, the thermal decomposition temperature, mechanical strength, and water resistance of the samples were also greatly improved. This study not only provides a new idea for the preparation of waterborne polyurethane coatings with excellent comprehensive properties but also expands the application of biomass material castor oil in the field of coating.

Keywords: waterborne polyurethane acrylate; comprehensive performance; castor oil; UV curing

1. Introduction

Polyurethane acrylate is a kind of cross-linked polymer with a carbamate structure that is synthesized by a polyaddition reaction of polyol and isocyanate. It is widely used in coatings, leather, adhesives, and sealants because of its excellent adhesion, flexibility, wear resistance, and weather resistance [1,2]. However, the viscosity of traditional polyurethane acrylate is relatively high, and active diluents are usually added to regulate it. The active diluents commonly used are mainly acrylates with low boiling point, volatile, irritating odor, and toxicity, which will bring great harm to the environment and human health. With the increase in people's awareness of environmental protection and the implementation of environmental protection policies, people have gradually turned their attention to pollution-free and green water. Its use as an active diluent in the polymerization process can not only reduce the viscosity of the polymer but also reduce the release of VOC [3–7] and enhance operational safety during use. In addition, the coating industry has begun to transform towards environmentally friendly products. Compared with traditional organic coatings, UV-cured coatings have advantages such as low VOC emissions, high curing efficiency, convenient operation, pollution-free properties, and low energy consumption [8–11]. In the field of photochemistry, UV curing has maintained an unprecedented position in the new generation of industry-related coating systems, and therefore, UV curing coatings have become one of the research hotspots.

With the rapid development of modern society, due to the increasing consumption of petrochemical products, resulting in the increasing shortage of petroleum raw materials, people gradually turn their attention to rich resources and low-cost, renewable, and biodegradable biomass materials [12–18]. Among numerous biomass materials, vegetable oil has received extensive attention from researchers due to its unique molecular structure, abundant content, suitable price, and biodegradability [19,20]. A variety of modified vegetable oil products can be prepared by using active sites such as C=C double bond, hydroxyl group, epoxy group, and ester group on vegetable oil, such as polyol materials for the synthesis of polyurethane acrylate [21,22]. For example, Wang et al. synthesized an oleic acid-based primary alcohol using methyl oleate as raw materials, which exhibited high reactivity as a polyurethane soft segment. At the same time, the prepared film material showed excellent tensile properties [23]. Li et al. synthesized epoxy soybean oil acrylate with multi-functional hydroxyl groups from epoxy soybean oil and introduced it into the main chain of waterborne polyurethane acrylate [24]. The results showed that the epoxy soybean oil-based solidified film had good mechanical and thermodynamic properties, and its material had been well used in the field of wood coatings. Moreover, Gaddam et al. used cottonseed oil to synthesize three kinds of phosphorylated polyols with different hydroxyl values and used them as soft segments of polyurethane to synthesize waterborne polyurethane dispersions (PUDs) without industrial hydrophilic chain extension and catalyst. The three types of PUDs exhibited excellent storage stability, and the tensile properties, glass transition temperature, thermal stability, hydrophobicity, and anti-corrosion properties of the coating all improved with the increase of the hydroxyl value of the phosphorylated polyols [25]. It can be concluded that the introduction of vegetable oil and its derivatives can effectively improve the mechanical properties and thermal stability of waterborne polyurethane acrylate coatings. However, few studies have reported that vegetable oil is used both as a raw material for polyurethane reaction and as a source of UV-curable double bond monomer.

In this study, castor oil was chemically modified by a two-step method with maleic anhydride, a biomass resource, to prepare multi-functional castor oil triacrylate (MACOG), and then different percentages of MACOG were introduced into waterborne polyurethane acrylate emulsion, and a series of coating materials were prepared by UV curing [26]. Subsequently, the thermodynamic properties, mechanical properties, thermal stability, gel content, and hydrophobic properties of the coating were tested. It was concluded that the vegetable oil derivatives could be used to prepare high-performance coating materials instead of petrochemical resources. Most importantly, this work not only made the raw materials sustainable but also made the preparation and curing processes green and pollution-free.

2. Experiment

2.1. Materials

Castor oil (CO) was sourced from Tianjin Fuyu Fine Chemical Reagent Co., Ltd. (Tianjin, China). Dibutyltin dilaurate (DBTDL) and maleic anhydride (MA) were purchased from Tianjin Fuchen Chemical Reagent Co., Ltd. (Tianjin, China). Isophorone diisocyanate (IPDI), glycidyl methacrylate (GMA), polybutanediol adipate (PBA, 1000 g/mol), hydroxyethyl acrylate (HEA), and dimethylol butyric acid (DMBA) were all derived from Shanghai Maclin Biochemical Co., Ltd. (Shanghai, China). Both hydroquinone and N,N-dimethylethanolamine (DMEA) were from Tianjin Damao Chemical Reagent Co., Ltd. (Tianjin, China). Triethylamine (TEA) was from Tianjin Yongda Chemical Reagent Co., Ltd. (Tianjin, China). 2-butyl ketone was purchased from Guangzhou Chemical Reagent Factory. 2-hydroxy-2-methyl-1-phenylacetone (PI-1173) was purchased from Tianjin Jiuri New Materials Co., Ltd. (Tianjin, China). The above pharmaceutical materials had not been further processed and had been directly used. The UV equipment (CH-UV06) used was from Shanghai Yuming Instrument Co., Ltd., Shanghai, China. The UV lamp had a power of 85 W and a wavelength of 395 nm.

2.2. Synthesis of Castor Oil with Ternary Carboxylic Acid

A total of 34.21 g of castor oil, 9.81 g of maleic anhydride, 0.09 g of hydroquinone, and 0.22 g of DMEA were added to the three-neck round bottling flask. The reaction mixture was then heated to 65 °C and stirred at 250 r/min until the MA was completely melted. The mixture was then heated to 105 °C and reacted for 3 h to obtain a yellow, transparent, and viscous liquid of castor oil-based terarboxylic acid, which was named MACO. The synthesis process is shown in Scheme 1.

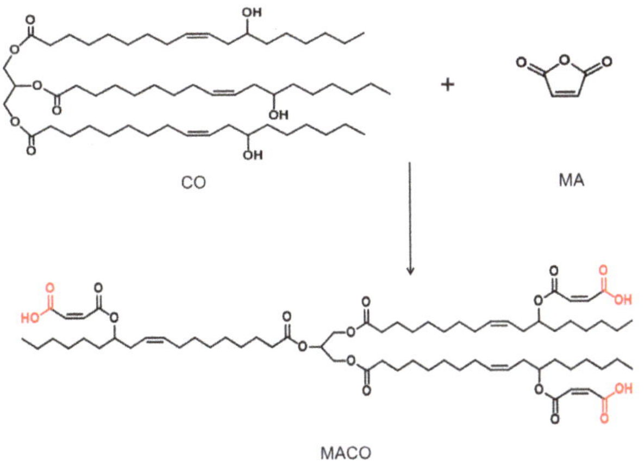

Scheme 1. The process of preparing MACO.

2.3. Synthesis of Castor Oil-Based Triacrylate

A total of 44.02 g of MACO and 0.29 g of DMEA were added to a 250 mL three-necked round-bottomed flask and heated to 90 °C. Then, the mixture of 14.66 g GMA and 0.15 g hydroquinone was added to the mixture by drops within 30 min and reacted for 4 h. Finally, a slightly orange viscous liquid of castor oil-based triacrylate was obtained and named MACOG. The synthesis route is shown in Scheme 2.

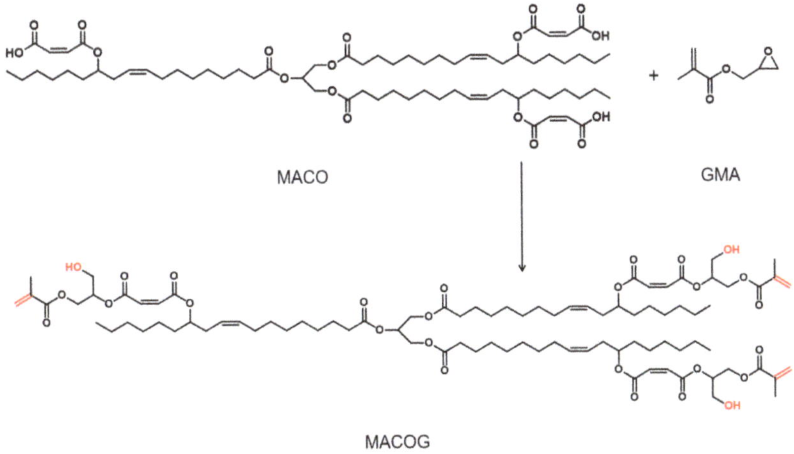

Scheme 2. The process of preparing MACOG.

2.4. Synthesis of Waterborne Polyurethane Acrylate Emulsion (WPUA)

First, appropriate amounts of DMBA, PBA, and IPDI were added to a 250 mL three-necked round-bottomed flask. Two drops of DBTDL and appropriate amounts of 2-butanone were added to the mixture. The mixture was stirred at 80 °C for 3 h in a nitrogen atmosphere. Subsequently, MACOG with different mass fractions was added to the mixture by drops within 30 min, and the detailed formula is shown in Table 1. After 4 h of reaction,

HEA containing 1 wt% hydroquinone was added to the flask. When NCO was completely consumed, the mixture was cooled to room temperature, and an appropriate amount of TEA was added and then stirred for 30 min to neutralize the carboxyl group. Finally, 30 wt% deionized water was added and stirred at high speed for 2 h, and 2-butanone was removed to obtain waterborne polyurethane acrylate dispersion (WPUA). The synthesis route is shown in Scheme 3.

Table 1. Formula of WPUA.

Samples	Mass Fraction of MACOG (wt%)	Raw Material Formula (g)					
		IPDI	PBA	MACOG	DMBA	HEA	TEA
S1	10	4.82	10	1.84	1.14	0.6	0.78
S2	20	5.30	10	4.72	1.14	0.6	0.78
S3	30	5.97	10	7.60	1.14	0.6	0.78
S4	40	6.95	10	12.47	1.14	0.6	0.78

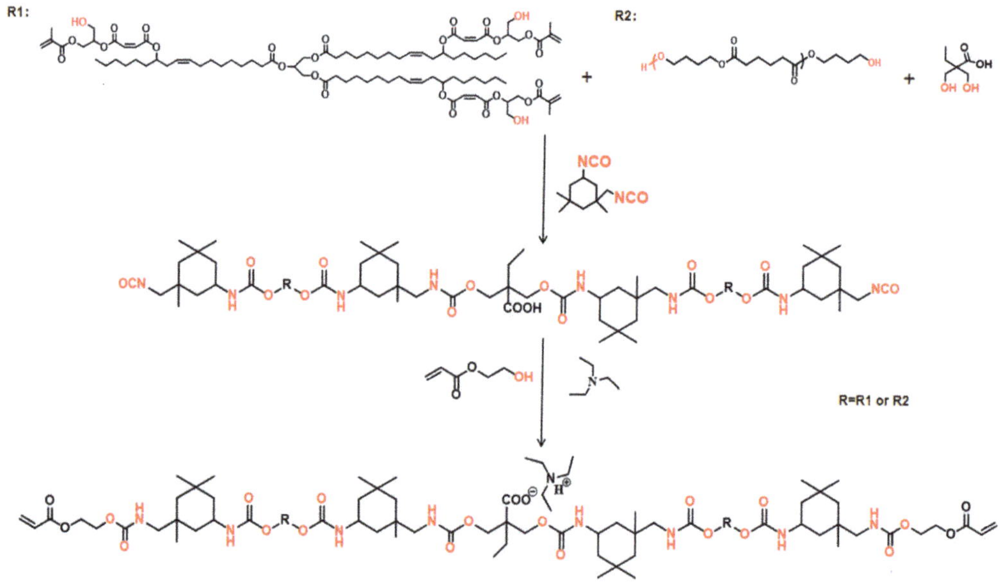

Scheme 3. The process of preparing WPUA.

2.5. Preparation of Waterborne Polyurethane Acrylate Photocurable Film

A certain amount of WPUA lotion and 4 wt% of PI-1173 were added into a 20 mL glass bottle and stirred evenly to form an aqueous dispersion. Subsequently, the dispersion was poured into a culture dish and dried at room temperature for 24 h before drying at 60 °C for 12 h. Finally, the cured film was obtained by irradiation for 30 s under a UV lamp with a distance of 2 cm and a light intensity of 387 mW/cm^2. The thickness of the obtained film was 0.4 mm.

2.6. Characterization

The liquid samples CO, MACO, and MACOG were coated on halide chips to form a liquid film and characterized using Fourier transform infrared spectroscopy (FTIR, Nicolet iS10, Thermo Fisher, Waltham, MA, USA), with a wavelength range of 500–4000 cm^{-1}. The proton shift of the product was characterized using a nuclear magnetic resonance spectrometer (Bruker AV 400, Bruker Biospin AG, Fällanden, Switzerland), and the solvent used in the test was CDCl$_3$. The dynamic mechanics of the sample were tested using the tensile mode of a dynamic thermomechanical analyzer (DMA 242E, German Naichi, Krefeld, Germany). The size of the test sample was 20.0 mm × 6.0 mm × 0.5 mm, with an oscillation frequency of 1 Hz, a testing temperature of −80~180 °C, and a heating rate of 3 K/min. Formula (1) was used to calculate the cross-linking density of the sample.

$$V_e = E'/3RT', \tag{1}$$

where T′ is the absolute temperature (T_g + 30 °C) in the rubber state, E′ is the storage modulus at T′, and R is the gas constant 8.314 J/(mol·K).

An electronic universal testing machine (UTM4204, Shenzhen Sansi Zongheng Technology Co., Ltd., Shenzhen, China) was used to test the tensile properties of the cured film according to the GB/T 1040.2-2006 standard. The tensile speed was set at 20 mm/min, and the sample size was 40.0 mm × 10.0 mm × 0.5 mm. The thermal stability of the sample in the nitrogen atmosphere was tested using a thermal analyzer (TG209F1LibraTM, Germany Nechi instrument manufacturing Co., Ltd. Shanghai, China). The heating rate of the test was 10 °C/min, and the temperature range was 35~800 °C. The particle size distribution and Zeta potential were measured using a laser particle size analyzer (Zetasizer Nano ZSE, Malvern, Shanghai, China). Before testing, the sample was diluted 100 times with distilled water. According to the GB/T 30693-2014 standard, a contact angle measuring instrument (OCA20, DATAPHYSICS, Shanghai, China) was used to test the water contact angle of the cured film surface. The mass of the sample before testing was m_0. The sample was taken out after soaking in water for 48 h, and the surface moisture was absorbed with filter paper. The mass of the sample was denoted as m_1. This process was repeated three times, and its average was finally taken. According to Formula (2), the water absorption rate of the solidified film was calculated.

$$\text{Water absorption rate} = (m_1 - m_0)/m_0 \times 100\% \tag{2}$$

The acetone extraction method was used to test the gel content of the solidified film. The steps were as follows: First, the solidified film with a mass of W_0 was accurately weighed at room temperature and then soaked in a sealed glass bottle containing acetone for 48 h. After that, the solidified film was taken out and dried in a vacuum oven at 60 °C to a constant weight, and its weight was weighed as W_1. The gel ratio was calculated according to Formula (3).

$$\text{Gel ratio} = (W_1/W_0) \times 100\% \tag{3}$$

The flexibility of the coating was tested according to GB/T 1731-1993. The pencil hardness of the coating was tested according to GB/T 6739-1996. According to GB/T 9274-1988, the acid and alkali resistance of the coating was tested.

3. Results and Discussion

3.1. Structural Characterization of Products

The infrared spectra of CO, MACO, and MACOG are shown in Figure 1. It could be seen that the characteristic hydroxyl peak of CO was found at 3421 cm^{-1}. However, in MACO, the characteristic peak at 3421 cm^{-1} disappeared, while a wide and strong carboxyl absorption peak appeared near 2500–3500 cm^{-1} [27]. This indicated that MACO had been successfully synthesized through the esterification reaction between CO and MA. In the infrared spectrum of MACOG, a new hydroxyl peak appeared at 3498 cm^{-1}, and the characteristic peaks at 1638 cm^{-1} and 813 cm^{-1} were attributed to the C=C stretching vibration absorption peak and the C=H bending vibration absorption peak, respectively. This indicated that MACOG had been successfully synthesized through the ring-opening esterification reaction between MACO and GMA. Subsequently, nuclear magnetic tests were used to further confirm that MACO and MACOG had been successfully synthesized, and the results are shown in Figure 2. MACO spectra showed that the hydroxyl proton peak disappeared at 3.63 ppm in CO, and a new chemical shift peak appeared at 5.00 ppm, which corresponded to the proton peak of newly formed ester-linked methylene [28]. In the MACOG spectra, the chemical shift at 3.82–4.54 ppm should be attributed to the methylene and methylene proton peaks in the structure of methacrylate. The chemical shift of the -CH$_2$- proton peak attached to the newly formed hydroxy-group appeared at 3.76 ppm, while the chemical shift of the -CH=CH$_2$ structure in GMA appeared at 5.57–6.18 ppm.

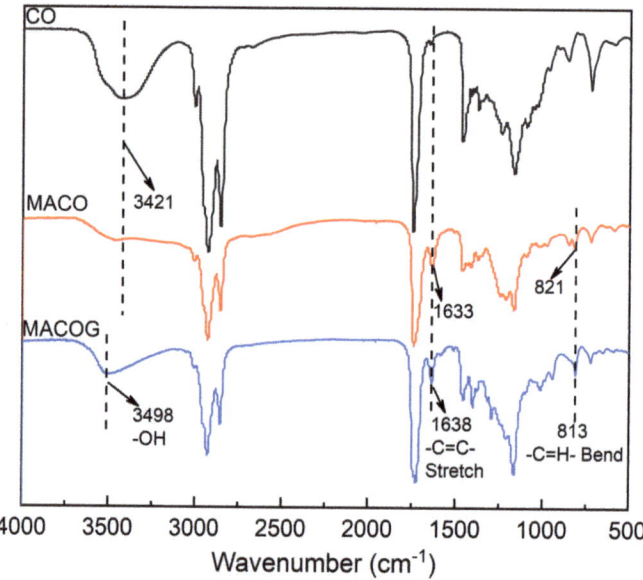

Figure 1. Infrared spectra of CO, MACO and MACOG.

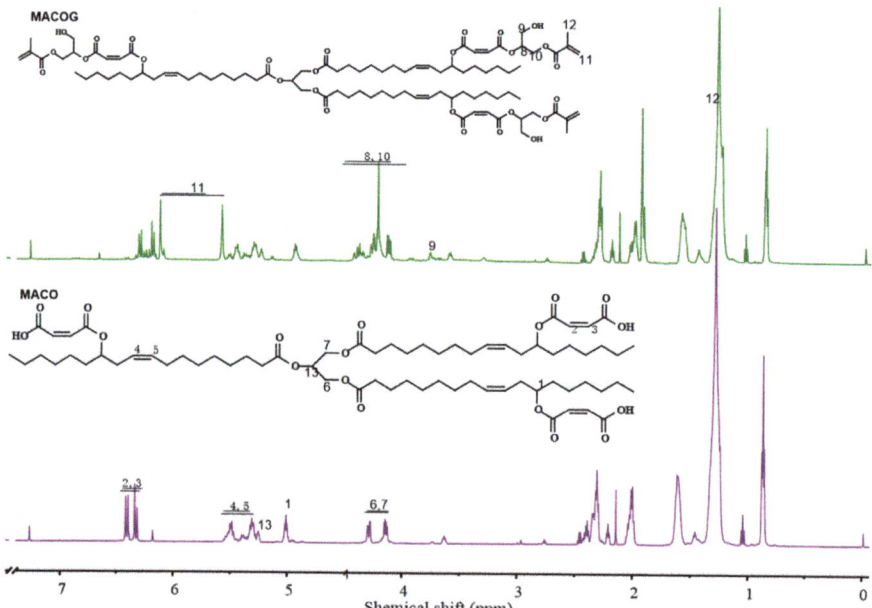

Figure 2. ^{1}H-NMR spectra of MACO and MACOG.

3.2. Particle Size Analysis of Emulsion

The storage stability of emulsion could be determined by measuring emulsion particle size distribution and Zeta potential [29]. The test results of the samples are shown in Figure 3 and Table 2. As can be seen from Figure 3, with the increase in MACOG content, the mean particle size (From 36.87 nm to 118.60 nm) and particle size distribution (From 0.136 to 0.165) of emulsion showed a trend of increasing and widening, respectively. This was mainly attributed to the increase in MACOG content, which reduced the relative content of DMBA in raw materials. After neutralization, the number of ionic groups was reduced, which would weaken the electrostatic repulsion between dispersed particles and improve the association of particles. In addition, the increase in MACOG would also introduce more hydrophobic chain segments, which would not only enhance the hydrophobicity of oligomers but also improve the cross-linking degree of prepolymers (as seen in Table 3). It can be seen from Table 2 that the PDI values of all emulsions were less than 0.3, indicating that they all had good dispersion. There was no obvious precipitation after 6 months at room temperature, and the Zeta potential was higher than 42.0 mV, indicating that the WPUA emulsion prepared in this work had good storage stability.

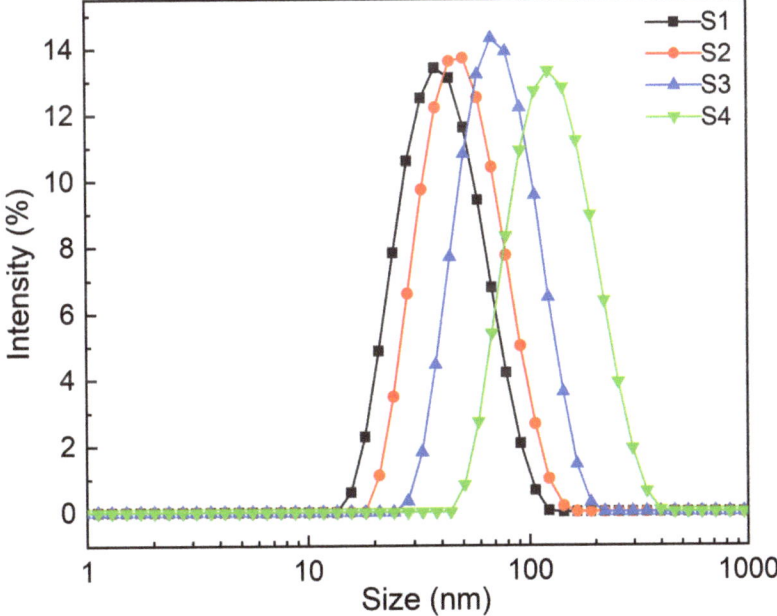

Figure 3. Particle size distribution curves of emulsion with different mass fractions of MACOG.

Table 2. Average particle size, PDI value, and Zeta potential of WPUA with different MACOG content.

Samples	DMBA Content (wt%)	Mean Particle Size (nm)	Dispersion Coefficient (PDI)	Zeta Potential (mV)
S1	6.20	36.87 ± 0.17	0.136 ± 0.009	44.43 ± 2.32
S2	5.24	45.63 ± 0.27	0.134 ± 0.010	42.73 ± 1.27
S3	4.50	66.37 ± 0.19	0.138 ± 0.004	45.30 ± 1.24
S4	3.66	118.60 ± 0.08	0.165 ± 0.003	55.17 ± 1.69

Table 3. DMA data and cross-linking density of cured films with different MACOG content.

Samples	E_{25} (MPa)	T_g (°C)	E' at T_g + 30 °C (MPa)	v_e (×10^3 mol/m^3)
S1	9.21	20.3	2.82	0.35
S2	22.89	22.1	6.39	0.79
S3	99.08	31.2	13.08	1.57
S4	298.59	46.6	23.74	2.72

3.3. Dynamic Thermomechanical Properties

The storage modulus (E'), glass transition temperature (T_g), and loss factor (Tanδ) of the cured film obtained through dynamic mechanical analysis are shown in Figure 4 and Table 3. As can be seen from Figure 4, the cured film exhibited a high storage modulus in the low-temperature region, and its value first increased and then decreased with the increase in MACOG content. At the same time, the storage modulus of all cured films was temperature-dependent, and its value decreased with the increase in heating temperature. The samples all showed a peak of the T_g on the Tan δ curve, which indicated that the cured film was homogeneous and the compatibility between the substances was good. At the same time, the peak shape decreased and widened with the increase in MACOG content, which was mainly related to the increase in the cross-linking density of the cured film [30]. The higher the cross-linking density, the weaker the mobility of the chain segment and the higher the glass transition temperature. As can be seen from Table 3, with the increase in MACOG content, the cross-linking density of samples increased, which was mainly related to the increased double-bond content in the system. The increase in cross-linking density would weaken the kinematic ability of chain segments in the system and then increase the T_g of the cured film, resulting in the T_g of the system increased from 20.3 °C to 46.6 °C.

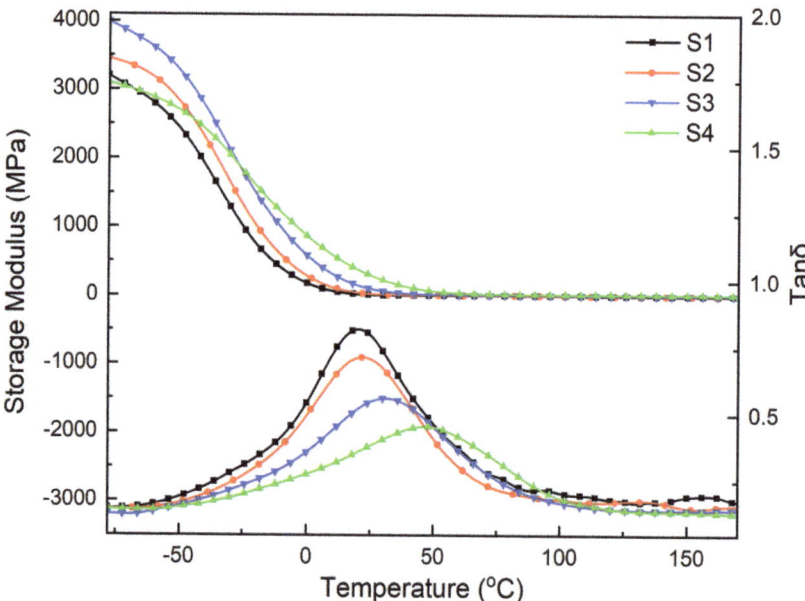

Figure 4. DMA curves of cured films with different MACOG content.

3.4. Mechanical Property

The test curve of the mechanical properties of the cured film is shown in Figure 5. It could be concluded that with the increase in MACOG content, the fracture strength and

Young's modulus of the cured film showed a trend of first increasing and then decreasing. The increase in fracture strength of cured film was mainly related to the increase in cross-linking density in the system (As shown in Table 3). However, when the content of MACOG reached 40 wt%, the fracture strength and Young's modulus of the sample were reduced. This was mainly because excessive cross-linking density in the system would not only lead to poor compatibility between hard and soft segments in the system but also cause an uneven distribution of cross-linking sites in the system [31]. When the content of MACOG was 30 wt%, the sample S3 had the largest breaking strength, which was about 23 MPa.

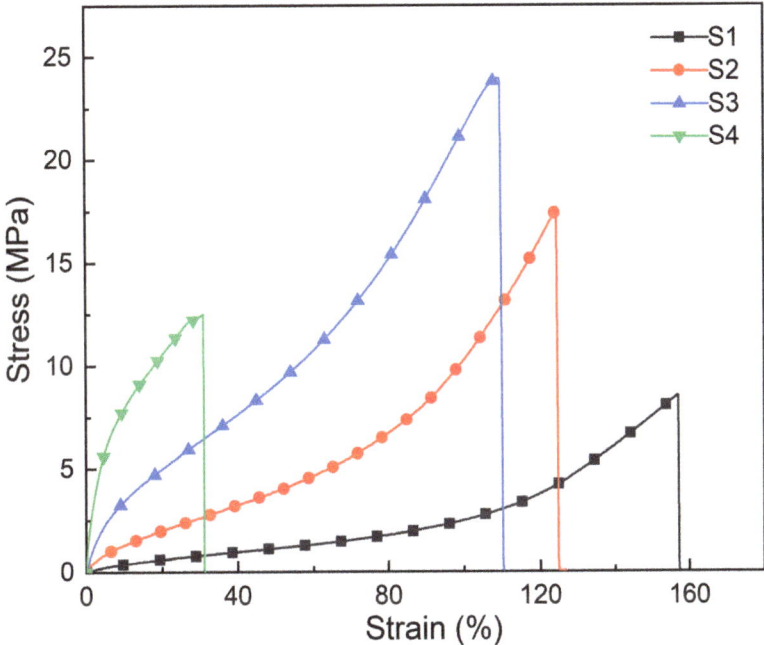

Figure 5. Stress-strain curves of cured films with different MACOG content.

3.5. Thermal Stability

The thermogravimetric method was used to analyze the thermal stability of the cured film, and its TGA and DTG curves are shown in Figure 6. It can be seen from the thermal decomposition curve that the cured film experienced three thermal degradation stages, and the degradation below 200 °C was mainly related to the residual moisture in the cured film. The degradation at 230 °C to 360 °C was mainly related to the decomposition of carbamate [32]. With the increase in MACOG content, the unstable content of carbamate in the system was reduced so that the thermal decomposition temperature was increased [33]. The decomposition between 360 °C and 500 °C was mainly related to the fracture of soft segments and cross-linking bonds in polyurethane. The temperature corresponding to a sample mass loss of 5 wt% ($T_{5\%}$) is generally used as the starting decomposition temperature of the sample. From Table 4, it could be seen that the temperature of sample S1 at $T_{5\%}$ was 231.9 °C. With the increase in MACOG content, the $T_{5\%}$ value of the sample further increased, reaching a maximum of 264.7 °C. In addition, when the mass loss of the sample was 30 wt% ($T_{30\%}$), the $T_{30\%}$ value of sample S1 was still the smallest, and its trend of change was consistent with $T_{5\%}$. After all samples were heated at high temperatures, the residual char rates of samples S1, S2, S3, and S4 at 790 °C were 2.5 wt%, 2.6 wt%, 3.2 wt%, and 3.8 wt%, respectively. This indicated that the residual char rate of samples at high temperatures increased with the increase in MACOG content. All samples in the

DTG curve exhibited two temperature peaks corresponding to the maximum thermal weight loss rate, which was caused by the thermodynamic incompatibility between the soft and hard segments of polyurethane. The thermal decomposition temperature of the hard segment was lower than that of the soft segment, so the maximum thermal decomposition rate at low temperatures was mainly related to the degradation of the hard segments. In addition, as MACOG increased, the cross-linking density of the sample increased, causing the maximum thermal weight loss rate temperature peak of the sample to shift toward higher temperatures. It could be concluded that the introduction of MACOG would increase the cross-linking density of the system, thereby improving the thermal stability of the system.

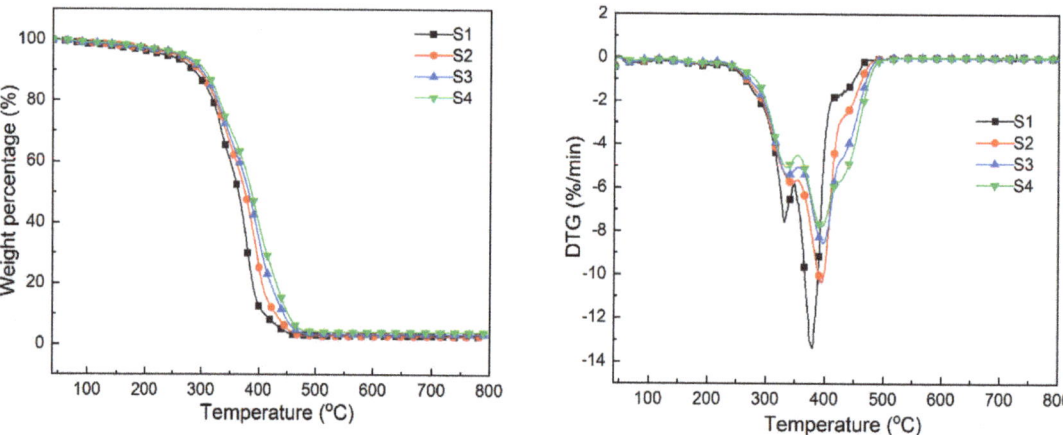

Figure 6. TGA and DTG curves of cured films with different MACOG content.

Table 4. Thermogravimetric data of cured films with different MACOG content.

Samples	$T_{5\%}$ (°C)	$T_{30\%}$ (°C)	Carbon Residue Rate (790 °C, wt%)
S1	231.9	332.9	2.5
S2	253.9	340.9	2.6
S3	262.0	344.5	3.2
S4	264.7	349.6	3.8

3.6. Contact Angle and Water Absorption

The test results of water absorption and surface water contact angle of the cured coating are shown in Figure 7. With the increase in MACOG content, the water absorption of the cured film decreased from 10.49 wt% to 6.36 wt%, and the water contact angle increased from 73.85° to 90.57°, which indicated that the introduction of MACOG could effectively improve the water-resistance of the cured film. On the one hand, it was related to the structure of hydrophobic long-chain fatty acids in MACOG [34]. On the other hand, it was related to the increased cross-linking density in the system, which enhanced the compactness of the coating [35].

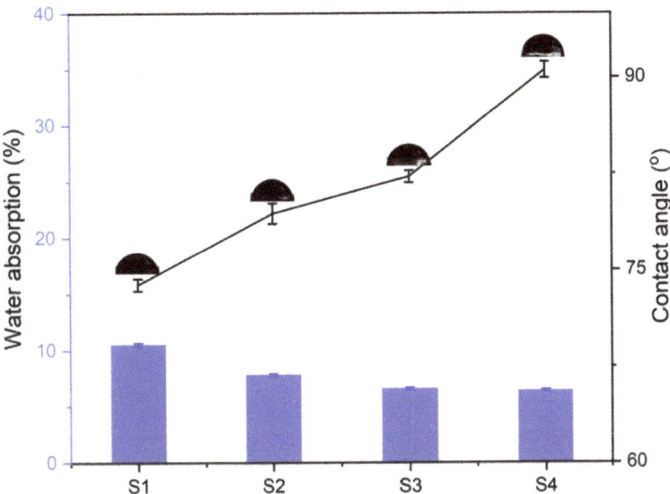

Figure 7. Water absorption and water contact angle of cured films with different MACOG contents.

3.7. General Performance of Curing Film

The general performance tests of the cured film are shown in Table 5. With the increase of the content of MACOG, the gel content and pencil hardness of the cured film showed the same trend of change. The gel ratio increased from 78.44 wt% to 94.71 wt%, and the pencil hardness increased from HB to 2H, which was mainly related to the increase in the cross-linking density of the system caused by the increase in MACOG. All cured films exhibited excellent flexibility, mainly due to the long-chain structure of PBA and the fatty acid long-chain structure of MACOG. By testing the acid and alkali resistance of the cured film, it was found that there was no significant change after immersion in HCl aqueous solution, indicating good acid resistance. However, after soaking in a NaOH solution, the surface of the cured film appeared white, which was mainly because the ester group contained in the structure of polyurethane was easy to hydrolysis in NaOH, leading to the destruction of its structure and weakening its alkali resistance [36].

Table 5. General properties of cured films with different MACOG contents.

Classification	S1	S2	S3	S4
Gel rate (wt%)	78.44	87.27	91.88	94.71
Pencil hardness	HB	H	2H	2H
Flexibility (mm)	2	2	2	2
Acid resistance (0.1 mol/L HCl)	Unchanged	Unchanged	Unchanged	Unchanged
Alkali resistance (0.1 mol/L NaOH)	Bleach	Bleach	Bleach	Bleach

4. Conclusions

In this work, a new type of multi-functional castor oil-based acrylate (MACOG) was successfully prepared using castor oil, maleic anhydride and glycidyl methacrylate as raw materials, and then it was synthesized into waterborne polyurethane acrylate emulsion, a variety of coating materials were prepared under UV, and the comprehensive performance of the cured film was evaluated. The results showed that the introduction of MACOG could effectively improve the thermal stability, glass transition temperature, acid resistance, gel rate, and breaking strength of cured films. The water contact angle of the coating surface increased from 73.85° to 90.57°, while the water absorption decreased from 10.49 wt% to 6.36 wt%. This was mainly related to the introduction of MACOG, which improved the cross-linking density in the system. In conclusion, this work not only prepared coatings with excellent comprehensive properties but also improved the added value of castor oil, which provided a new research idea for using vegetable oil instead of petrochemical resources to prepare coating products.

Author Contributions: Conceptualization, Y.T. and C.-A.X.; Methodology, Y.T., X.L., Y.X. and T.Y.; Software, Y.T. and C.-A.X.; Validation, Y.X.; Formal analysis, X.L. and T.Y.; Investigation, Y.T. and C.-A.X.; Resources, Y.X. and C.-A.X.; Data curation, X.L. and C.-A.X.; Writing—original draft, Y.T.; Writing—review & editing, C.-A.X. and T.Y.; Project administration, T.Y.; Funding acquisition, T.Y. All authors have read and agreed to the published version of the manuscript.

Funding: This work was supported by the National Natural Science Foundation of China (22178138, 21808070), Guangdong Basic and Applied Basic Research Foundation (2020B1515120099).

Institutional Review Board Statement: Not applicable.

Data Availability Statement: Data are contained within the article.

Conflicts of Interest: The authors declare no conflicts of interest.

References

1. Kang, S.Y.; Ji, Z.; Tseng, L.F.; Turner, S.A.; Villanueva, D.A.; Johnson, R.; Albano, A.; Langer, R. Design and Synthesis of Waterborne Polyurethanes. *Adv. Mater.* **2018**, *30*, e1706237. [CrossRef]
2. Shen, Y.; Liu, J.; Li, Z.; Luo, J.; Wang, S.; Tang, J.; Wang, P.; Wang, D.; Wang, X.; Hu, X.; et al. UV-curable organic silicone grafting modified waterborne polyurethane acrylate: Preparation and properties. *Int. J. Adhes. Adhes.* **2024**, *129*, 103583. [CrossRef]
3. Xiao, Y.; Fu, X.; Zhang, Y.; Liu, Z.; Jiang, L.; Lei, J. Preparation of waterborne polyurethanes based on the organic solvent-free process. *Green Chem.* **2016**, *18*, 412–416. [CrossRef]
4. Wang, J.; Zhang, H.; Miao, Y.; Qiao, L.; Wang, X. A whole-procedure solvent-free route to CO_2-based waterborne polyurethane by an elevated-temperature dispersing strategy. *Green Chem.* **2017**, *19*, 2194–2200. [CrossRef]
5. Zhang, C.; Wang, H.; Zhou, Q. Waterborne isocyanate-free polyurethane epoxy hybrid coatings synthesized from sustainable fatty acid diamine. *Green Chem.* **2020**, *22*, 1329–1337. [CrossRef]
6. Zou, C.; Zhang, H.; Qiao, L.; Wang, X.; Wang, F. Near neutral waterborne cationic polyurethane from CO_2-polyol, a compatible binder to aqueous conducting polyaniline for eco-friendly anti-corrosion purposes. *Green Chem.* **2020**, *22*, 7823–7831. [CrossRef]
7. Peng, Z.; Zhu, A. The novel preparation of waterborne acrylic polyurethane-silica organic-inorganic interpenetrating network coatings. *Prog. Org. Coat.* **2024**, *187*, 108157. [CrossRef]
8. Xu, C.-A.; Qu, Z.; Lu, M.; Meng, H.; Zhan, Y.; Chen, B.; Wu, K.; Shi, J. Effect of rosin on the antibacterial activity against S.aureus and adhesion properties of UV-curable polyurethane/polysiloxane pressure-sensitive adhesive. *Colloids Surf. A Physicochem. Eng. Asp.* **2021**, *614*, 126146. [CrossRef]
9. Zhang, J.; Shang, Q.; Hu, Y.; Zhu, G.; Huang, J.; Yu, X.; Cheng, J.; Liu, C.; Chen, J.; Feng, G.; et al. Castor-oil-based UV-curable hybrid coatings with self-healing, recyclability, removability, and hydrophobicity. *Prog. Org. Coat.* **2022**, *165*, 106742. [CrossRef]
10. Patil, R.S.; Thomas, J.; Patil, M.; John, J. To shed light on the UV curable coating technology: Current state of the art and perspectives. *J. Compos. Sci.* **2023**, *7*, 513. [CrossRef]
11. Tong, J.; Xie, S.; Miao, J.-T.; Luo, J.; Liu, R. Preparation of UV-cured polyurethane-urea acrylate coatings with high hardness and toughness. *Prog. Org. Coat.* **2024**, *186*, 107969. [CrossRef]
12. Zhang, J.; Wu, Y.-M.; Zhang, H.-L.; Yan, T.-H.; Huang, Y.-Z.; Jiang, J.-X.; Tang, J.-J. Castor oil-glycerol-based waterborne polyurethane dispersions. *Prog. Org. Coat.* **2021**, *157*, 106333. [CrossRef]
13. Wu, Q.; Hu, Y.; Tang, J.; Zhang, J.; Wang, C.; Shang, Q.; Feng, G.; Liu, C.; Zhou, Y.; Lei, W. High-performance soybean-oil-based epoxy acrylate resins: "Green" synthesis and application in UV-curable coatings. *ACS Sustain. Chem. Eng.* **2018**, *6*, 8340–8349. [CrossRef]
14. Yang, X.; Li, S.; Xia, J.; Song, J.; Huang, K.; Li, M. Novel renewable resource-based UV-curable copolymers derived from myrcene and tung oil: Preparation, characterization and properties. *Ind. Crops Prod.* **2015**, *63*, 17–25. [CrossRef]
15. Huang, Y.; Pang, L.; Wang, H.; Zhong, R.; Zeng, Z.; Yang, J. Synthesis and properties of UV-curable tung oil based resins via modification of Diels–Alder reaction, nonisocyanate polyurethane and acrylates. *Prog. Org. Coat.* **2013**, *76*, 654–661. [CrossRef]
16. Huang, J.; Zhang, J.; Zhu, G.; Yu, X.; Hu, Y.; Shang, Q.; Chen, J.; Hu, L.; Zhou, Y.; Liu, C. Self-healing, high-performance, and high-biobased-content UV-curable coatings derived from rubber seed oil and itaconic acid. *Prog. Org. Coat.* **2021**, *159*, 106391. [CrossRef]
17. Liu, R.; Luo, J.; Ariyasivam, S.; Liu, X.; Chen, Z. High biocontent natural plant oil based UV-curable branched oligomers. *Prog. Org. Coat.* **2017**, *105*, 143–148. [CrossRef]
18. Su, Y.; Zhang, S.; Chen, Y.; Yuan, T.; Yang, Z. One-step synthesis of novel renewable multi-functional linseed oil-based acrylate prepolymers and its application in UV-curable coatings. *Prog. Org. Coat.* **2020**, *148*, 105820. [CrossRef]
19. Pezzana, L.; Wolff, R.; Stampfl, J.; Liska, R.; Sangermano, M. High temperature vat photopolymerization 3D printing of fully bio-based composites: Green vegetable oil epoxy matrix & bio-derived filler powder. *Addit. Manuf.* **2024**, *79*, 103929.

20. Tran, K.; Liu, Y.; Soleimani, M.; Lucas, F.; Winnik, M.A. Waterborne 2-component polyurethane coatings based on acrylic polyols with secondary alcohols. *Prog. Org. Coat.* **2024**, *190*, 108374. [CrossRef]
21. Wang, D.; Huang, Z.; Shi, S.; Ren, J.; Dong, X. Environmentally friendly plant-based waterborne polyurethane for hydrophobic and heat-resistant films. *J. Appl. Polym. Sci.* **2022**, *139*, e52437. [CrossRef]
22. Zhang, C.; Garrison, T.F.; Madbouly, S.A.; Kessler, M.R. Recent advances in vegetable oil-based polymers and their composites. *Prog. Polym. Sci.* **2017**, *71*, 91–143. [CrossRef]
23. Wang, L.; Xiang, J.; Wang, S.; Sun, Z.; Wen, J.; Li, J.; Zheng, Z.; Fan, H. Synthesis of oleic-based primary glycol with high molecular weight for bio-based waterborne polyurethane. *Ind. Crops Prod.* **2022**, *176*, 114276. [CrossRef]
24. Li, X.; Wang, D.; Zhao, L.; Hou, X.; Liu, L.; Feng, B.; Li, M.; Zheng, P.; Zhao, X.; Wei, S. UV LED curable epoxy soybean-oil-based waterborne PUA resin for wood coatings. *Prog. Org. Coat.* **2021**, *151*, 105942. [CrossRef]
25. Gaddam, S.K.; Kutcherlapati, S.N.R.; Palanisamy, A. Self-cross-linkable anionic waterborne polyurethane–silanol dispersions from cottonseed-oil-based phosphorylated polyol as ionic soft segment. *ACS Sustain. Chem. Eng.* **2017**, *5*, 6447–6455. [CrossRef]
26. Hermens, J.G.H.; Jensma, A.; Feringa, B.L. Highly efficient biobased synthesis of acrylic acid. *Angew. Chem. Int. Ed.* **2021**, *61*, e202112618. [CrossRef] [PubMed]
27. Liang, B.; Kuang, S.; Huang, J.; Man, L.; Yang, Z.; Yuan, T. Synthesis and characterization of novel renewable tung oil-based UV-curable active monomers and bio-based copolymers. *Prog. Org. Coat.* **2019**, *129*, 116–124. [CrossRef]
28. Mistri, E.; Routh, S.; Ray, D.; Sahoo, S.; Misra, M. Green composites from maleated castor oil and jute fibres. *Ind. Crops Prod.* **2011**, *34*, 900–906. [CrossRef]
29. Lei, W.; Zhou, X.; Fang, C.; Song, Y.; Li, Y. Eco-friendly waterborne polyurethane reinforced with cellulose nanocrystal from office waste paper by two different methods. *Carbohydr. Polym.* **2019**, *209*, 299–309. [CrossRef]
30. Man, L.; Feng, Y.; Hu, Y.; Yuan, T.; Yang, Z. A renewable and multifunctional eco-friendly coating from novel tung oil-based cationic waterborne polyurethane dispersions. *J. Clean. Prod.* **2019**, *241*, 118341. [CrossRef]
31. Hong, C.; Zhou, X.; Ye, Y.; Li, W. Synthesis and characterization of UV-curable waterborne Polyurethane–acrylate modified with hydroxyl-terminated polydimethylsiloxane: UV-cured film with excellent water resistance. *Prog. Org. Coat.* **2021**, *156*, 106251. [CrossRef]
32. Liang, H.; Lu, Q.; Liu, M.; Ou, R.; Wang, Q.; Quirino, R.L.; Luo, Y.; Zhang, C. UV absorption; anticorrosion, and long-term antibacterial performance of vegetable oil based cationic waterborne polyurethanes enabled by amino acids. *Chem. Eng. J.* **2021**, *421*, 127774. [CrossRef]
33. Chandra, S.; Karak, N. Environmentally friendly polyurethane dispersion derived from dimer acid and citric acid. *ACS Sustain. Chem. Eng.* **2018**, *6*, 16412–16423. [CrossRef]
34. Lin, S.; Huang, J.; Chang, P.R.; Wei, S.; Xu, Y.; Zhang, Q. Structure and mechanical properties of new biomass-based nanocomposite: Castor oil-based polyurethane reinforced with acetylated cellulose nanocrystal. *Carbohydr. Polym.* **2013**, *95*, 91–99. [CrossRef] [PubMed]
35. Bai, C.Y.; Zhang, X.Y.; Dai, J.B.; Zhang, C.Y. Water resistance of the membranes for UV curable waterborne polyurethane dispersions. *Prog. Org. Coat.* **2007**, *59*, 331–336. [CrossRef]
36. Dai, Z.; Jiang, P.; Lou, W.; Zhang, P.; Bao, Y.; Gao, X.; Xia, J.; Haryono, A. Preparation of degradable vegetable oil-based waterborne polyurethane with tunable mechanical and thermal properties. *Eur. Polym. J.* **2020**, *139*, 109994. [CrossRef]

Disclaimer/Publisher's Note: The statements, opinions and data contained in all publications are solely those of the individual author(s) and contributor(s) and not of MDPI and/or the editor(s). MDPI and/or the editor(s) disclaim responsibility for any injury to people or property resulting from any ideas, methods, instructions or products referred to in the content.

Article

Development and Application of a Lignin-Based Polyol for Sustainable Reactive Polyurethane Adhesives Synthesis

Víctor M. Serrano-Martínez [1,*], Carlota Hernández-Fernández [1,*], Henoc Pérez-Aguilar [1], María Pilar Carbonell-Blasco [1], Avelina García-García [2] and Elena Orgilés-Calpena [1]

[1] Footwear Technology Centre, Campo Alto Campground, 03600 Alicante, Spain; pcarbonell@inescop.es (M.P.C.-B.); eorgiles@inescop.es (E.O.-C.)
[2] MCMA Group, Department of Inorganic Chemistry and Institute of Materials, University of Alicante, Ap. 99, 03080 Alicante, Spain; a.garcia@ua.es
* Correspondence: vmserrano@inescop.es (V.M.S.-M.); chernandez@inescop.es (C.H.-F.); Tel.: +34-965-395-213 (V.M.S.-M. & C.H.-F.)

Abstract: In response to the environmental impacts of conventional polyurethane adhesives derived from fossil fuels, this study introduces a sustainable alternative utilizing lignin-based polyols extracted from rice straw through a process developed at INESCOP. This research explores the partial substitution of traditional polyols with lignin-based equivalents in the synthesis of reactive hot melt polyurethane adhesives (HMPUR) for the footwear industry. The performance of these eco-friendly adhesives was rigorously assessed through Thermogravimetric Analysis (TGA), Differential Scanning Calorimetry (DSC), rheological analysis, and T-peel tests to ensure their compliance with relevant industry standards. Preliminary results demonstrate that lignin-based polyols can effectively replace a significant portion of fossil-derived polyols, maintaining essential adhesive properties and marking a significant step towards more sustainable adhesive solutions. This study not only highlights the potential of lignin in the realm of sustainable adhesive production but also emphasises the valorisation of agricultural by-products, thus aligning with the principles of green chemistry and sustainability objectives in the polymer industry.

Keywords: polyurethane adhesives; lignin-based polyols; sustainable synthesis; rice straw valorisation; hot melt adhesives (HMPUR); biopolymers; renewable resources

Citation: Serrano-Martínez, V.M.; Hernández-Fernández, C.; Pérez-Aguilar, H.; Carbonell-Blasco, M.P.; García-García, A.; Orgilés-Calpena, E. Development and Application of a Lignin-Based Polyol for Sustainable Reactive Polyurethane Adhesives Synthesis. *Polymers* **2024**, *16*, 1928. https://doi.org/10.3390/polym16131928

Academic Editors: Zhuohong Yang and Chang-An Xu

Received: 7 June 2024
Revised: 1 July 2024
Accepted: 3 July 2024
Published: 6 July 2024

Copyright: © 2024 by the authors. Licensee MDPI, Basel, Switzerland. This article is an open access article distributed under the terms and conditions of the Creative Commons Attribution (CC BY) license (https://creativecommons.org/licenses/by/4.0/).

1. Introduction

Traditional polyurethane adhesives are prevalent in various industries, including footwear, and are predominantly synthesised from polyols and isocyanates derived from fossil sources. These adhesives are non-renewable, contributing to a significant environmental impact due to their carbon footprint and sustainability issues [1–3]. The reliance on petrochemical sources to produce polyurethane highlights the urgent need for sustainable alternatives in adhesive synthesis, driven by global initiatives and regulations aimed at reducing environmental harm and promoting sustainability [4–7]. The environmental impacts of traditional polyurethane adhesives stem from their petrochemical origin, which is associated with greenhouse gas emissions and depletion of non-renewable resources. This has led to a growing interest in developing greener adhesives that minimise the use or generation of hazardous substances, decrease environmental impacts, and reduce energy consumption throughout their life cycle [8]. Efforts to develop more sustainable adhesives have included the exploration of bio-based polyols, such as those derived from lignin, vegetable oils, and other renewable resources; however, these efforts face challenges, including the need to match or exceed the performance of traditional adhesives and to ensure the economic viability of the greener alternatives [9–11]. Lignin-based polyols represent a promising avenue for adhesive innovation, due to lignin's abundance as a by-product of the pulp and paper industry and its unique properties. Lignin is a natural aromatic polymer rich in functional hydroxyl groups, making it an attractive bio-based alternative to fossil-based polyols for the synthesis of polyurethanes. Despite its potential, lignin remains under-utilised in industrial applications, partly due to challenges in processing and modifying lignin to achieve the desired reactivity and performance in adhesive formulations [12–14]. Previous works have demonstrated the viability of lignin-based polyols in producing adhesives with comparable or even superior properties to their petroleum-based counterparts, addressing both performance and environmental sustainability concerns [5]. The exploration of lignin in adhesive innovation is not new, but its application has gained momentum due to the pressing need for sustainable material sources. The development of lignin-based polyols for PU adhesives represents a significant step towards reducing the environmental impact of adhesives, thus aligning with global sustainability goals and addressing the urgent need for renewable and less harmful materials in the adhesive industry [11].

Expanding on these developments, reactive polyurethane hot melt adhesives (HMPUR) stand out as a versatile and robust option that is gaining traction across various industrial and commercial sectors. At room temperature, these adhesives maintain a solid state and undergo curing upon exposure to moisture, which is facilitated by the reaction between the polyols and an excess of diisocyanates. This characteristic not only accelerates the setting process but also enhances the bond strength, making HMPUR ideal for applications demanding quick and durable adhesion. Adopting bio-based polyols, such as those derived from lignin, further reduces the environmental burden of these adhesives, supporting the transition towards more sustainable manufacturing practices. In addition to utilizing bio-based resources, recent research has also emphasised the integration of recycled polyols into HMPUR formulations. Existing studies have explored the feasibility of using bio-based polyols from vegetable sources, confirming their capability to meet the stringent quality standards for footwear applications while integrating a higher percentage of sustainable materials into adhesive formulations [14]. Moreover, a broader review of sustainable practices highlighted that materials such as lignin provide a significant hydroxyl functionality, which is crucial for polyurethane production, fostering the move towards a more circular economy through reducing reliance on petrochemicals [15]. Such advancements underscore the potential for integrating innovative and environmentally friendly materials into the industries traditionally dominated by petrochemical products.

Building upon this background, our previous study specifically introduced a novel process developed at INESCOP, employing a steam explosion method to extract lignin from rice straw waste—a significantly abundant agricultural by-product. This process is not only innovative, but also critical in addressing the urgent need for more sustainable waste management practices. The extracted lignin will be then transformed into polyols for the synthesis of HMPUR, particularly for use in the footwear industry. Our previous work demonstrated the feasibility of this method, which significantly enhances the accessibility and reactivity of lignin, facilitating its incorporation into polyol formulations with improved performance characteristics [16].

The objectives of this study are multifaceted. We aim to demonstrate that lignin-based polyols can effectively replace a substantial portion of fossil-based polyols without compromising the essential properties required for high-quality adhesives. This research also aligns with broader sustainability and green chemistry goals through valorising agricultural waste and reducing reliance on non-renewable resources [17–20]. Furthermore, the potential applications of these bio-based polyols in creating adhesives that meet industry standards for performance and environmental impact are explored.

In conclusion, the application of lignin-based polyols in the synthesis of reactive polyurethane adhesives presents a tangible step towards more sustainable material production. Through rigorous assessments, including Thermogravimetric Analysis (TGA), Differential Scanning Calorimetry (DSC), rheological analysis, and T-peel tests, this study aims to establish a new benchmark for the adhesive industry, contributing to its evolution towards sustainability.

2. Materials and Methods

2.1. Materials

For the synthesis of the lignin-based polyol, pure absolute ethanol (Panreac Appliquem, Barcelona, Spain; purity ≥ 99.8%, CAS 64-17-5), ethyl acetate (Quimidroga SA, Barcelona, Spain; CAS 141-78-6), and hydrochloric acid solution (ITW Reagents, Barcelona, Spain; 35%, CAS 7647-01-0) were used.

Regarding the synthesis of adhesives, conventional polyols as well as the more sustainable polyol obtained from lignin were used. This lignin-based polyol is denoted as LIGNOC (Mw = 929 g/mol, IOH = 182.25 mg kOH/g), which was developed at INESCOP; its properties were measured at the University of the Basque Country UPV/EHU. The 1,4-butanediol polyadipate (Hoopol F-580 Synthesia Technology, Barcelona, Spain; Mw = 3000 g/mol, IOH = 37–40 mgKOH/g) and polypropylene glycol (Quimidroga SA, Barcelona, Spain; Mw = 425 g/mol, IOH = 250–270 mgKOH/g) were used as polyols from fossil resources, and 4,4-diphenylmethane diisocyanate (MDI) (98% purity, Sigma Aldrich, Barcelona, Spain) was used as isocyanate.

2.2. Lignin-Based Polyol Synthesis Method from Rice Straw

Building on our prior work, this study utilised lignin derived from the supernatant of a steam explosion process designed for cellulose extraction from rice straw, as comprehensively detailed in [16]. This foundational process involves treating rice straw at 200 °C to facilitate the separation of cellulose and lignin, capturing the lignin in the supernatant.

For the current study, 20 g of this lignin (step I) was dissolved in a 50/50 mixture of ethanol and water (60 mL of each) to promote solubilisation. The lignin raw material and subsequent processing steps are illustrated in Figure 1. After mixing, the lignin solution was subjected to organosolvent fractionation (step II) at 200 °C for 75 min to further break down its complex structure, called fractionation. The treated mixture was then vacuum filtered to separate the liquid fraction (black liquor) from the solids. The filtrate was processed through ultrasonication for one hour (step III), utilizing 35% of the power and 10 s on/off cycles, which helped to further degrade the lignin into smaller components. This step was crucial for preparing the lignin for subsequent chemical manipulation [21–23].

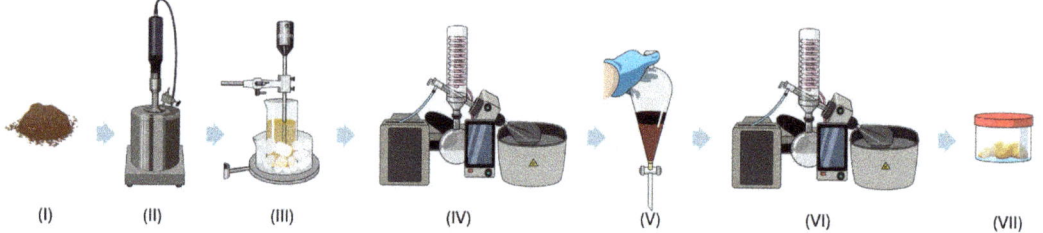

Figure 1. Lignin-based polyol synthesis process.

The black liquor was then subjected to rotary evaporation (step IV) in order to remove the ethanol, thus concentrating the lignin derivatives. The pH of this concentrated solution was adjusted to 2.5 using 6M hydrochloric acid. To enhance the separation of lignin derivatives, 80 mL of ethyl acetate was added. Phase separation was conducted using a separatory funnel (step V), where the ethyl acetate layer containing the lignin-derived polyol was carefully isolated.

The isolated ethyl acetate phase underwent a second rotary evaporation (step VI) to remove any residual solvent. The resultant polyol (step VII) was then purified by agitation in a washing bath to eliminate traces of ethyl acetate, thus ensuring its purity [24,25].

This refined method transforms lignin from rice straw into a high-value polyol suitable for producing HMPUR. The approach not only leverages rice straw, an agricultural by-product, as a feedstock for advanced material applications, but also exemplifies the scalability of our initial extraction process.

2.3. Reactive Polyurethane Hot Melt Adhesives (HMPURs) Synthesis Process

The synthesis process for reactive hot melt polyurethane adhesives can be affected by several factors, such as the molecular weight of the polyols used, the functionality of both polyols and isocyanate, or the stoichiometric ratio between the raw materials (NCO/OH).

Reactive polyurethane hot melt adhesives were synthesised using the prepolymer method [1,26], with an optimal NCO/OH index of 1.5.

The procedure followed for the synthesis and application of the adhesives is shown in Figure 2. First, the polyols were melted and mixed at 90 °C in a glass reactor jacketed with thermal oil (step VIII), subjected to constant mechanical stirring of 300 rpm (Heidolph RZR 2021, Kel-heim, Germany) in a nitrogen atmosphere, according to previous research [14]. The isocyanate was then added. The progress of the reaction was monitored by determining the percentage of free NCO through the dibutylamine titration method [27]. Once the desired percentage of free isocyanate had been reached (Table 1), the reaction was stopped and the HMPUR adhesive was transferred in a melted state to a hermetically sealed cartridge, where the reaction continued for 18 h (annealing process). The airtight cartridge allowed for the subsequent application of the adhesive by means of a manual heat gun (step IX).

(VIII)   (IX)

Figure 2. HMPURs synthesis process (VIII) and application of the adhesive (IX).

Table 1. Nomenclature and formulation of synthesised adhesives with PPG and lignin-based polyol percentages and desired free NCO%.

Formulation	Lignin-Based Polyol wt.%	PPG wt.%	Desired Free NCO (%)
HMPUR-REF	-	100.0	3.75
LIGNOC-2.5%	2.5	97.5	3.70
LIGNOC-5%	5	95.0	3.67
LIGNOC-7.5%	7.5	92.5	3.63

In the synthesis of HMPUR, the reference formula employed an equal mixture of 1,4-butanediol polyadipate and polypropylene glycol (PPG). A key component of this study involves conducting multiple syntheses in which this PPG is partially substituted with the polyol synthesised from rice straw in proportions of 2.5%, 5%, and 7.5%, as detailed in Table 1.

2.4. Characterisation Techniques

2.4.1. Fourier Transform Infrared Spectroscopy (FTIR)

The chemical structure was studied using a Varian 660-IR (Varian Australia PTY LTD, Mulgrave, Australia) with a diamond prism. A total of 16 scans were averaged, with a resolution of 4 cm^{-1}, the range analysed being from 500 to 4000 cm^{-1} with attenuated total reflection (ATR) technology [28].

2.4.2. Thermogravimetric Analysis (TGA)

The thermal stability was evaluated using a TGA 2 STARe System thermal balance, equipped with the STARe v16.4 software from Mettler-Toledo, Switzerland. A sample size of approximately 7 to 10 mg of the adhesive was placed into an alumina crucible. The sample was then heated from 30 to 600 °C at a rate of 10 °C/min under an inert nitrogen atmosphere, with a nitrogen flow rate of 30 mL/min [29,30].

2.4.3. Differential Scanning Calorimetry (DSC)

The thermal behaviour was studied using a DSC3 + STARe Systems calorimeter from Mettler-Toledo AG, Schwerzenbach, Switzerland. The experiments used samples weighing between 9 and 12 mg placed in aluminium pans, and were carried out in an inert nitrogen atmosphere with a flow rate of 30 mL/min and a temperature change rate of 10 °C/min. The analysis included two sequential runs: (i) Initial heating from −15 °C to 110 °C, followed by isothermal maintenance at 110 °C for three minutes to remove the sample's thermal history, and (ii) a second heating phase from −65 °C to 100 °C, succeeded by isothermal cooling at 25 °C for 45 min. The optimal conditions for these DSC experiments were previously determined by the authors in earlier research [31].

2.4.4. Rheological Analysis

The viscoelastic properties were determined on a Kinexus Pro+ rheometer (Malvern Panalytical, Malvern, UK). A 25 mm diameter aluminium top plate was used, on which 50 mL of the sample was placed.

For the polyol, a gap of 0.1 mm was set between the plates, and the tests were carried out with a shear rate sweep from 0.01 to 1000 s^{-1} (50 °C). For the adhesives, a gap of 0.4 mm was set, and a cooling sweep was performed from 160 to 25 °C, maintaining a cooling rate of 5 °C/min and using a constant frequency of 1 Hz at a deformation equal to 1%.

2.4.5. T-Peel Strength Test

The adhesion properties were assessed using the method outlined in the EN 1392:2007 standard [32]. For this evaluation, in addition to the reference adhesive and the newly synthesised adhesives, reference materials commonly used in footwear were used as substrates for the adhesive bonds. A medium-hard vulcanised styrene–butadiene rubber SBR-2 was used as the reference sole material, while chrome-tanned split leather was used as the upper reference material. Both materials were supplied by Proyección Europlan XXI S.L. and were die-cut into 150 × 30 mm specimens.

Prior to forming the joints, each material underwent specific surface treatments. The split leather samples were roughened using a P100 aluminium oxide abrasive cloth (Due Emme Abrasivi, Pavia, Italy) at 2800 rpm on a roughing machine from Superlema S.A. (Zaragoza, Spain). The SBR-2 rubber was roughened and then halogenated using a 2 wt% trichloroisocyanuric acid solution in ethyl acetate. The upper to sole adhesive joints, measuring 150 × 30 mm, were then prepared by bonding leather/HMPUR/SBR rubber.

The joint formation process began 30 min after applying the adhesive, where both adhesive films were activated by exposing them to infrared radiation at 80 °C in a CAN 02/01 temperature-controlled heater from AC&N (Elda, Spain). Immediately after, the materials were pressed together under a pressure of 1.8 bar for 10 s to ensure a strong bond. The adhesive joints were then conditioned at 23 °C and 50% relative humidity for 72 h. T-peel strength was finally evaluated using an Instron 34TM-10 universal testing machine (Instron Ltd., Buckinghamshire, UK) at a cross-head speed of 100 mm/min [33]. The bond strength values were calculated based on the width and scale of the test pieces, leading to an average and typical deviation calculated from five samples per joint. Failure modes were assessed according to ISO standard 17708 [34].

3. Results and Discussion

Throughout this section, the experimental results obtained are precisely described, and their interpretation and discussion are presented. With reference to the characterisation, the lignin polyol obtained was analysed through DSC, TGA, and FTIR, while the adhesives were characterised using FTIR, DSC, TGA, rheology, and T-peel strength tests.

3.1. Polyol Characterisation

After the polyol synthesis, an experimental characterisation of the product was carried out. The chemical composition was first studied using FTIR, with the spectra obtained shown in Figure 3.

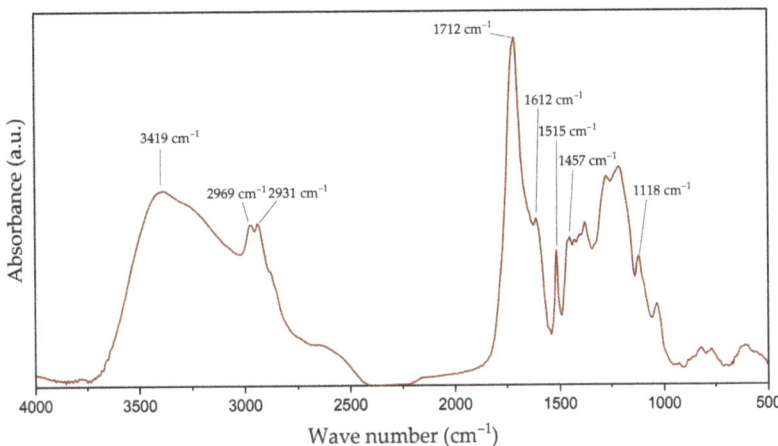

Figure 3. Infrared spectra of the polyol based on lignin obtained from rice straw.

The FTIR-ATR spectra illustrated in Figure 3 reveal distinct absorption bands that are indicative of various functional groups inherent to the lignin structure and the modifications involved in polyol synthesis. The spectrum displayed a broad absorption band in the region of 3500–3300 cm^{-1}, related to the stretching vibration O-H of the hydroxyl group, common in polyols. Furthermore, 2969 and 2931 cm^{-1} peaks were assigned to the vibrations of aliphatic C-H bonds of methylene and methyl groups, indicating the presence of saturated hydrocarbon chains [35].

The absorption peak at 1712 cm^{-1} was identified as corresponding to carbonyl (C=O) stretching in unconjugated ketonic or ester groups, which are often introduced in lignin during the oxidative modification processes [36].

Finally, the phenylpropane skeleton of lignin—a fundamental component of its macromolecular structure—was evidenced by sharp and distinct bands at 1612, 1515, and 1457 cm^{-1}. These bands are typically assigned to the vibrations of C=C bonds within the aromatic rings, confirming the retention of aromatic structures within the polyol. Furthermore, the stretching vibrations of C-O bonds with a specific peak at 1118 cm^{-1} is associated with C-O in secondary alcohols and aliphatic ethers [23]. The FTIR-ATR spectral data corroborate the chemical structure of the lignin-derived polyol, providing insight into the functional groups that contribute to its application potential in various polymeric and material science contexts.

The thermal stability of the polyol was then evaluated through TGA. Figure 4a shows the weight loss obtained, and its derivate is shown in Figure 4b; the results are collected in Table 2.

Table 2. Decomposition temperatures and weight loss data obtained from TGA for the developed lignin-based polyol.

Sample	T1 (°C)	Δw1 (wt%)	T2 (°C)	Δw2 (wt%)	Residue (%)
Lignin-based polyol LIGNOC	140	25	302	37	28

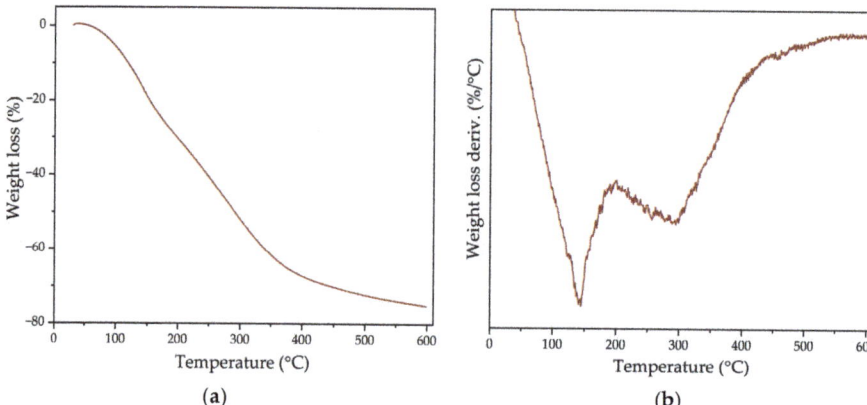

Figure 4. TGA results for the developed lignin-based polyol: (**a**) weight loss; (**b**) weight loss derivate.

Regarding the weight loss derivate, it is possible to understand how the thermal degradation takes place in two different areas. The first area—between 100–200 °C—shows a maximum degradation peak at 140 °C, which could be assigned to the most volatile compounds present in the polyol. Meanwhile, the second degradation area appears between 200–500 °C, with a maximum peak at 302 °C associated with the degradation of lignin, as well as other oligomers present in the structure of the polyol [21].

In this way, the thermal properties of the polyol were also evaluated by DSC. The DSC curve obtained in the second heating sweep is shown in Figure 5, from which it can be observed that the polyol presents an amorphous state, with a single glass transition temperature of −40 °C [37].

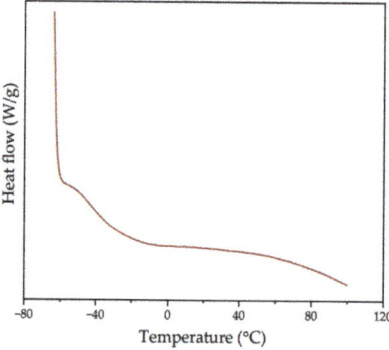

Figure 5. DSC curve of the lignin-based polyol during the second heating sweep.

Finally, plate–plate rheometry was used to determine the viscosity of the polyol—a parameter that represents its flow resistance. Figure 6 shows the result obtained in the shear rate sweep, from which it can be observed that the polyol presents a non-Newtonian pseudoplastic behaviour, decreasing its viscosity as the shear stress increases [38–40].

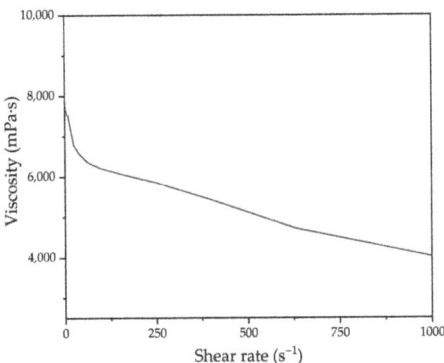

Figure 6. Viscosity of the lignin-based polyol obtained by plate–plate rheometric analysis.

3.2. HMPURs Synthesis and Characterisation

The reaction time and free isocyanate (%NCO) measurement are crucial aspects in the synthesis of polyurethane adhesives, as detailed in Table 3, which presents the reaction conditions. Accurately measuring the reaction time is vital as it directly influences the final properties of the adhesive, such as viscosity, strength, and flexibility. In addition, the amount of free NCO at the end of the process is a key indicator of the reactivity and complete conversion of the precursors into the final product. Monitoring these parameters allows for the optimisation of the synthesis process to obtain polyurethane adhesives with desirable characteristics, ensuring consistent, high-quality performance in the final applications.

Table 3. Obtained conditions of the synthesised adhesives.

Adhesive	Reaction Time (min)	Free Isocyanate (%NCO)
HMPUR-REF	60	3.87
LIGNOC-2.5%	26	3.85
LIGNOC-5%	16	3.68
LIGNOC-7.5%	9	3.35

The values shown in Table 3 reflect the high reactivity of the developed lignin polyol, obtaining a reduction in reaction time of more than 50% for the synthesis with the lowest substitution and reaching a reduction of 85% in the synthesis time for the highest PPG substitution. Otherwise, the final free isocyanate obtained was almost constant in all of the adhesives.

Regarding the adhesive characterisation, first, the chemical composition of the adhesives was studied using FTIR. Figure 7a shows the spectra obtained for the recently applied adhesive, from which it can be seen that the band corresponding to free -NCO appears in all cases for a wavelength of ~2250 cm^{-1}. This band is characteristic of uncured polyurethane adhesives, and it was observed that the incorporation of the lignin-based polyol did not produce the appearance of any new bands in the spectra analysed, when compared with the reference HMPUR. Figure 7b shows the spectra obtained after 24 h of adhesive application, with a decrease in the band corresponding to the free -NCO in the case of adhesives with lignin-based polyols, corresponding to the consumption of free -NCO during the curing reaction with ambient moisture and, thus, showing much faster curing compared with the reference adhesive.

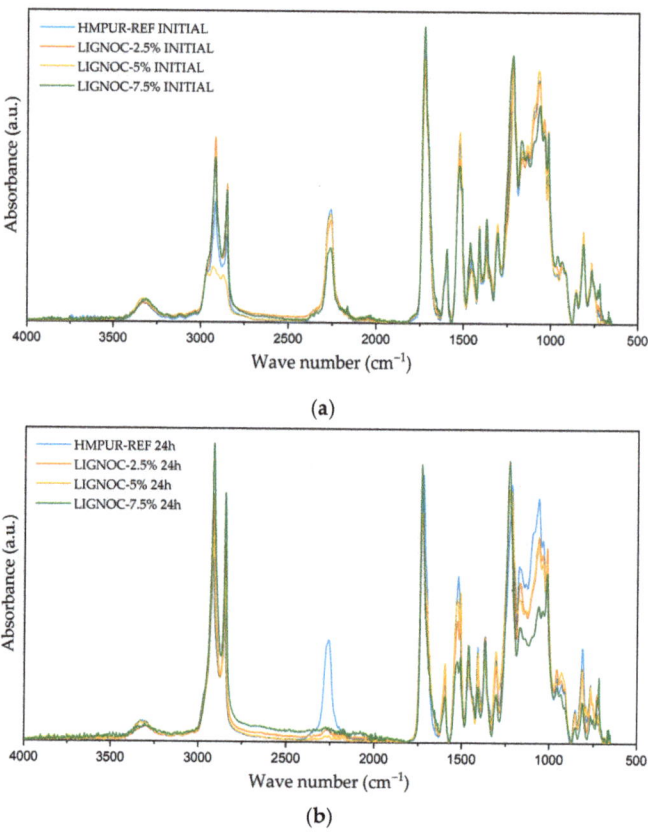

Figure 7. Comparison of the infrared spectra of applied adhesives: (**a**) infrared spectrum obtained immediately after application; and (**b**) infrared spectrum obtained after 24 h from application.

The thermal stability of the adhesives was then evaluated through TGA. Figure 8a,b show the comparison of the mass loss obtained for each sample and its derivate, respectively. All of the numerical results are collected in Table 4.

The data collected in Table 4 suggest that the addition of lignin-based polyol to the adhesives did not affect the thermal stability of the formulations with lower concentrations (LIGNOC-2.5% and LIGNOC-5%), maintaining thermal decomposition parameters comparable to the reference adhesive. However, the formulation with the highest substitution (LIGNOC-7.5%) demonstrated a marked improvement in thermal stability, evidenced by an increase in the first decomposition temperature, as well as a higher residue content. This increase in thermal stability and post-decomposition residue percentage can be attributed to the aromatic structure of lignin and its inherent thermal resistance, which provides greater structural integrity to the adhesive under high-temperature conditions [39,41,42]. Furthermore, observing Figure 8b, it is possible to verify that the ratio between hard segments—coming from isocyanate and low molecular weight polyols (decomposition above 338–400 °C)—and soft segments—coming from high molecular weight polyols (decomposition between 400–416 °C)—was maintained.

Table 4. Decomposition temperatures and weight loss data obtained from TGA for the HMPUR adhesives.

Adhesive	T1 (°C)	Δw1 (wt%)	T2 (°C)	Δw2 (wt%)	T3 (°C)	Δw3 (wt%)	Residue (%)
HMPUR-REF	334	62	399	33	-	-	7
LIGNOC-2.5%	333	60	400	34	-	-	7
LIGNOC-5%	334	58	399	34	-	-	9
LIGNOC-7.5%	341	50	375	32	438	6	12

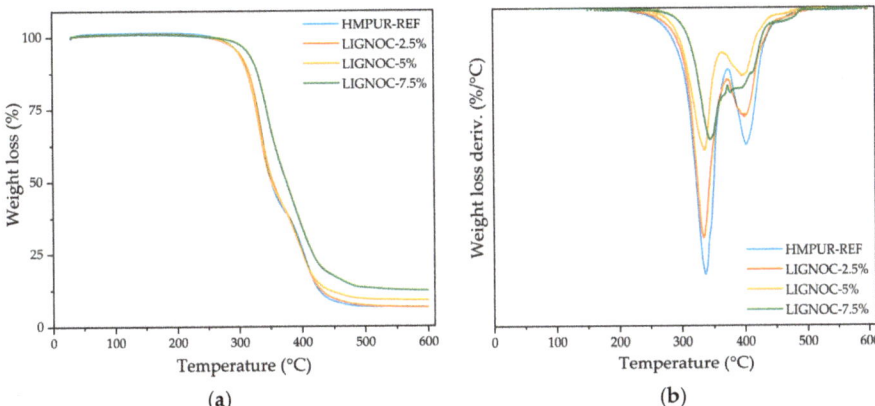

Figure 8. Comparison of the TGA results for the adhesives: (**a**) comparison of the samples' weight loss; and (**b**) comparison of the samples' weight loss derivate.

Thus, a slower decomposition rate is observed as the concentration of incorporated lignin-based polyol increases, suggesting that a higher content of it in the soft segments promotes a more stable matrix at high temperatures, which is critical for applications demanding high heat resistance in polyurethane adhesives [43,44].

Thereafter, the thermal behaviour of the adhesives was evaluated using DSC. Figure 9 shows the DSC curve corresponding to the second heating sweep, obtained after eliminating the thermal history of the adhesives during the first temperature sweep. The obtained glass transition temperature (T_g) values are included in Table 5.

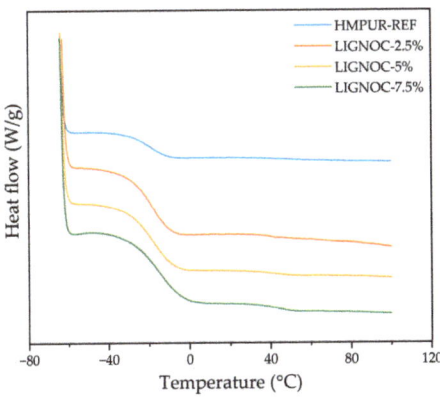

Figure 9. Comparison of the DSC curves of the HMPUR adhesives during the second heating sweep.

Table 5. Glass transition temperatures obtained from DSC thermograms of the HMPUR adhesives.

Adhesive	T_g (°C)
HMPUR-REF	−18.86
LIGNOC-2.5%	−17.02
LIGNOC-5%	−18.06
LIGNOC-7.5%	−14.21

The reference adhesive shows a glass transition temperature around −19 °C, while the substitution of polypropylene glycol with the lignin-based polyol resulted in an increase in Tg for all cases. These results imply that even a small amount of lignin-based polyol acts to stiffen the adhesive matrix. This effect becomes more pronounced with higher concentrations of lignin-based polyol, observing that it contributes significantly to the rigidity of the adhesive, increasing the glass transition temperature even more at the highest substitution. Overall, these observations indicate that the polyol used predominantly influences the amorphous regions of the polymer, leading to progressively higher glass transition temperatures as the concentration of lignin increases, potentially due to increased interactions between polymer chains or alterations in the polymer structure introduced by the lignin-based polyol [7,45].

The complex viscosity of the adhesives was also evaluated through plate–plate rheology. As shown in Figure 10, in all cases, the complex viscosities of the new adhesives were higher than that of the reference adhesive.

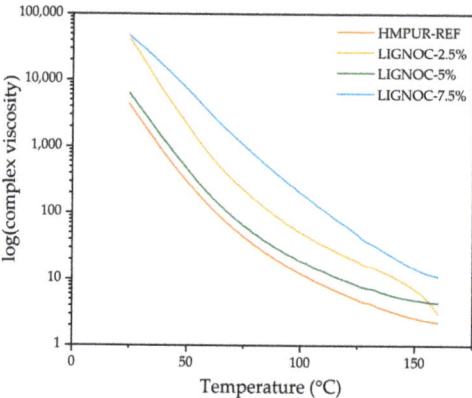

Figure 10. Complex viscosity of the HMPUR adhesives obtained through plate–plate rheometric analysis.

From Figure 10, it is possible to notice how, in all cases, the complex viscosity decreased logarithmically with temperature, which is typical for polymers as their mobility increases [46,47]. The modification of the reference adhesive with different concentrations of lignin-based polyol increased the complex viscosity in all cases for the temperature range studied. This increase can be attributed to a higher rate of chain interactions due to the lignin-based polyol restricting the mobility of the polymer. It should be pointed out that the LIGNOC-7.5% adhesive had a markedly higher viscosity than the other adhesives throughout the entire range, which could result in difficulties in application using a heated manual gun at lower temperatures, where the difference in viscosities is higher.

Finally, T-peel tests were performed to evaluate the adhesion in upper to sole split/adhesive HMPUR/SBR rubber joints—the most demanding test in the footwear sector. Figure 11 shows the results obtained after the peel test.

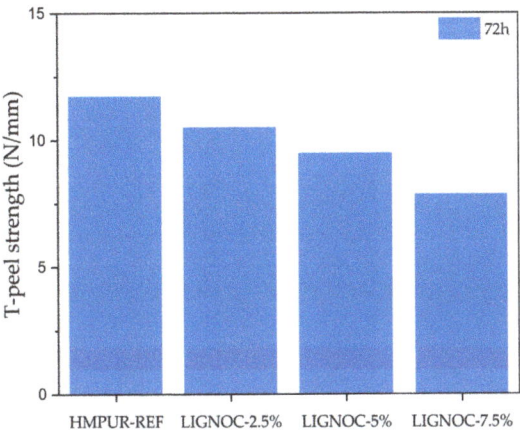

Figure 11. T-peel test results after 72 h from the bonding time.

All adhesives showed a lower adhesive strength than the reference adhesive. However, all of them met the minimum requirements for footwear, according to the standard EN 15307:2015 [48], which indicates a peel strength ≥3.5 N/mm for moderate-/high-demand footwear. Additionally, it is crucial to evaluate not only the peel resistance value but also the nature of the bond; specifically, the appearance of the surface when peeled off, as shown in Figure 12.

Figure 12. Appearance of samples after T-peel test. From left to right: HMPUR-REF, LIGNOC-2.5%, LIGNOC-5%, and LIGNOC-7.5%.

Typically, failures in adhesion can be categorised as either adhesion failures, where the separation occurs between two different materials, cohesive failures, where the separation happens within the same material, or a mixture of both types [14]. In this case, both bonding performed with the reference adhesive and the bio-adhesives showed adhesion failure, as determined through the visual checking of the samples after the test, where the adhesive remained on one of the substrates after the separation, concretely separating from the SBR rubber.

4. Conclusions

This study provided compelling evidence supporting the advancement of sustainable adhesive production through the use of lignin-based polyols derived from agricultural by-products, such as rice straw. The exploration and utilisation of lignin—an abundant and under-utilised natural resource—in the synthesis of HMPUR has significant environmental and industrial implications.

Our research established that lignin-based polyols can effectively replace a substantial portion of fossil-based polyols without compromising the essential adhesive properties. After successful conversion from rice straw-derived lignin to polyol, the experimental outcomes revealed that the adhesives synthesised with up to 7.5% lignin-derived polyol not only maintained comparable thermal and mechanical properties but, in some instances, showed an enhanced thermal stability and higher residue content after decomposition. This indicates an improvement in the structural integrity and resistance of the adhesives at elevated temperatures. Moreover, the integration of lignin-based polyols accelerated the curing process and increased the glass transition temperature, suggesting the potential for higher performance in applications demanding robust thermal characteristics.

The successful incorporation of lignin-based polyols in HMPUR adhesives represents a significant stride towards reducing the carbon footprint of polyurethane adhesives. Through leveraging bio-based resources, the adhesive industry can diminish its reliance on non-renewable petrochemicals, thus aligning with global sustainability goals. Furthermore, the use of agricultural waste not only helps in waste reduction but also adds value to otherwise low value by-products. The environmental benefits, coupled with the retention and even enhancement of adhesive properties, underscore the potential of lignin-based polyols as a pivotal component in the future of green chemistry.

Looking forward, it is imperative to expand the scope of research to optimise the extraction and modification processes of lignin, in order to enhance its compatibility and performance in adhesive formulations. Additionally, exploring other types of agricultural and industrial wastes as potential sources of lignin could further broaden the resource base. Potential applications could extend beyond the footwear industry, into areas such as the automotive, construction, and packaging industries, in which the demand for strong, durable, and environmentally friendly adhesives has been increasing. Moreover, advancing our understanding of the interactions between lignin-based polyols and other polymer components could lead to innovations in other types of polymers and composites, potentially opening new avenues for the use of sustainable materials in various high-performance applications.

In summary, the adoption of lignin-based polyols in the synthesis of polyurethane adhesives both supports environmental sustainability and presents an opportunity for the adhesive industry to contribute actively to a more circular economy. Continued research and development in this field are crucial to overcoming existing challenges and unlocking the full potential of bio-based polyols in commercial adhesive applications.

Author Contributions: Conceptualisation and investigation, V.M.S.-M., C.H.-F., H.P.-A. and M.P.C.-B.; methodology, V.M.S.-M., C.H.-F., H.P.-A. and M.P.C.-B.; validation and formal analysis, V.M.S.-M., C.H.-F., H.P.-A. and M.P.C.-B.; writing—original draft preparation, V.M.S.-M. and C.H.-F.; writing—review and editing, V.M.S.-M., C.H.-F., H.P.-A. and M.P.C.-B.; supervision, A.G.-G. and E.O.-C. All authors have read and agreed to the published version of the manuscript.

Funding: This research was funded by the Ministry of Science and Innovation (MCIN) and the State Research Agency (AEI) of Spain through the grant number TED2021-129932B-C22, BioTech-RICE.

Institutional Review Board Statement: Not applicable.

Informed Consent Statement: Not applicable.

Data Availability Statement: Data are contained within the article.

Conflicts of Interest: The authors declare no conflicts of interest.

References

1. Orgilés-Calpena, E.; Arán-Aís, F.; Torró-Palau, A.M.; Orgilés-Barceló, C. Novel Polyurethane Reactive Hot Melt Adhesives Based on Polycarbonate Polyols Derived from CO_2 for the Footwear Industry. *Int. J. Adhes. Adhes.* **2016**, *70*, 218–224. [CrossRef]
2. Orgilés-Calpena, E.; Arán-Aís, F.; Torró-Palau, A.M.; Montiel-Parreño, E.; Orgilés-Barceló, C. Synthesis of Polyurethanes from CO_2-Based Polyols: A Challenge for Sustainable Adhesives. *Int. J. Adhes. Adhes.* **2016**, *67*, 63–68. [CrossRef]

3. Carbonell-Blasco, M.P.; Pérez-Limiñana, M.A.; Ruzafa-Silvestre, C.; Arán-Ais, F.; Orgilés-Calpena, E. Influence of Biobased Polyol Type on the Properties of Polyurethane Hotmelt Adhesives for Footwear Joints. *Appl. Adhes. Sci.* **2021**, *9*, 8. [CrossRef]
4. Dorn, L.; Thirion, A.; Ghorbani, M.; Olaechea, L.M.; Mayer, I. Exploring Fully Biobased Adhesives: Sustainable Kraft Lignin and 5-HMF Adhesive for Particleboards. *Polymers* **2023**, *15*, 2668. [CrossRef] [PubMed]
5. Mary, A.; Blanchet, P.; Landry, V. Polyurethane Wood Adhesives from Microbrewery Spent Grains. In *Bio-Based Building Materials*; Springer: Cham, Switzerland, 2023; pp. 14–28.
6. Hernández-Ramos, F.; de Hoyos-Martínez, P.L.; Barriga, S.; Erdocia, X.; Labidi, J. Current Approaches for Polyurethane Production from Lignin. In *Biorefinery: A Sustainable Approach for the Production of Biomaterials, Biochemicals and Biofuels*; Springer Nature: Singapore, 2023; pp. 153–202.
7. Alinejad, M.; Henry, C.; Nikafshar, S.; Gondaliya, A.; Bagheri, S.; Chen, N.; Singh, S.; Hodge, D.; Nejad, M. Lignin-Based Polyurethanes: Opportunities for Bio-Based Foams, Elastomers, Coatings and Adhesives. *Polymers* **2019**, *11*, 1202. [CrossRef] [PubMed]
8. Vieira, F.R.; Gama, N.; Magina, S.; Barros-Timmons, A.; Evtuguin, D.V.; Pinto, P.C.O.R. Polyurethane Adhesives Based on Oxyalkylated Kraft Lignin. *Polymers* **2022**, *14*, 5305. [CrossRef]
9. Rao, Y.; Wan, G. Sustainable Adhesives: Bioadhesives, Chemistries, Recyclability, and Reversibility. In *Advances in Structural Adhesive Bonding*; Elsevier: Amsterdam, The Netherlands, 2023; pp. 953–985.
10. Bhakri, S.; Ghozali, M.; Cahyono, E.; Triwulandari, E.; Kartika Restu, W.; Nurfajrin Solihat, N.; Heri Iswanto, A.; Antov, P.; Savov, V.; Seng Hua, L.; et al. Development and Characterization of Eco-Friendly Non-Isocyanate Urethane Monomer from Jatropha Curcas Oil for Wood Composite Applications. *J. Renew. Mater.* **2023**, *11*, 41–59. [CrossRef]
11. Asare, M.A.; de Souza, F.M.; Gupta, R.K. Natural Resources for Polyurethanes Industries. In *Specialty Polymers*; CRC Press: Boca Raton, FL, USA, 2022; pp. 29–46.
12. Vieira, F.R.; Magina, S.; Evtuguin, D.V.; Barros-Timmons, A. Lignin as a Renewable Building Block for Sustainable Polyurethanes. *Materials* **2022**, *15*, 6182. [CrossRef] [PubMed]
13. Ma, Y.; Xiao, Y.; Zhao, Y.; Bei, Y.; Hu, L.; Zhou, Y.; Jia, P. Biomass Based Polyols and Biomass Based Polyurethane Materials as a Route towards Sustainability. *React. Funct. Polym.* **2022**, *175*, 105285. [CrossRef]
14. Blasco, M.P.C.; Limiñana, M.Á.P.; Silvestre, C.R.; Calpena, E.O.; Aís, F.A. Sustainable Reactive Polyurethane Hot Melt Adhesives Based on Vegetable Polyols for Footwear Industry. *Polymers* **2022**, *14*, 284. [CrossRef]
15. Tenorio-Alfonso, A.; Sánchez, M.C.; Franco, J.M. A Review of the Sustainable Approaches in the Production of Bio-Based Polyurethanes and Their Applications in the Adhesive Field. *J. Polym. Environ.* **2020**, *28*, 749–774. [CrossRef]
16. Serrano-Martínez, V.M.; Pérez-Aguilar, H.; Carbonell-Blasco, M.P.; Arán-Ais, F.; Orgilés-Calpena, E. Steam Explosion-Based Method for the Extraction of Cellulose and Lignin from Rice Straw Waste. *Appl. Sci.* **2024**, *14*, 2059. [CrossRef]
17. Heinrich, L.A. Future Opportunities for Bio-Based Adhesives—Advantages beyond Renewability. *Green Chem.* **2019**, *21*, 1866–1888. [CrossRef]
18. Khongchamnan, P.; Wanmolee, W.; Laosiripojana, N.; Champreda, V.; Suriyachai, N.; Kreetachat, T.; Sakulthaew, C.; Chokejaroenrat, C.; Imman, S. Solvothermal-Based Lignin Fractionation From Corn Stover: Process Optimization and Product Characteristics. *Front. Chem.* **2021**, *9*, 697237. [CrossRef] [PubMed]
19. Ouyang, X. Effect of Simultaneous Steam Explosion and Alkaline Depolymerization on Corncob Lignin and Cellulose Structure. *Chem. Biochem. Eng. Q.* **2018**, *32*, 177–189. [CrossRef]
20. Pan, C.; Liu, Z.; Bai, X.; Hui, L. Structural Changes of Lignin from Wheat Straw by Steam Explosion and Ethanol Pretreatments. *Bioresources* **2016**, *11*, 6477–6488. [CrossRef]
21. Hernández-Ramos, F.; Alriols, M.G.; Calvo-Correas, T.; Labidi, J.; Erdocia, X. Renewable Biopolyols from Residual Aqueous Phase Resulting after Lignin Precipitation. *ACS Sustain. Chem. Eng.* **2021**, *9*, 3608–3615. [CrossRef]
22. Hernández-Ramos, F.; Alriols, M.G.; Antxustegi, M.M.; Labidi, J.; Erdocia, X. Valorisation of Crude Glycerol in the Production of Liquefied Lignin Bio-Polyols for Polyurethane Formulations. *Int. J. Biol. Macromol.* **2023**, *247*, 125855. [CrossRef] [PubMed]
23. Perez-Arce, J.; Centeno-Pedrazo, A.; Labidi, J.; Ochoa-Gómez, J.R.; Garcia-Suarez, E.J. A Novel and Efficient Approach to Obtain Lignin-Based Polyols with Potential Industrial Applications. *Polym. Chem.* **2020**, *11*, 7362–7369. [CrossRef]
24. Wells, T.; Kosa, M.; Ragauskas, A.J. Polymerization of Kraft Lignin via Ultrasonication for High-Molecular-Weight Applications. *Ultrason. Sonochem.* **2013**, *20*, 1463–1469. [CrossRef]
25. Kai, D.; Tan, M.J.; Chee, P.L.; Chua, Y.K.; Yap, Y.L.; Loh, X.J. Towards Lignin-Based Functional Materials in a Sustainable World. *Green Chem.* **2016**, *18*, 1175–1200. [CrossRef]
26. Orgilés Calpena, E.; Carbonell Blasco, P.; Pérez Limiñana, M.A. Sustainable Polyurethane Adhesives for Footwear Based on an Algal Biomass Co-Product as Macroglycol. In Proceedings of the 5th International Conference on Structural Adhesive Bonding, Porto, Portugal, 11–12 July 2019; pp. 11–12.
27. EN 1242:2013; Adhesives—Determination of Isocyanate Content. CEN: Bruxelles, Belgium, 2013. Available online: https://standards.iteh.ai/catalog/standards/cen/7167d460-c11a-4522-b759-1364d3bd90f6/en-1242-2013 (accessed on 17 May 2024).
28. Pérez-Limiñana, M.A.; Pérez-Aguilar, H.; Ruzafa-Silvestre, C.; Orgilés-Calpena, E.; Arán-Ais, F. Effect of Processing Time of Steam-Explosion for the Extraction of Cellulose Fibers from Phoenix Canariensis Palm Leaves as Potential Renewable Feedstock for Materials. *Polymers* **2022**, *14*, 5206. [CrossRef] [PubMed]

29. Carbonell-Blasco, M.P.; Moyano, M.A.; Hernández-Fernández, C.; Sierra-Molero, F.J.; Pastor, I.M.; Alonso, D.A.; Arán-Aís, F.; Orgilés-Calpena, E. Polyurethane Adhesives with Chemically Debondable Properties via Diels–Alder Bonds. *Polymers* **2023**, *16*, 21. [CrossRef] [PubMed]
30. Ferrández-García, A.; Ferrández-Villena, M.; Ferrández-García, C.E.; García-Ortuño, T.; Ferrández-García, M.T. Potential Use of Phoenix Canariensis Biomass in Binderless Particleboards at Low Temperature and Pressure. *Bioresources* **2017**, *12*, 6698–6712. [CrossRef]
31. Orgilés-Calpena, E.; Arán-Aís, F.; Torró-Palau, A.; Orgilés-Barceló, C. Biodegradable Polyurethane Adhesives Based on Polyols Derived from Renewable Resources. *Proc. Inst. Mech. Eng. Part L J. Mater. Des. Appl.* **2014**, *228*, 125–136. [CrossRef]
32. EN 1392.2007; Adhesives for Leather and Materials for Footwear. Solvent-Based and Dis-Persion Adhesives. Test Methods to Measure Bond Strength under Specific Conditions. CEN: Bruxelles, Belgium, 2007. Available online: https://standards.iteh.ai/catalog/standards/cen/f4e2c6c0-9846-43b0-bcef-adda17102931/en-1392-1998 (accessed on 17 May 2024).
33. Arán-Ais, F.; Ruzafa-Silvestre, C.; Carbonell-Blasco, M.; Pérez-Limiñana, M.; Orgilés-Calpena, E. Sustainable Adhesives and Adhesion Processes for the Footwear Industry. *Proc. Inst. Mech. Eng. C J. Mech. Eng. Sci.* **2021**, *235*, 585–596. [CrossRef]
34. UNE-EN ISO 17708:2019; Footwear-Test Methods for Whole Shoe-Upper Sole Adhesion. ISO: Geneva, Switzerland, 2019.
35. da Silva, S.H.F.; dos Santos, P.S.B.; Thomas da Silva, D.; Briones, R.; Gatto, D.A.; Labidi, J. Kraft Lignin-Based Polyols by Microwave: Optimizing Reaction Conditions. *J. Wood Chem. Technol.* **2017**, *37*, 343–358. [CrossRef]
36. Saffar, T.; Bouafif, H.; Braghiroli, F.L.; Magdouli, S.; Langlois, A.; Koubaa, A. Production of Bio-Based Polyol from Oxypropylated Pyrolytic Lignin for Rigid Polyurethane Foam Application. *Waste Biomass Valorization* **2020**, *11*, 6411–6427. [CrossRef]
37. Cruz-Aldaco, K.; Flores-Loyola, E.; Aguilar-González, C.N.; Burgos, N.; Jiménez, A. Synthesis and Thermal Characterization of Polyurethanes Obtained from Cottonseed and Corn Oil-Based Polyols. *J. Renew. Mater.* **2016**, *4*, 178–184. [CrossRef]
38. Li, M.-C.; Wu, Q.; Song, K.; De Hoop, C.F.; Lee, S.; Qing, Y.; Wu, Y. Cellulose Nanocrystals and Polyanionic Cellulose as Additives in Bentonite Water-Based Drilling Fluids: Rheological Modeling and Filtration Mechanisms. *Ind. Eng. Chem. Res.* **2016**, *55*, 133–143. [CrossRef]
39. Li, F.; Qiu, C.-P.; Zhang, X.-L.; Tan, R.-W.; de Hoop, C.F.; Curole, J.P.; Qi, J.-Q.; Xiao, H.; Chen, Y.-Z.; Xie, J.-L.; et al. Effect of Biomass Source on the Physico-Mechanical Properties of Polyurethane Foam Produced by Microwave-Assisted Liquefaction. *Bioresources* **2020**, *15*, 7034–7047. [CrossRef]
40. Ivdre, A.; Abolins, A.; Volkovs, N.; Vevere, L.; Paze, A.; Makars, R.; Godina, D.; Rizikovs, J. Rigid Polyurethane Foams as Thermal Insulation Material from Novel Suberinic Acid-Based Polyols. *Polymers* **2023**, *15*, 3124. [CrossRef] [PubMed]
41. Kaur, R.; Singh, P.; Tanwar, S.; Varshney, G.; Yadav, S. Assessment of Bio-Based Polyurethanes: Perspective on Applications and Bio-Degradation. *Macromol* **2022**, *2*, 284–314. [CrossRef]
42. Lu, W.; Li, Q.; Zhang, Y.; Yu, H.; Hirose, S.; Hatakeyama, H.; Matsumoto, Y.; Jin, Z. Lignosulfonate/APP IFR and Its Flame Retardancy in Lignosulfonate-Based Rigid Polyurethane Foams. *J. Wood Sci.* **2018**, *64*, 287–293. [CrossRef]
43. Chahar, S.; Dastidar, M.G.; Choudhary, V.; Sharma, D.K. Synthesis and Characterisation of Polyurethanes Derived from Waste Black Liquor Lignin. *J. Adhes. Sci. Technol.* **2004**, *18*, 169–179. [CrossRef]
44. Tavares, L.B.; Boas, C.V.; Schleder, G.R.; Nacas, A.M.; Rosa, D.S.; Santos, D.J. Bio-Based Polyurethane Prepared from Kraft Lignin and Modified Castor Oil. *Express Polym. Lett.* **2016**, *10*, 927–940. [CrossRef]
45. Pregi, E.; Romsics, I.; Várdai, R.; Pukánszky, B. Interactions, Structure and Properties of PLA/Lignin/PBAT Hybrid Blends. *Polymers* **2023**, *15*, 3237. [CrossRef]
46. Dobrescu, V.E.; Radovici, C. Temperature Dependence of Melt Viscosity of Polymers. *Polym. Bull.* **1983**, *10*, 134–140. [CrossRef]
47. Zlatkevich, L. Molecular Mobility and Transitions in Polymers. In *Radiothermoluminescence and Transitions in Polymers*; Springer: New York, NY, USA, 1987; pp. 55–80.
48. EN 15307:2015; Adhesives for Leather and Footwear Materials—Sole-Upper Bonds—Minimum Strength Requirements. CEN: Bruxelles, Belgium, 2015. Available online: https://standards.iteh.ai/catalog/standards/cen/42392010-5d08-4985-9d1f-ce74820be5cf/en-15307-2014 (accessed on 17 May 2024).

Disclaimer/Publisher's Note: The statements, opinions and data contained in all publications are solely those of the individual author(s) and contributor(s) and not of MDPI and/or the editor(s). MDPI and/or the editor(s) disclaim responsibility for any injury to people or property resulting from any ideas, methods, instructions or products referred to in the content.

Article

Advanced Anticorrosive Graphene Oxide-Doped Organic-Inorganic Hybrid Nanocomposite Coating Derived from *Leucaena leucocephala* Oil

Wejdan Al-otaibi, Naser M. Alandis, Yasser M. Al-Mohammad and Manawwer Alam *

Department of Chemistry, College of Science, King Saud University, P.O. Box 2455, Riyadh 11451, Saudi Arabia; wejdanotaiby@gmail.com (W.A.-o.); nandis@ksu.edu.sa (N.M.A.)
* Correspondence: maalam@ksu.edu.sa

Abstract: Metal corrosion poses a substantial economic challenge in a technologically advanced world. In this study, novel environmentally friendly anticorrosive graphene oxide (GO)-doped organic-inorganic hybrid polyurethane (LFAOIH@GO-PU) nanocomposite coatings were developed from *Leucaena leucocephala* oil (LLO). The formulation was produced by the amidation reaction of LLO to form diol fatty amide followed by the reaction of tetraethoxysilane (TEOS) and a dispersion of GO_x (X = 0.25, 0.50, and 0.75 wt%) along with the reaction of isophorane diisocyanate (IPDI) (25–40 wt%) to form LFAOIH@GO_x-PU_{35} nanocomposites. The synthesized materials were characterized by Fourier transform infrared spectroscopy (FTIR); ^1H, ^{13}C, and ^{29}Si nuclear magnetic resonance; and X-ray photoelectron spectroscopy. A detailed examination of LFAOIH@$GO_{0.5}$-PU_{35} morphology was conducted using X-ray diffraction, scanning electron microscopy, energy-dispersive X-ray spectroscopy, and transmission electron microscopy. These studies revealed distinctive surface roughness features along with a contact angle of around 88 G.U preserving their structural integrity at temperatures of up to 235 °C with minimal loading of GO. Additionally, improved mechanical properties, including scratch hardness (3 kg), pencil hardness (5H), impact resistance, bending, gloss value (79), crosshatch adhesion, and thickness were evaluated with the dispersion of GO. Electrochemical corrosion studies, involving Nyquist, Bode, and Tafel plots, provided clear evidence of the outstanding anticorrosion performance of the coatings.

Keywords: bio-resource; polyurethanamide; organic-inorganic hybrid; nanocomposite; coating; corrosion inhibition

Citation: Al-otaibi, W.; Alandis, N.M.; Al-Mohammad, Y.M.; Alam, M. Advanced Anticorrosive Graphene Oxide-Doped Organic-Inorganic Hybrid Nanocomposite Coating Derived from *Leucaena leucocephala* Oil. *Polymers* **2023**, *15*, 4390. https://doi.org/10.3390/polym15224390

Academic Editors: Chang-An Xu and Zhuohong Yang

Received: 2 October 2023
Revised: 19 October 2023
Accepted: 25 October 2023
Published: 12 November 2023

Copyright: © 2023 by the authors. Licensee MDPI, Basel, Switzerland. This article is an open access article distributed under the terms and conditions of the Creative Commons Attribution (CC BY) license (https://creativecommons.org/licenses/by/4.0/).

1. Introduction

Corrosion is a process that diminishes the strength and integrity of metallic materials [1]. It can be initiated by both natural processes and human activities and is influenced by a range of factors, including humidity, gaseous surroundings, electrolytes, temperature, and pH variations [2,3]. The economic consequences associated with corrosion are notably substantial, and the corrosion protection coating market is anticipated to witness robust expansion, progressing from an initial valuation of USD 10.4 billion in 2023 to a projected sum of USD 12.4 billion by the year 2028 [4]. This expense increases in parallel with technological advancement, affecting the worldwide economy and human welfare. The utilization of polymeric coatings emerges as a prime strategy to protect metallic materials against corrosive environments. Amongst the various polymeric materials employed as coatings, organic-inorganic hybrid (OIH) coatings stand out for their extensive utilization for advanced applications [5,6]. The utilization of OIH materials has garnered considerable attention in the research field due to their capacity to systematically amalgamate and regulate properties stemming from both organic and inorganic components. These hybrid materials are distinguished by their exceptional mechanical, thermal, cryogenic, and dynamic attributes, rendering them highly advantageous for a wide spectrum of challenging applications [5]. Significantly, these applications encompass the development of coatings for automotive, aircraft, aerospace, optically active films, contact lenses, packaging materials, abrasion-resistant paints, and protective coatings [7].

The escalating environmental apprehensions encompassing issues like toxicity, high cost, gas flaring, and industrial/domestic waste, coupled with the volatile and unpredictable pricing dynamics of petroleum-derived feedstocks, bear significant emphasis. The global economy's dependence on this exhaustible, environmentally non-benign resource underscores the necessity of replacing this indispensable energy and feedstock source with sustainable, eco-friendly materials [8]. Hence, in the development of OIH materials, the substitution of petroleum-derived organic components with bio-based alternatives such as polysaccharides, cellulose, proteins, cashew nut shell liquid, and vegetable seed oils (VSOs) offers advantageous prospects [9]. VSOs have gained notable recognition as a promising source of bio-polyols, largely due to their abundant accessibility, uncomplicated processing, and inherent chemical versatility [10]. Organic domains derived from vegetable oils exhibit significant promise as viable substitutes of petroleum products resulting in improved performance attributes with substantial ecological and economic implications. Various materials obtained from VSOs, such as soybean, rapeseed, castor, and *Leucaena leucocephala* (*LL*), have been meticulously designed to cater to specific application requirements [9]. Among different VSOs, *LL* oil (*LLO*), produced from LL, a member of the Fabaceae family, is an agro-industrial product with diverse applications in the realm of bioenergy. LLO derivatives like polyetheramide, polyesteramide, and polyurethane have been utilized in the field of anticorrosive coatings. But there is a further need to improve their mechanical and anticorrosive performance.

Integrating graphene oxide (GO) into polymers as nanocomposites offers notable benefits, including improved flexibility and electrical conductivity through its enhanced electron mobility and surface area, as shown by graphene's versatile use in various applications [11]. In recent times, there has been a surge of interest in GO-doped OIH materials among researchers due to their strong thermal and electrical stability and mechanical strength, with graphene emerging as a pivotal dispersion component. Al Rashed et al. [12] synthesized polyurethane/polysiloxane OIH hybrid coatings filled with GO, which revealed good mechanical and optical properties. The results demonstrated that GO, along with isocyanate treatment, reduced oxygen and moisture penetration in the coatings. The effect of GO facilitated polysiloxane condensation on their surfaces, enhancing corrosion protection on aluminum alloy surfaces. Harb et al. [13] designed GO-doped OIH coatings for protecting steel surfaces from corrosion by benzoyl peroxide-induced polymerization of methyl methacrylate covalently bonded through 3-(trimethoxysilyl)propyl methacrylate to silica domains formed by the hydrolytic condensation of tetraethoxysilane. Ahmad et al. [5] fabricated OIH coatings composed of polyurethane fatty amide and silica components under ambient conditions. The synthesized material had high physicomechanical, chemical, and thermal stability and was produced with the objective of addressing corrosion issues in mild steel. A literature survey reveals that no work has been reported on the proposed research work.

The aim of the current study is to achieve a sustainable and environmentally friendly synthesis of GO-doped *LLO*-based organic-inorganic hybrid (LFAOIH) polyurethane (PU) material (LFAOIH@GO-PU) for advanced anticorrosive application. The LFAOIH@GO-PU formulation was produced in two simple steps: (i) preparation of *LLO*-based diol fatty amide (LLFAD) and (ii) reaction of freshly prepared LLFAD with tetraethoxy silane (TEOS) to form an organic-inorganic hybrid (LFAOIH), followed by in situ dispersion of GO (different percentages) to form LFAOIH @GO, followed by the reaction of isophorane diisocyanate (IPDI) to form ambient cured LFAOIH@GO-PU coating material. The present work explores the utility of synthesized nanocomposite material in mechanically robust chemically resistant and anticorrosive coatings. Notably, the study pioneers the application of Electrochemical Impedance Spectroscopy (EIS) to elucidate the corrosion protection in the presence of 3.5% NaCl, employing Nyquist and Bode plots over for 12 days along with the potentiodynamic (Tafel) analysis. The investigation also encompasses the structural validation through Fourier transform infrared (FTIR), $^1H/^{13}C/^{29}Si$ Nuclear magnetic resonance (NMR) and X-ray photoelectron spectroscopy (XPS). Additionally, the morphology is scrutinized using X-ray diffraction (XRD), scanning electron microscopy (SEM), and transmission electron microscopy (TEM). The thermal stability aspect is addressed through thermogravimetric and differential scanning calorimetry analysis (TGA/DTG and DSC). Furthermore, the introduced one-pot synthesis process from LLFAD contributes significantly to the fabrication of eco-friendly polymeric protective coatings.

2. Experimental

2.1. Materials and Methods

The study involved the extraction of *Leucaena leucocephala* seed oil from mature legumes collected at the University Campus of King Saud University. The seeds were subsequently ground into a powdered form, and the oil was then extracted from the powdered seeds utilizing a Soxhlet apparatus with petroleum ether as the chosen solvent. The elimination of ether was achieved through the use of a vacuum rotary evaporator [14]. Diethanolamine was procured from AppliChem GmbH in Darmstadt, Germany, while sodium metal, sodium chloride, and toluene were acquired from Winlab in the United Kingdom. Additionally, the reagents TEOS (98%) and IPDI (98%) were sourced from Acros Organics in NJ, USA. Notably. GO was obtained from Grafen Chemical Industries Co in Ankara, Turkey. Diethyl ether, methanol, and dibutyltin dilaurate were purchased from Sigma-Aldrich in St. Louis, NJ, USA.

2.2. Synthesis of Leucaena leucocephala Fattyamide Diol (LLFAD)

LLFAD was prepared as per the reported method [5].

2.3. Synthesis of LFAOIH and Its PU (LFAOIH-PU_{30-40})

Freshly prepared LLFAD (14 g, 0.04 mole) was introduced into a three-necked conical flask equipped with essential apparatuses, including a dropping funnel, thermometer, and condenser. Sequentially, 8.32 g (0.04 moles) of TEOS was added in a controlled manner over a 30 min interval under continuous agitation at room temperature. Upon complete incorporation of TEOS, the reaction mixture was subjected to an elevated temperature of 80 ± 5 °C for 1 h. The contents were further heated at 120 ± 5 °C for 1 h to complete the reaction. The progression of the reaction was diligently monitored via appearance of contents (conversion of fogginess to clear) and FTIR. Subsequently, the reaction mixture underwent gradual cooling to ambient temperature while maintained in continuous agitation. Within the same experimental context, incremental additions of IPDI were meticulously introduced, each at varying proportions (30, 35, and 40% with regard to LLFAD) over 30 min duration with a drop of DBTL and maintained the temperature at 60 °C for 3 h. A minute volume (8–12%) of toluene was incorporated to sustain the desired viscosity. The trajectory of the reaction was monitored by FTIR. The resulting LFAOIH-PU products were denoted as LFAOIH-PU_{30}, LFAOIH-PU_{35}, and LFAOIH-PU_{40}, the subscript numeral indicating the % of IPDI.

2.4. Synthesis of LFAOIH@GO and Their PU Nanocomposite (LFAOIH@GO_x-PU_{35})

LFAOIH was synthesized as the aforementioned setup with an equimolar amount of LLFAD and TEOS (0.04 moles). GO_X (X = 0.25%, 0.50% and 0.75% w.r.t. LLFAD), initially dispersed (by sonication for 10 min) in minimal amounts of toluene, was added in the same pot at the same temperature. The temperature was maintained for 1 h. The reaction progress was tracked by using viscosity and FTIR. Afterward, the mixture was slowly cooled with continuous agitation. In the same vessel, an optimized 35% IPDI, along with a drop of DBTL, was added over 30 min and at 60 °C for 3 h. A small volume of toluene (8–12%) was used to maintain the desired viscosity. The trajectory of the reaction was checked by viscosity and FTIR analysis. The resulting products were denoted as LFAOIH@$GO_{0.25}$-PU_{35}, LFAOIH@$GO_{0.5}$-PU_{35}, and LFAOIH@$GO_{0.75}$-PU_{35}.

2.5. Characterizations

Various spectroscopic techniques were employed to characterize the prepared samples. FTIR measurements were conducted using FTIR spectrophotometer; Spectrum 100, Perkin Elmer Cetus Instrument, Norwalk, CT, USA spanning a range of 4000 to 400 cm^{-1} at room temperature (28–30 °C) with a resolution of 4 cm^{-1}. Nuclear Magnetic Resonance (NMR) spectra, specifically ^{1}H, ^{13}C, and ^{29}Si NMR, were obtained using a Jeol DPX400 MHz instrument, Tokyo, Japan. Deuterated dimethyl sulfoxide (DMSO) served as the solvent, while tetramethyl silane was used as an internal standard for calibration. The sample was analyzed by X-ray photoelectron spectroscopy (XPS) by using a JPS-9030; JEOL, Japan MgKα (1253.6 eV). The solubility assessment of the prepared sample involved dissolving a 20 mg sample in 10 mL of both polar and nonpolar solvents (dimethyl sulfoxide, pyridine, benzene, toluene, ethyl-methyl ketone, dichloromethane, butanol, chloroform, diethyl ether, benzyl alcohol, carbon tetrachloride, N,N-dimethyl formamide, isoamyl alcohol, petroleum ether, butan-1-ol, 1-hexanol, cyclohexene, furan, propanol, 1,4-dioxane, ethanol, acetone, and ethyl acetate). The resulting solution was allowed to stand undisturbed for a duration of 24 h. Various physico-mechanical tests were conducted, including thickness measurements using ASTM D1186-B, Elcometer Model 345 from Elcometer Instrument Ltd. in Manchester, UK, scratch hardness following BS 3900 standards, pencil hardness in accordance with ASTM D3363-05, crosshatch testing following ASTM D3359-02, impact testing as per IS 101:1988, bend testing based on ASTM D3281-84, and gloss measurement using a Gloss meter Model KSJ MG6-F1 from KSJ Photoelectrical Instruments Co., Ltd. in Quanzhou, China. To investigate the composition and morphology of the film, a scanning electron microscope (SEM), namely the JSM 7600F by JEOL in Japan, was utilized. For elemental analysis, Energy Dispersive X-ray spectroscopy (EDX) coupled to the SEM was employed. To facilitate observation under the microscope, the sample was sputter-coated with Pt and the microscope operated at 15 kV. Transmission electron microscopy (TEM) was conducted using a JEM-2100F transmission electron microscope by JEOL, Japan. For checking the wettability of the surface, a drop of double distilled water was deposited on the surface of the coating and photographed with contact angle measurements (CAM200 Attention goniometer). Electrochemical impedance spectroscopy (EIS) was employed to assess the corrosion resistance of the prepared sample in a 3.5% NaCl solution. This analysis was conducted using a three-electrode flat glass cell, alongside an Autolab potentiostat/galvanostat (PGSTAT204-FRA32) and NOVA 2.1 software. The equipment used for this purpose was Metrohom Autolab B.V., located at Kanaalweg 29-G, 3526 KM, Utrecht, Netherlands. The experimental setup involved three electrodes: (I) Ag/AgCl utilized as a reference electrode, (II) platinum serving as an auxiliary electrode, and (III) the test specimen functioning as the working electrode. The solution was applied to a 1.0 cm^2 area of the working electrode, adhering to the ASTM G59-97 standard. The purpose was to gauge the performance of the coated surface in terms of its impedance behavior against corrosion in the specified environment. The electrochemical parameters such as corrosion potential (Ecorr) and current density (Icorr) were calculated from Tafel extrapolation method and the corrosion inhibition efficiency can be calculated form Equation (1)

$$IE\% = \frac{corrosion\ current\ density\ of\ (bare\ metal - coated\ surface)}{corrosion\ current\ density\ of\ coated\ surface} \times 100 \quad (1)$$

3. Results and Discussion

Schemes 1 and 2 show the synthesis of LFAOIH-PU$_{35}$ and LFAOIH@GO$_x$-PU$_{35}$ nanocomposite material from LLO. It reveals the formulation carried out in two different steps. The first step of reaction involves the conversion of LLO into LLFAD by the base catalyzed reaction of LLO with diethanolamine to form LLFAD. Then, the purified LLFAD reacted with the equimolar amount of TEOS to form LFAOIH at 100 ± 5 °C. When TEOS was introduced to LLFAD, it was noticed that the reaction mixture became cloudy at 80 ± 5 °C; however, this cloudiness vanished after 1 h of heating at 120 ± 5 °C. This is connected to the chemical reaction between the hydroxyl groups of LLFAD and the alkoxy group of TEOS, which generates O-Si-O bonds as a result of the reaction [5]. After that dispersion of GOx (X = 0.25%, 0.50% and 0.75% w.r.t. LLFAD) followed by the reaction of the content with an optimized percentage of IPDI (35%) at 60 ± 5 °C temperature to form the crosslinked material for coating. IPDI crosslinked the materials through the reaction of –NCO with the residual functionality of LFAOIH@GO$_{0.5}$ to form LFAOIH@GO$_{0.5}$-PU$_{35}$. The solubility behavior of the LFAOIH@GO$_{0.5}$-PU$_{35}$ nanocomposite material is influenced by the polarities, functional groups, and molecular structures of both the nanocomposite itself and the solvents. Solubility in dimethyl sulfoxide, pyridine, benzene, toluene, ethyl-methyl ketone, dichloromethane, butanol, and chloroform may be due to the presence of compatible functional groups and molecular arrangements [15,16]. The interactions between these components likely facilitate dissolution. The insolubility in diethyl ether, benzyl alcohol, carbon tetrachloride, isoamyl alcohol, petroleum ether, butan-1-ol, 1-hexanol, and cyclohexane may result from differences in polarity, hydrogen bonding capabilities, and molecular sizes between both the nanocomposite and these solvents. These factors hinder the establishment of strong enough interactions to enable dissolution [9]. The materials were found to be partially soluble in cyclohexene, furan, propanol, 1,4-dioxane, ethanol, acetone, and ethyl acetate, which may be due to the compatibility between the nanocomposite and these solvents, allowing for some degree of dissolution due to similar chemical properties.

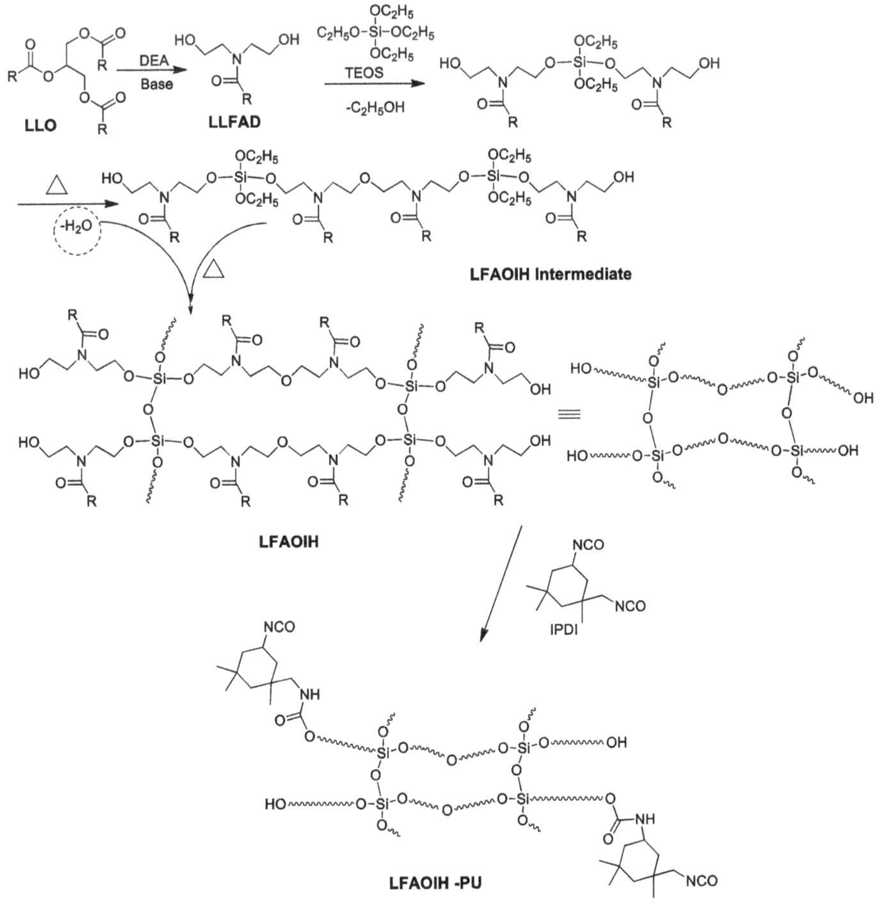

Scheme 1. Synthesis of LFAOIH-PU.

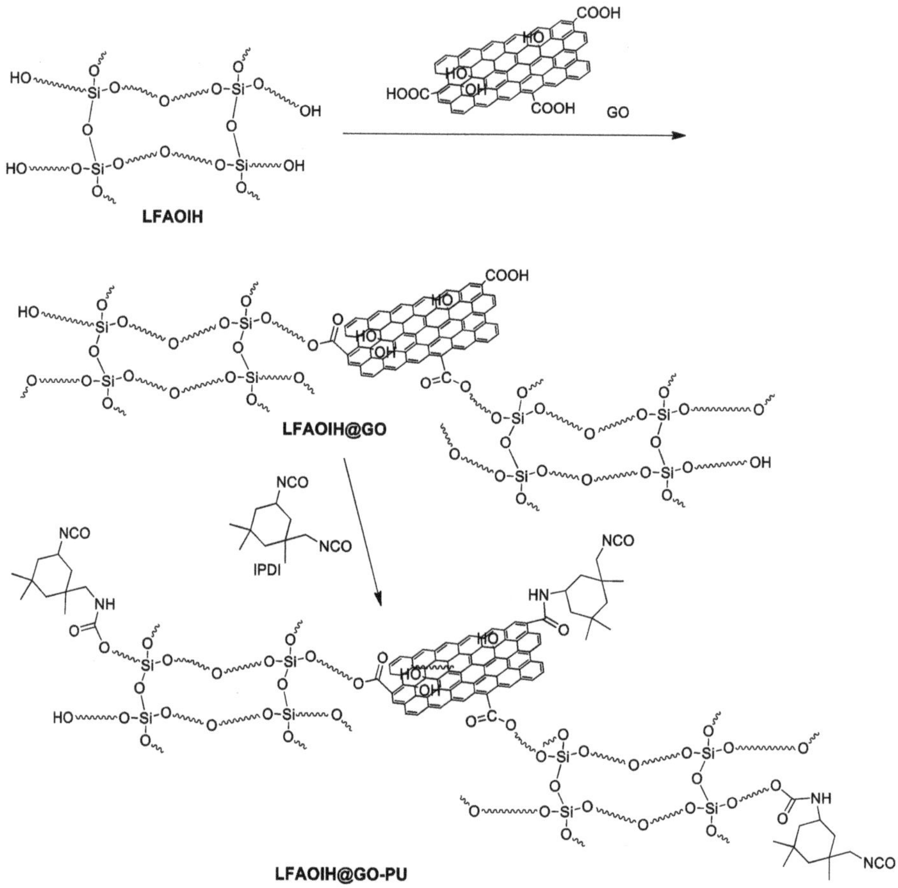

Scheme 2. Synthesis of LFAOIH@GO-PU.

3.1. Spectral Analysis

3.1.1. FTIR Analysis

The formation of LFAOIH-PU$_{35}$ and LFAOIH@GO$_{0.5}$-PU$_{35}$ was confirmed through FTIR spectrum analysis. To validate this confirmation, a comparison was carried out by analyzing the FTIR spectra of LLFAD, LFAOIH-PU$_{35}$ and LFAOIH@GO$_{0.5}$-PU$_{35}$, as shown in Figure 1. The characteristics peaks are as:

LLFAD: In the FTIR analysis of LLFAD, distinct absorption bands were observed corresponding to specific functional groups and molecular vibrations [Figure 1(1)]. These include an absorption band at 3359 cm^{-1} attributed to hydroxyl groups (-OH), indicating the presence of alcohols. The band at 3008 cm^{-1} signifies the presence of C=C double bonds. Bands at 2853 cm^{-1} and 2924 cm^{-1} correspond to alkyl stretching vibrations, indicating the presence of methyl (-CH$_3$) and methylene (-CH$_2$) groups [14]. The absorption band at 1618 cm^{-1} is indicative of the amide carbonyl group (C=O) stretching vibration. The region between 1466 cm^{-1} and 1384 cm^{-1} represents the bending vibrations of methyl and methylene groups (-CH$_3$, -CH$_2$). The bands at 1210 cm^{-1} and 1052 cm^{-1} are characteristic of the ether linkage (-(C=O)-O-C) vibration [16], and the presence of -C-O- bonds, respectively [14].

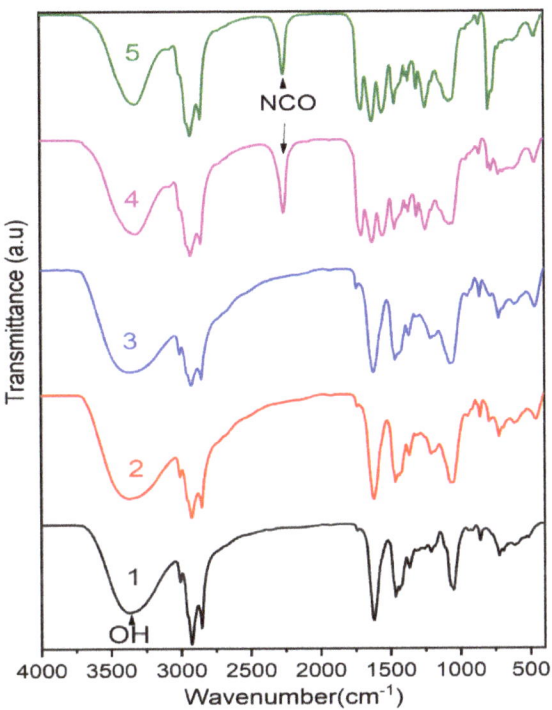

Figure 1. FTIR spectra of (**1**) LLFAD, (**2**) LFAOIH, (**3**) LFAOIH@GO, (**4**) LFAOIH−PU$_{35}$ and (**5**) LFAOIH@GO$_{0.5}$-PU$_{35}$.

LFAOIH: In the FTIR analysis of LFAOIH, the observed absorption bands are consistent with those in LLFAD, indicating the presence of similar functional groups. Notably, additional peaks corresponding to silicon oxide interactions are also evident [14]. Furthermore, distinctive peaks associated with silicon oxide interactions emerge, encompassing a symmetrical stretch at 614 cm^{-1} (Si-O-Si), an asymmetrical stretch at 1069 cm^{-1} (Si-O-Si), and a bending vibration at 461 cm^{-1} (Si-O-Si) [5]. These absorptions signify the involvement of silicon oxide moieties within the LLFAD-TEOS compound.

LFAOIH-PU: The hydroxyl group (-OH) peak, previously at 3368 cm^{-1}, shifts to 3323 cm^{-1}, which may be due to hydrogen bonding [Figure 1]. The peak appears at 1704 cm^{-1} related to the urethane carbonyl{-O-C(=O)-NH-} [5,14]. Other peaks include 1626 cm^{-1}, 793 cm^{-1} (ArC=C- bending), 468 cm^{-1} (Si-O-Si bending), and 1074 cm^{-1} (asymmetrical Si-O-Si stretch) [5], with notable peaks at 2259 cm^{-1} (-NCO group). This analysis provides insights into molecular interactions and the polyurethane's structure.

LFAOIH@GO$_x$: The detection of GO incorporation is not primarily determined by the presence of GO's absorption bands, as GO encompasses numerous -O-containing functional groups that can overlap with those present in the LLFAD (Figure 1). Instead, the indication of GO's presence is derived from shifts in absorption values. LFAOIH@GO$_{0.5}$ has all the characteristic peaks as found in LFAOIH along with broadening of the peak at 3368 due to hydrogen bonding and additional peaks at 1738 cm^{-1}, which is due to ester C=O that formed by the reaction of the hydroxyl group of LFAOIH with -COOH group GO [16,17]. The detection of GO's presence may not be evident solely through the observation of absorption bands associated with GO because it possesses numerous -O- functional groups that often overlap with functional groups found. Instead, one can identify the presence of GO by noting changes in the absorption values of specific functional groups like -OH, >C=O ester and >C=O amide in comparison to those observed in virgin LFAOIH. These alterations in absorption values are attributed to the interactions occurring between LFAOIH and GO [18–20].

LFAOIH@GO$_{0.5}$-PU$_{35}$: The emergence of an additional peak at 2259 cm^{-1} and 1704 cm^{-1} (overlap with >C=O ester peak of LFAOIH@GO$_x$) can be attributed to the presence of the urethane linkage formed between the isocyanate (-NCO) and hydroxyl (-OH) groups during the process of polymerization along with the same peaks found in LFAOIH@GO$_{0.5}$, which confirmed LFAOIH@GO$_{0.5}$-PU$_{35}$ formulation (Figure 1) [16,17].

The broad hump in the FTIR spectrum in the 3200–3500 cm^{-1} region is a distinctive indication of the presence of hydrogen bonding. It results from the broadening of the O-H stretching vibration peak due to the dynamic and variable nature of hydrogen bond interactions within the sample [18]. The comparative FTIR data of LFAOIH-PU$_{35}$ and LFAOIH@GO$_{0.5}$-PU$_{35}$ [Figure 1] indicate that the peaks of LFAOIH@GO$_{0.5}$-PU$_{35}$ shows a broad hump as compared to LFAOIH-PU$_{35}$, which revealed the characteristic of hydrogen bonding and indicates that multiple O-H groups in the sample are engaged in hydrogen bond interactions. Hydrogen bonding exerts a discernible influence on the enhancement of mechanical properties, including scratch hardness, pencil hardness, bending and gloss values [19]. Robust hydrogen bonding fosters enhanced adhesion between coatings and substrates, and bolsters material toughness. It also contributes to resistance against wear and abrasion and augments ductility, mitigating the susceptibility to deformation-induced cracking [20].

3.1.2. NMR Analysis

^{1}H NMR: In the ^{1}H NMR spectrum of LFAOIH, distinct peaks were observed (Figure 2a), such as 0.715 ppm for -CH$_3$ groups, 1.08–1.21 ppm for -CH$_2$ groups, and 1.329 ppm for >N-CO-CH$_2$-CH$_2$ moieties. Other notable peaks included 1.87 ppm (-CH$_2$-CH$_2$-CH=), 2.16 ppm (>N-CO-CH$_2$-), 2.59 ppm (=CH-CH$_2$-N), and broad 3.5 ppm (-CH-O-Si-), as well as 5.23 ppm (-OH) [21], whereas in the formulation of LFAOIH-PU$_{35}$, the same characteristic peaks were observed, with a sharper and quadruple 3.5 ppm (-CH-O-Si-) peak due to its involvement in the reaction. Additionally, a minor peak at ~4 ppm, attributed to (-O-CH$_2$-) from the incorporation of IPDI, and a sharp peak around 5.2 ppm confirmed polymerization via IPDI insertion.

^{13}C NMR: In the ^{13}C NMR spectrum of LFAOIH, specific resonance peaks were identified, including 13.96 ppm (attributed to -CH$_3$), a range spanning 21.14–33.90 ppm (associated with the CH$_2$ chain), and a prominent peak at 40.1 ppm indicative of dimethyl sulfoxide (DMSO) (Figure 2b) [21]. Additionally, resonances were observed within the range of 49.01–59.12 ppm, corresponding to the >N-CO-CH$_2$CH$_2$- moiety. Further distinctive signals were detected at 129–130.01 ppm (-CH=CH-) and 171 ppm (-CH-O-Si-) [22]. In the LFAOIH-PU$_{35}$ formulation, all these characteristic peaks were retained; however, a noteworthy observation was the diminished intensity of the peak around 171 ppm. This attenuation suggests a plausible interaction involving this carbon in subsequent reactions with IPDI and the formulation of LFAOIH-PU$_{35}$.

^{29}Si NMR: In the ^{29}Si NMR spectrum, a sharp chemical shift around -105 ppm has been found, which confirmed the presence of Si-O-Si atoms (Q4species) bonded to four oxygen atoms in siloxane (-Si-O-) within the compound (Figure 2c) [22]. In the case of LFAOIH, the peak around -100 is found to be sharp whereas in the exact peak it was found to be a little broad. The broadening of the same (-105 ppm) peak in the spectrum when reacting with IPDI is a result of the chemical changes and increased structural complexity causing dynamic exchange between silicon species and yielding a highly crosslinked structure of LFAOIH-PU$_{35}$ [22].

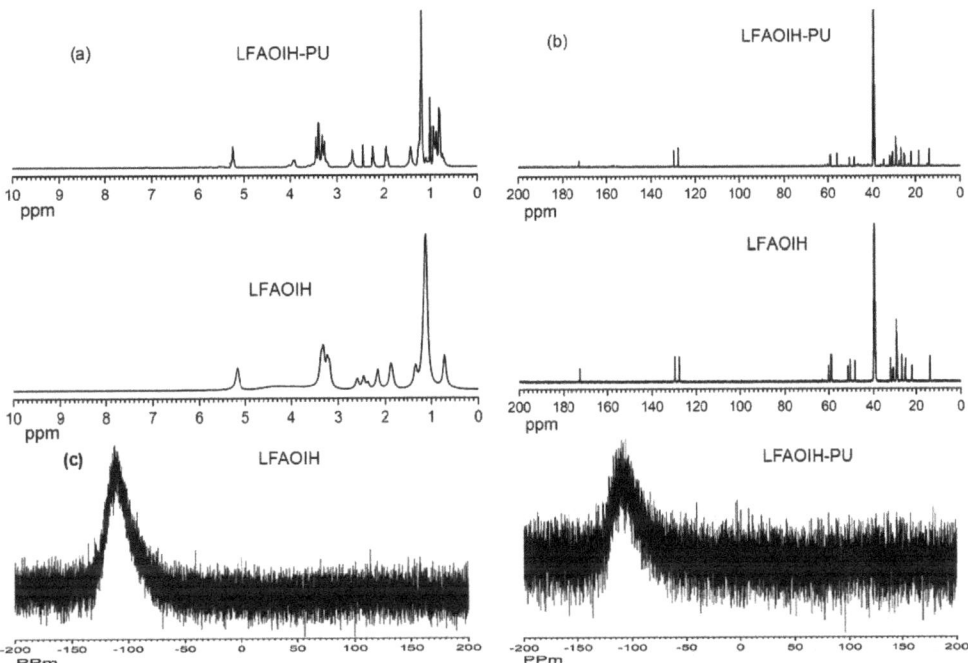

Figure 2. (a) ^{1}H, (b) ^{13}C and (c) ^{29}Si NMR spectra of LFAOIH and LFAOIH-PU, respectively.

3.1.3. XPS Analysis

The chemical composition of LFAOIH@GO$_{0.5}$-PU$_{35}$ was thoroughly investigated using XPS analysis (Figure 3). The resulting deconvoluted spectra, depicted in Figure 3 for their respective elements (C 1s, N 1s, O 1s, and Si 2p$_{3/2}$), offer detailed insights into the material's composition. In the deconvoluted C1s spectrum, distinctive peaks were identified at four different binding energies: 283.6 eV, 285.2 eV, 286.5 eV, and 289 eV [23]. These peaks are associated with specific carbon bonding configurations, namely C-C, C-N, C-O, and C=O bonds, respectively [23]. Within the deconvoluted N1s spectrum, two prominent peaks were discerned at 399.6 eV and 400.25 eV, revealing the presence of nitrogen atoms engaged in N-C and N-H bonds [24]. In the deconvoluted O1s spectrum, two well-defined peaks were observed at 532.5 eV and 533.6 eV, indicating the binding energies of oxygen atoms in C-O/C-OH and C-O bonds [24]. Additionally, the characteristic Si 2p3/2 peak at 102.1 eV, as observed, signifies the existence of Si-O-Si and Si-OH bonds within the analyzed material [25]. This XPS analysis provides valuable insights into the chemical composition of LFAOIH@GO$_{0.5}$-PU$_{35}$, revealing the nature of carbon, nitrogen, oxygen, and silicon bonds present within the material.

3.1.4. Physicomechanical Properties

Table 1 presents the physicomechanical impacts from LFAOIH-PU$_{35}$ to LFAOIH@GO$_X$-PU$_{35}$. The study contrasts the physical and mechanical attributes of GO-infused PU with those of pure PU. The results indicate that the inclusion of GO enhances the scratch hardness, pencil hardness, cross hatch and gloss values as observed from LFAOIH-PU$_{35}$ to LFAOIH@GO$_X$-PU$_{35}$. The enhancement in mechanical properties is also correlated to the presence of hydrogen bonding that is confirmed by the FTIR analysis of the samples. This improvement can be attributed to effective crosslinking through them [5]. Notably, the cross-hatch adhesion findings substantiate heightened adherence to mild steel strips.

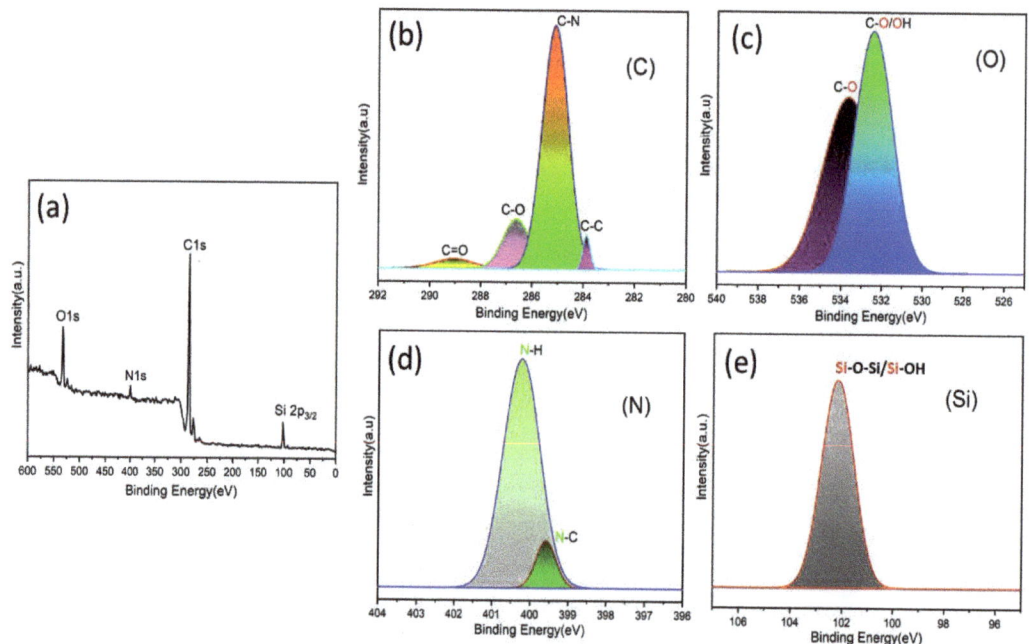

Figure 3. XPS analysis of LFAOIH@GO$_{0.5}$–PU$_{35}$, (**a**) Survey spectrum, (**b**) C 1s peaks, (**c**) O1s peaks and (**d**) N1s 1s peaks and (**e**) Si 2p.

Table 1. Various mechanical properties of LFAOIH-PU$_{35}$ and LFAOIH@GO$_x$-PU$_{35}$.

Properties	LFAOIH-PU$_{35}$	LFAOIH@GO$_{0.25}$-PU$_{35}$	LFAOIH@GO$_{0.5}$-PU$_{35}$	LFAOIH@GO$_{0.75}$-PU$_{35}$
Scratch hardness (kg)	2	2.4	3.0	2.8
Impact (lb/inch) 150	Pass	Pass	Pass	fail
Bending (1/8)	pass	pass	pass	pass
Pencil hardness	3H	4H	5H	4H
Cross Hatch (%)	100	100	100	98
Gloss at 60°	69	75	79	82
Thickness (micron)	88	129	135	140

Also, the incorporation of GO positively influences the satisfactory results observed in the physico-mechanical tests of all coating systems. This enhancement may be attributed to the presence of functional groups such as —OH and >C=O, along with double bonds (the saturation levels can differ based on the fatty acid composition found in the parent seed oil) and extended hydrocarbon chains [14,21]. These constituents contribute to the plasticizing effect on the coatings, thereby ensuring enhanced flexibility. However, LFAOIH@GO$_{0.5}$-PU$_{35}$ demonstrates brittleness during bend tests as the value of GO also affects the impact resistance, gloss values and cross hatch and other values decrease as GO value increases as shown in Table 1. It may be because the loading of GO increases; these attributes exhibit a declining trend [21,26]. Elevated GO concentrations lead to the formation of agglomerates, resulting in adversely affecting the overall coating performance [21]. Hence, out of all, LFAOIH@GO$_{0.5}$-PU$_{35}$ showed the best physicomechanical results and was optimized for further studies.

3.1.5. XRD

XRD analyses were performed on both LFAOIH-PU$_{35}$ and the doped variant with GO LFAOIH@GO$_{0.5}$-PU$_{35}$ to investigate possible modifications in the crystalline characteristics of the LFAOIH-PU$_{35}$ matrix after introducing inorganic components and displayed in Figure 4. The XRD profiles of the materials demonstrated a consistent characteristic pattern, featuring a broad hump at $2\theta = 21°$, indicative of the amorphous nature of the matrix for both LFAOIH-PU$_{35}$ and LFAOIH@GO$_{0.5}$-PU$_{35}$ due to the presence of silicon. Notably, smaller peaks were also observed around 32° and 45°. Upon doping with GO, the diffraction pattern exhibited similarities, with an additional peak emerging at 26° [27]. This emergence could potentially be attributed to the presence of graphite within the polymer matrix. Overall, both the LFAOIH-PU$_{35}$ to LFAOIH@GO$_{0.5}$-PU$_{35}$ showed a semicrystalline behavior as shown in Figure 4.

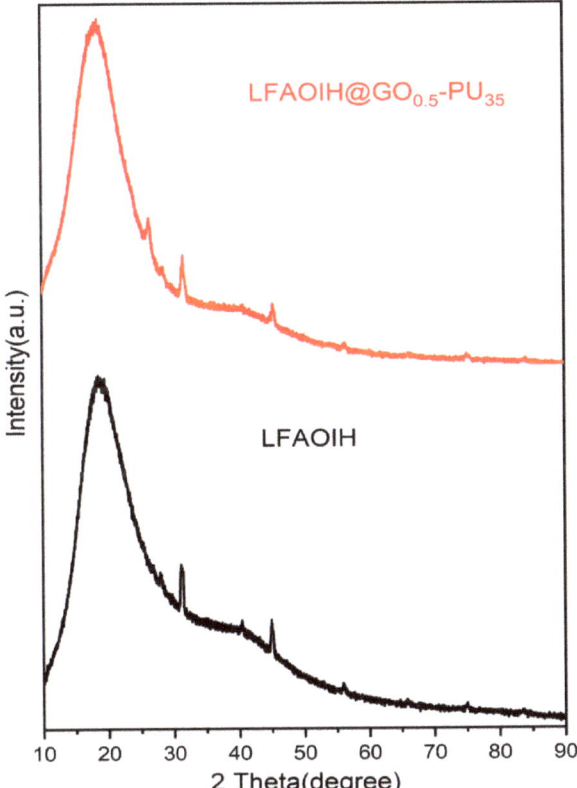

Figure 4. XRD thermograms of LFAOIH and LFAOIH@GO$_{0.5}$-PU$_{35}$.

3.1.6. Thermal Analysis

The DSC thermogram analysis revealed the presence of two distinct endothermic transitions (Figure 5a). For LFAOIH-PU$_{35}$, the initial endotherm occurs in the temperature range of 239 to 280 °C, with a peak centered at 263 °C. The subsequent endotherm takes place between 400 and 475 °C, with its peak at 438 °C. In the case of the LFAOIH@GO$_{0.5}$-PU$_{35}$ nanocomposite, the first endotherm transpires from 240 to 312 °C and is centered at 270 °C, while the second endotherm spans the range of 401 to 477 °C, with a peak at 440 °C. The first endothermic transition is associated with a phase transition, while the second is attributed to melting phenomena [14]. The observed shifts in these endotherms signify enhanced thermal properties brought about by the incorporation of GO. These shifts underline the improvement in thermal stability of the LFAOIH@GO$_{0.5}$-PU$_{35}$ matrix through the incorporation of GO, indicative of effective intermolecular interactions within the structure [21].

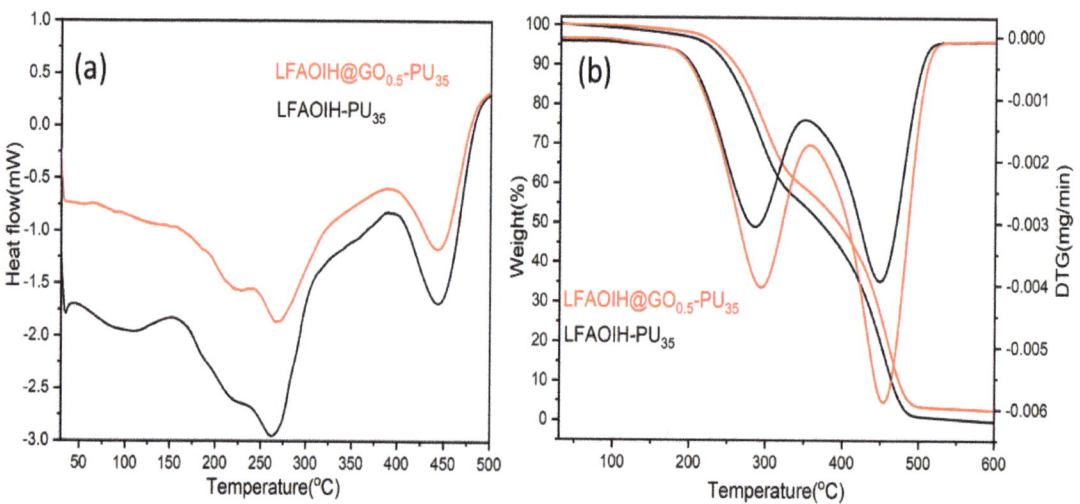

Figure 5. (a) DSC and (b) TGA/DTG thermograms of LFAOIH and LFAOIH@GO$_{0.5}$-PU$_{35}$.

The TGA thermograms of LFAOIH-PU$_{35}$ and LFAOIH@GO$_{0.5}$-PU$_{35}$ exhibit a biphasic degradation pattern, which is also evident in the corresponding DTG thermogram (Figure 5b). Up to 225 °C and 235 °C for LFAOIH-PU$_{35}$ and LFAOIH@GO$_{0.5}$-PU$_{35}$, respectively, a mass loss of 5 wt% is observed due to the evaporation of moisture and volatile impurities. Subsequently, at 475 °C and 480 °C, a significant 90 wt% loss signifies the comprehensive degradation of the oil component, marked by the breakdown of the ester, amide and hydrocarbon chain in LFAOIH-PU$_{35}$ and LFAOIH@GO$_{0.5}$-PU$_{35}$, respectively [14]. These degradation trends are supported by the presence of two endothermic peaks in the DTG thermograms, observed at 275 and 445 °C for LFAOIH-PU$_{35}$ and at 300 and 455 °C for LFAOIH@GO$_{0.5}$-PU$_{35}$ [14]. This improved stability is attributed to the effective dispersion of GO within the matrix, promoting favorable interactions that contribute to the enhanced thermal properties [21].

3.1.7. Contact Angle

Surface wettability was assessed through contact angle (CA) studies, revealing the hydrophilic or hydrophobic nature of the OIH coating. The CA signifies the molecular interaction strength between liquids and solids. By depositing a 0.1 mL double-distilled water droplet onto the pristine interfaces of LFAOIH-PU$_{35}$ and LFAOIH@GO$_{0.5}$-PU$_{35}$ the CA was measured. The outcomes of CA measurements (as depicted in Figure 6) unveiled a rise in the contact angle 73.25 G.U (73.45°–73.04°) in LFAOIH-PU$_{35}$ to 88.64 G.U (88.79°–88.48°) in LFAOIH@GO$_{0.5}$-PU$_{35}$. This shift indicates a subtle enhancement in hydrophobicity following the incorporation of GO [21]. The observed augmentation in hydrophobicity could be attributed to the presence of hydrocarbon chains within the IOH coating [28]. This enhancement in the contact angle value is unequivocally linked to the integration of GO, fostering the development of a nanostructured assembly that augments surface roughness π-π stacking interactions, and the existence of hydrophobic domains. Consequently, such structural enhancement offers promising prospects for leveraging these coatings in humid environments [29].

Figure 6. Contact angle analysis of (**a**) LFAOIH and (**b**) LFAOIH@GO$_{0.5}$-PU$_{35}$.

3.1.8. Morphology

SEM: SEM analysis was systematically executed on LFAOIH@GO$_{0.5}$-PU$_{35}$ in order to comprehensively evaluate its surface morphology and assess the consequential effects stemming from the dispersion of GO within the OIH matrix (Figure 7) [21]. The analytical results unveiled a discernible roughness on the surface, which was attributed to the precise incorporation of GO at the nanoscale. Furthermore, the analysis detected the presence of rod-like or spherical-shaped features, augmenting the overall textural roughness of the surface. The identification of Si is discernible from the observed peaks at 1.7 KeV. The EDX spectra signify the existence of several elements, encompassing C, N, O, and Si. The absence of any other elements further suggests the purity of the polymer matrix.

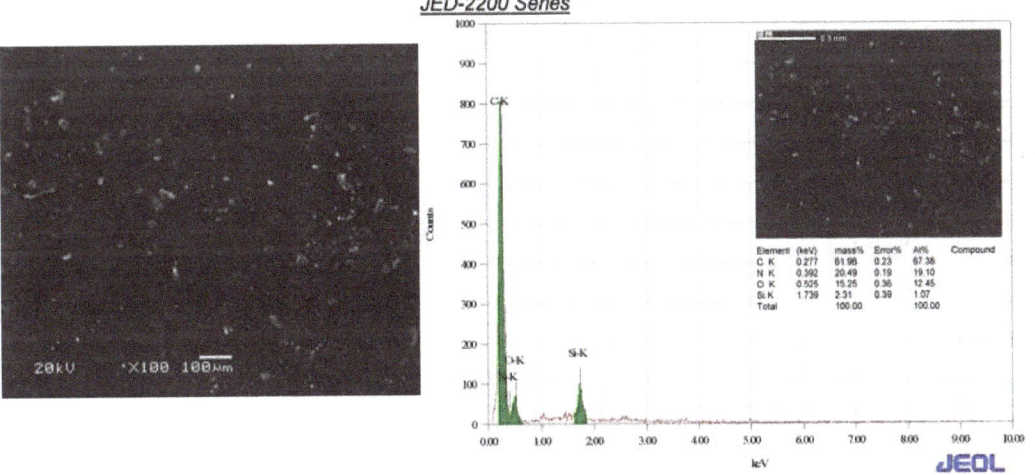

Figure 7. SEM and EDX micrograph of LFAOIH@GO$_{0.5}$-PU$_{35}$.

TEM: TEM studies have elucidated a conspicuous aggregation of particulate material, distinctly distributed across the surface (Figure 8). The densely concentrated dark regions observed in the micrograph are indicative of the uniform dispersion of GO nanoparticles within the LFAOIH@GO$_{0.5}$-PU$_{35}$ matrix [21]. These particulates represent these entities generated through the hydrolytic condensation of TEOS. This phenomenon suggests the occurrence of SiO$_2$ grafting onto the surface of GO [30].

3.1.9. Anticorrosion Study

Electrochemical impedance spectroscopic (EIS) analysis

To assess the anti-corrosive performance of coatings comprising LFAOIH-PU$_{35}$ and LFAOIH@GO$_{0.5}$-PU$_{35}$ on mild steel strips (measuring 1.0 cm^2), EIS measures have been conducted at various time intervals (1, 3, 6, 9, and 12 days) subsequent to exposure to a 3.5% NaCl environment. Figures 9 and 10 present the Nyquist plots for LFAOIH-PU$_{35}$ and LFAOIH@GO$_{0.5}$-PU$_{35}$ over a 12-day exposure period. In the analysis of these Nyquist plots, diverse equivalent circuit models are employed to effectively fit the experimental data. These Nyquist plots serve as valuable tools for unraveling the intricate interactions transpiring at the interface between the metallic substrate and the surrounding solution. They enable us to discern the onset of corrosion, the ingress of water through the coating-metal interface, and the subsequent formation of corrosion byproducts [9]. Furthermore, they facilitate the determination of critical parameters, including solution resistance (Rs), coating resistance (Rc comprising Rp1 and Rp2), and open circuit potential (OCP), as presented in the accompanying Table 2.

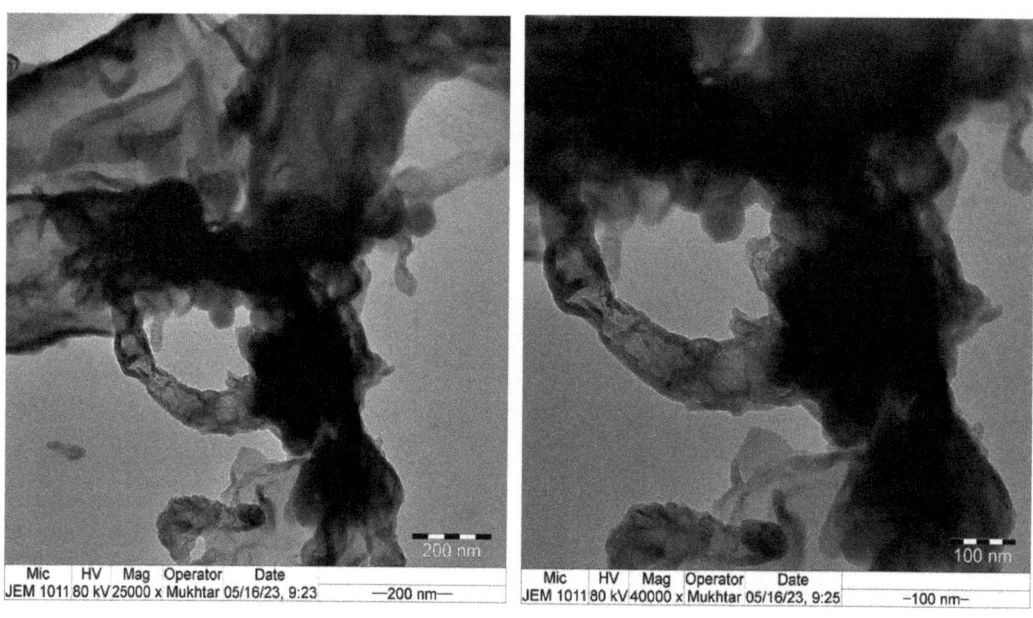

Figure 8. TEM micrograph of LFAOIH@GO$_{0.5}$-PU$_{35}$ at 100 nm and 200 nm.

To gain a more comprehensive understanding of the interaction transpiring at the metal–solution interface, it is imperative to consider the values associated with the constant phase elements (CPE$_1$ and CPE$_2$). The CPE serves as an indicator of the extent to which practical behavior deviates from the ideal capacitive behavior within the framework of the equivalent circuit model. Its inclusion within this model is intended to mitigate systematic errors, ultimately contributing to the precision of the fitting results. The CPE is characterized by an exponent denoted as "n" ($0 \leq n \leq 1$), which quantifies the degree of deviation from ideal dielectric behavior. For LFAOIH-PU$_{35}$, CPE$_1$ signifies a typical double-layered capacitor, characterized by significant porosity, as indicated by its n$_1$ values approximating 1 (0.97), while CPE$_2$ exhibits Warburg-like attributes, with n$_2$ values hovering around 0.56. This suggests a notable level of corrosion resistance for the surface of the manufactured coatings. Conversely, for LFAOIH@GO$_{0.5}$-PU$_{35}$, a single circuit with a CPE value within the range of 0.781 is observed at the 12th day of exposure, indicating that the dispersion of GO contributes to enhanced corrosion resistance [21]. Rc values, directly linked to the anti-corrosion effectiveness of coatings, reveal higher values associated with superior corrosion resistance. Notably, Rp1 represents the charge transfer resistance at the interface between LFAOIH-PU$_{35}$ coatings and the solution, while Rp2 denotes the polarisation resistance at the interface between a corrosion product layer and the solution [21]. The overall polarisation resistance is a result of the parallel combination of Rp1 and Rp2. Conversely, Rs, signifying solution resistance, demonstrates a gradual decline as the exposure period lengthens. This suggests that, over time, the surface integrity of the coated mild steel strips weakens, enabling the easier diffusion of corrosive ions. In cases where coatings lack corrosion resistance, a substantial and rapid decrease in Rs (solution resistance) may occur over time. However, in the present study, we observe a gradual reduction in Rs, with the minimum value reached after a 12 day exposure period. Notably, the data reveal that, even after 12 days of exposure, LFAOIH@GO$_{0.5}$-PU$_{35}$ exhibits a higher Rs value compared to LFAOIH-PU$_{35}$, highlighting the role of GO in providing an effective barrier against the diffusion of corrosive ions and preserving the integrity of the underlying metal surface [21]. Additionally, we observe that the Open Circuit Potential (OCP) values for LFAOIH@GO$_{0.5}$-PU$_{35}$ are initially higher (more positive) on the first day of immersion, but they progressively become more negative as the exposure time lengthens, especially when compared to LFAOIH-PU$_{35}$. This behavior underscores the dynamic response of LFAOIH@GO$_{0.5}$-PU$_{35}$ to the corrosive environment over time.

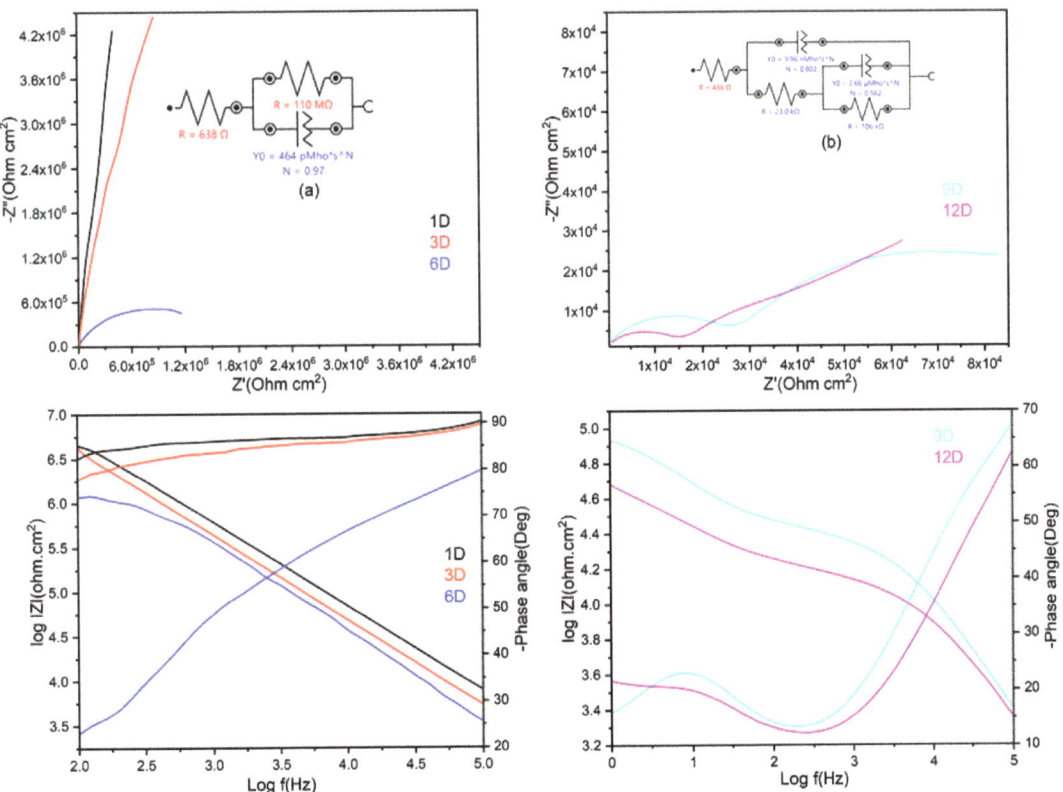

Figure 9. EIS and Bode theta analysis for LFAOIH-PU$_{35}$ for various (1, 3, 9, 12) days.

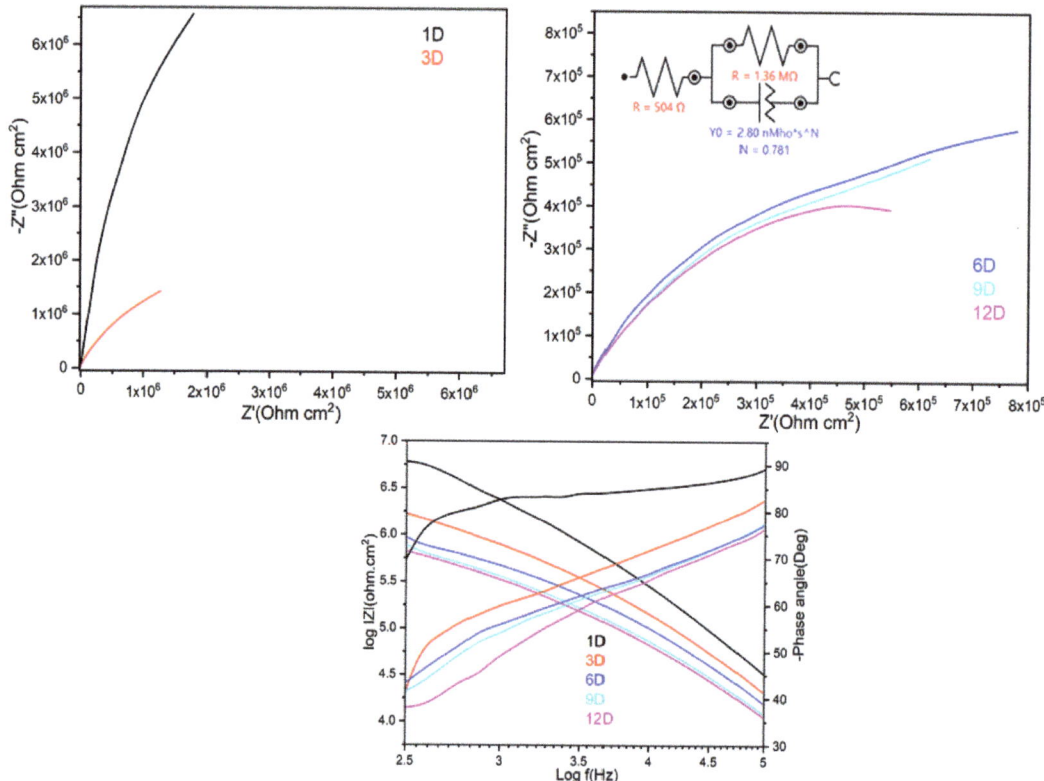

Figure 10. EIS and Bode theta analysis for LFAOIH@GO$_{0.5}$-PU$_{35}$ for various (1, 3, 9, 12) days.

Table 2. EIS studies of LFAOIH-PU$_{35}$ and LFAOIH@GO$_{0.5}$-PU$_{35}$ coatings.

LFAOIH-PU$_{35}$											
Time (Day)	OCP (V)	χ^2	Rs (Ω/cm^2)	Rc ($M\Omega/m^2$)	CPE$_1$						
					Y$_0$, pMho*s^n	n					
1	0.183	0.090	638	110	464	0.97					
3	−0.281	0.243	622	58	443	0.95					
6	−0.409	0.114	615	25	391	0.79				CPE$_2$	
				Rp1 (KΩ/cm^2)	Y$_0$, nMho*s^n	n$_1$		Rp2 (KΩ/cm^2)	Y$_0$, μMho*s^n	n$_2$	
9	−0.473	0.078	436	23	9.96	0.80		106	2.66	0.56	
12	−0.478	0.074	425	11.1	10.2	0.79		297	12.4	0.365	
LFAOIH@GO$_{0.5}$-PU$_{35}$											
Time (Day)	OCP (V)	χ^2	Rs (Ω/cm^2)	Rc ($M\Omega/cm^2$)	CPE						
					Y$_0$, nMho*s^n	n					
1	0.211	0.708	616	53.4	3.58	0.911					
3	−0.311	0.192	572	14.5	3.43	0.801					
6	−0.325	0.157	527	12.5	3.67	0.782					
9	−0.356	0.134	526	1.50	2.97	0.782					
12	−0.402	0.152	504	1.36	2.80	0.781					

The impedance behavior, following different immersion durations, is visually represented in Figure 10 through a Nyquist and Bode plot. The Nyquist plot displays the real impedance against imaginary impedance for pure LFAOIH-PU$_{35}$ and GO dispersed LFAOIH@GO$_{0.5}$-PU$_{35}$ after 1, 3, 6, 9, and 12 days of immersion. The corrosion rate is discernible from the size of the Nyquist plot, where a larger diameter corresponds to a slower corrosion rate or a heightened resistance to corrosive ions [31]. In the case of both LFAOIH-PU$_{35}$ and LFAOIH@GO$_{0.5}$-PU$_{35}$ coatings, an initial phase featuring a linear trend was noted on the first day of exposure. This occurrence can be attributed to the fact that not all organic coatings exhibit long-term impermeability. Consequently, ions and water have the potential to infiltrate through the capillary channels and pores inherent in the coating structure, leading to a reduction in resistance (Rc). With prolonged exposure of a coated surface to a corrosive environment, any voids or capillary channels within the coating permit the ingress of water molecules and corrosive ions through the coating–metal interface. This ingress subsequently results in a reduction in resistance over extended periods of immersion. Notably, an arc emerges at the initiation of the first semicircle in the EIS spectrum after 3 days of submersion for both LFAOIH-PU$_{35}$ and LFAOIH@GO$_{0.5}$-PU$_{35}$ coatings [9]. Furthermore, it is worth emphasizing that the radii of these semicircles exhibit a gradual reduction over time. This trend indicates a potential acceleration in the corrosion rate, which can be attributed to various factors such as the formation of fresh coating pores, an expansion in the exposed surface area of pre-existing pores, or the existence of structural irregularities and imperfections within the coating matrix. After 9 days of exposure, a semicircle with a straight line was observed for LFAOIH-PU$_{35}$, which indicated a combination of two impedance components. The semicircle represents a capacitive or diffusive process, while the straight line corresponds to a purely resistive process. This combination often indicates a complex electrochemical behavior at the studied interface. The semicircle portion may be associated with charge transfer resistance or the presence of a protective barrier, while the straight line suggests a more simplistic resistive behavior, possibly related to ionic conduction or diffusion through the coating with the value of 6×10^4 ohm-cm^2. Whereas, for LFAOIH@GO$_{0.5}$-PU$_{35}$, the arc of the semicircle has been decreased, which is found in the range of 6×10^5 ohm-cm^2, which is higher as compared to the case of LFAOIH-PU$_{35}$.

Moreover, the assessment of corrosion resistance was corroborated through the utilization of a Bode plot (Figure 9). Bode impedance and phase angle graphs were generated as a function of frequency subsequent to the immersion of the coatings in 3.5% NaCl solutions. The logarithm of impedance (Log Z) exhibits a consistent monotonic decrease across all frequencies as the coatings undergo immersion, indicative of capacitive behavior for both LFAOIH-PU$_{35}$ and LFAOIH@GO$_{0.5}$-PU$_{35}$ [9,32]. The impedance modulus (|Z| value), which serves as a semiquantitative indicator of the coating's potential as a barrier, exhibits a direct correlation with the coating's resistance and capacitance. Figure 10 illustrates higher phase angles (37°) and elevated impedance values than LFAOIH@GO$_{0.5}$-PU$_{35}$, persisting even after exposure over 12 days. These observations show that both the coatings show good accounts for corrosion resistive behavior, but LFAOIH@GO$_{0.5}$-PU$_{35}$ has higher anticorrosion properties, reinforcing its efficacy in mitigating corrosion.

Tafel Analysis

Tafel slopes are determined by extrapolating data from the Tafel region within polarization curves to examine whether the coatings are cathodic or anodic or both (Figure 11). A Tafel plot is employed to visually examine electrochemical characteristics, portraying the X-axis as potential and the Y-axis as the logarithm of current density. As evident in Figure 11, treating a mild steel substrate with LFAOIH-PU$_{35}$ and LFAOIH@GO$_{0.5}$-PU$_{35}$ coatings leads to a shift in both the anodic and cathodic branches towards lower current densities showing the corrosion inhibition. In Table 3, we present critical electrochemical parameters for LFAOIH-PU$_{35}$ and LFAOIH@GO$_{0.5}$-PU$_{35}$, E_{corr}, I_{corr}, corrosion rate, Open Circuit Potential (OCP), and IE values. E_{corr} and I_{corr} are interconnected through fundamental electrochemical principles, enabling the assessment and comparison of corrosion resistance in diverse materials and environments. A more positive E_{corr} signifies superior corrosion resistance, while a lower I_{corr} value suggests enhanced corrosion resistance. Analysis of Table 3 reveals that the E_{corr} value for LFAOIH@GO$_{0.5}$-PU$_{35}$ remains more positive ($-0.424 > -0.557$) in comparison to LFAOIH-PU$_{35}$, even after 12 days. Conversely, Icorr values increase for both materials with prolonged exposure. The decrease in IE% over time is minimal, even after 12 days of immersion. These findings underscore the superior corrosion resistance of LFAOIH@GO$_{0.5}$-PU$_{35}$ nanocomposites over LFAOIH-PU$_{35}$. This enhanced corrosion resistance is attributed to the incorporation of GO, which significantly augments the barrier effect, thereby suppressing the rate of the anodic reaction (primarily the oxidation of iron) and retarding the corrosion process. Consequently, LFAOIH@GO$_{0.5}$-PU$_{35}$ exhibits notable anticorrosive properties as a result of these aforementioned mechanisms.

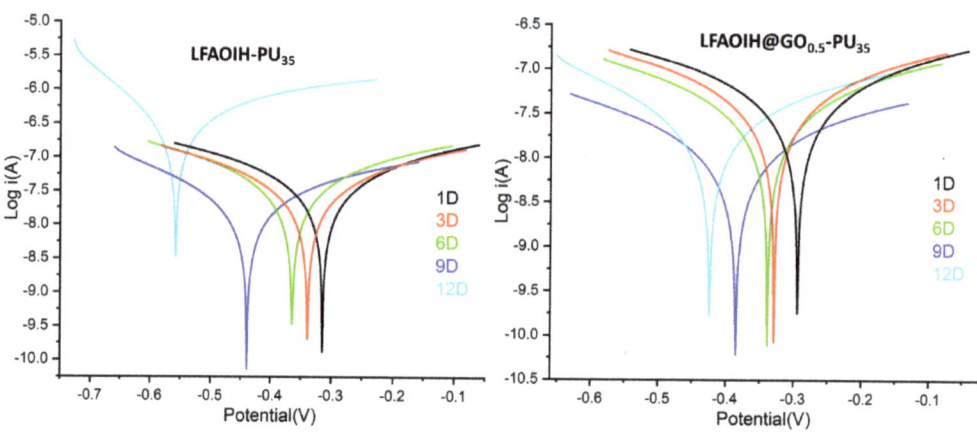

Figure 11. Tafel analysis for LFAOIH-PU$_{35}$ and LFAOIH@GO$_{0.5}$-PU$_{35}$ for various (1, 3, 9, 12) days.

Table 3. Corrosion Potentiodynamic of LFAOIH-PU$_{35}$ and LFAOIH@GO$_{0.5}$-PU$_{35}$ coatings.

Immersion Time (Day)	E_{corr} (V)	i_{corr} (A/cm^2)	Corrosion Rate (mm/Year)	LPR (Ω)	OCP(V)	IE(%)
			LFAOIH-PU$_{35}$			
1	−0.314	1.282×10^{-8}	2.430×10^{-4}	3.895×10^{6}	−0.311	99.97
3	−0.338	2.100×10^{-8}	2.689×10^{-4}	2.745×10^{6}	−0.333	99.95
6	−0.364	2.314×10^{-8}	2.930×10^{-4}	2.683×10^{6}	−0.355	99.94
9	−0.440	2.521×10^{-8}	3.091×10^{-4}	2.304×10^{6}	−0.410	99.94
12	−0.557	1.591×10^{-7}	1.845×10^{-3}	1.928×10^{5}	−0.478	99.96
			LFAOIH@GO$_{0.5}$-PU$_{35}$			
1	−0.303	2.170×10^{-8}	2.249×10^{-4}	3.846×10^{6}	−0.300	99.95
3	−0.328	2.780×10^{-8}	2.521×10^{-4}	2.501×10^{6}	−0.325	99.34
6	−0.338	2.936×10^{-8}	3.245×10^{-4}	2.315×10^{6}	−0.333	99.93
9	−0.385	2.350×10^{-8}	5.556×10^{-4}	2.024×10^{6}	−0.382	99.94
12	−0.424	7.016×10^{-7}	8.153×10^{-3}	7.909×10^{5}	−0.402	99.83

4. Conclusions

An environmentally friendly and efficient synthesis method has been developed for the production of LFAOIH@GO$_{0.5}$-PU$_{35}$ using *Leucaena leucocephala*, a bio-resource. This approach involves the synthesis of LFAOIH-PU with 30–40% IPDI and the addition of a small quantity (0.25–0.75 wt%) of GO to create an LFAOIH@GO$_{0.5}$-PU$_{35}$ nanocomposite. The formulations selected for further investigation were LFAOIH-PU$_{35}$ and LFAOIH@GO$_{0.5}$-PU$_{35}$. The chemical structure of these nanocomposites was elucidated using various analytical techniques, including FTIR, ^{1}H, ^{13}C, ^{29}Si NMR, and XPS analysis. The study comprehensively explored the influence of different GO loadings on the nanocomposites structural, morphological, thermal stability, mechanical properties, and corrosion resistance. Several mechanical properties, including scratch hardness, pencil hardness, impact resistance, bending, gloss value, crosshatch adhesion, and thickness, were meticulously examined. Morphological characteristics were investigated through XRD, FE-SEM, EDX, and TEM analyses, revealing distinct surface roughness features. Remarkably, the resulting nanocomposites, particularly LFAOIH@GO$_{0.5}$-PU$_{35}$, displayed substantial improvements in mechanical properties, including significantly enhanced scratch hardness (3 Kg), pencil hardness (5H), and increased gloss values (79). Moreover, these nanocomposites exhibited exceptional thermal stability, retaining their structural integrity at temperatures up to 480 °C. Additionally, they demonstrated enhanced surface hydrophobicity, as indicated by a contact angle of 88.79°/88.48°, achieved with a minimal (0.50%) loading of GO. Crucially, EIS studies, including Nyquist, Bode, and Tafel plots, provided compelling evidence of the outstanding anti-corrosion performance of these nanocomposites. Consequently, these studied nanocomposites hold significant promise for the large-scale production of robust anticorrosive coatings with enhanced mechanical properties and durability.

The future competitiveness of OIH nanocomposite anticorrosive coatings is contingent on their capacity to address the mounting demand for corrosion protection in diverse industrial sectors. Ongoing innovations in nanotechnology hold the potential to enhance performance through the development of advanced nanomaterials. The ability to tailor these coatings to meet the specific corrosion challenges of various industries, coupled with their long-term durability and cost-effectiveness, will be pivotal for maintaining a competitive edge. These coatings are well-positioned to not only provide corrosion resistance, but can also offer supplementary functionalities such as self-healing and antimicrobial properties, rendering them multifaceted. Furthermore, ensuring compatibility with emerging materials and substrates, expanding their global reach, substantial investments in research and development, and fostering collaboration with scientists and industry stakeholders are integral to their sustained competitiveness within a dynamic and burgeoning market.

Author Contributions: W.A.-o.: methodology, data, generation, drafting; N.M.A.: formal analysis, project administration: Y.M.A.-M.: formal analysis; M.A.: formal analysis, funding acquisition, writing—review and editing, manuscript handling. All authors have read and agreed to the published version of the manuscript.

Funding: Researchers Supporting Project (RSP2023R113), King Saud University, Riyadh, Saudi Arabia.

Data Availability Statement: The data are available from the corresponding author on reasonable request.

Acknowledgments: The authors are grateful to the Researchers Supporting Project number (RSP2023R113), King Saud University, Riyadh, Saudi Arabia for the support.

Conflicts of Interest: The authors state no conflict of interest.

References

1. Bhardwaj, N.; Sharma, P.; Kumar, V. Phytochemicals as Steel Corrosion Inhibitor: An Insight into Mechanism. *Corros. Rev.* **2021**, *39*, 27–41. [CrossRef]
2. Eliaz, N. Corrosion of Metallic Biomaterials: A Review. *Materials* **2019**, *12*, 407. [CrossRef] [PubMed]
3. Types of Corrosive Environments. In *Corrosion: Understanding the Basics*; ASM International: New York, NY, USA, 2000; pp. 193–236.
4. Available online: https://www.marketsandmarkets.com/Market-Reports/corrosion-protection-coating-market-150815310.html (accessed on 18 October 2023).

5. Ahmad, S.; Zafar, F.; Sharmin, E.; Garg, N.; Kashif, M. Synthesis and Characterization of Corrosion Protective Polyurethane-fattyamide/Silica Hybrid Coating Material. *Prog. Org. Coat.* **2012**, *73*, 112–117. [CrossRef]
6. Hammer, P.; Uvida, M.C.; Trentin, A. Self-Healing Organic-Inorganic Coatings. *Coatings* **2022**, *12*, 1668. [CrossRef]
7. Vijayan, J.G.; Chandrashekar, A.; AG, J.; Prabhu, T.N.; Kalappa, P. Polyurethane and Its Composites Derived from Bio-Sources: Synthesis, Characterization and Adsorption Studies. *Polym. Polym. Compos.* **2022**, *30*, 09673911221110347. [CrossRef]
8. Siyanbola, T.O.; Adebowale, A.D.; Adeboye, S.A.; Rao, S.J.V.; Ndukwe, N.A.; Sodiya, E.F.; Ajayi, A.A.; Akintayo, E.T.; Basak, P.; Narayan, R. Development of Functional Polyurethane-Cenosphere Hybrid Composite Coatings from Ricinus Communis Seed Oil. *Sci. Afr.* **2023**, *20*, e01711. [CrossRef]
9. Shaily; Shahzaib, A.; Zafar, F.; Khan, S.; Kaur, B.; Ghosal, A.; Alam, M.; Azam, M.; Haq, Q.M.R.; Nishat, N. Superhydrophobic Mn(II)-Coordinated Technical Cashew Nut Shell Liquid-Based Bactericidal and Corrosion-Resistant Advanced Polyurethane Coatings. *Mater. Today Commun.* **2023**, *35*, 105947. [CrossRef]
10. Ghasemlou, M.; Daver, F.; Ivanova, E.P.; Adhikari, B. Polyurethanes from Seed Oil-Based Polyols: A Review of Synthesis, Mechanical and Thermal Properties. *Ind. Crops Prod.* **2019**, *142*, 111841. [CrossRef]
11. He, S.; Wei, G.; Zhang, Z.; Yang, L.; Lin, Y.; Du, L.; Du, X. Incorporation of Graphene Oxide Modified with Polyamide Curing Agent into the Epoxy–Zinc Composite Coating for Promoting Its Corrosion Resistance. *Polymers* **2023**, *15*, 1873. [CrossRef]
12. Alrashed, M.M.; Soucek, M.D.; Jana, S.C. Role of Graphene Oxide and Functionalized Graphene Oxide in Protective Hybrid Coatings. *Prog. Org. Coat.* **2019**, *134*, 197–208. [CrossRef]
13. Harb, S.V.; Pulcinelli, S.H.; Santilli, C.V.; Knowles, K.M.; Hammer, P. A Comparative Study on Graphene Oxide and Carbon Nanotube Reinforcement of PMMA-Siloxane-Silica Anticorrosive Coatings. *ACS Appl. Mater. Interfaces* **2016**, *8*, 16339–16350. [CrossRef] [PubMed]
14. Alam, M.; Alandis, N.M.; Sharmin, E.; Ahmad, N.; Alrayes, B.F.; Ali, D. Characterization of Leucaena (*Leucaena leucephala*) Oil by Direct Analysis in Real Time (DART) Ion Source and Gas Chromatography. *Grasas Y Aceites* **2017**, *68*, 190. [CrossRef]
15. Zafar, F.; Zafar, H.; Sharmin, E.; Ashraf, S.M.; Ahmad, S. Studies on Ambient Cured Biobased Mn(II), Co(II) and Cu(II) Containing Metallopolyesteramides. *J. Inorg. Organomet. Polym. Mater.* **2011**, *21*, 646–654. [CrossRef]
16. Alam, M.; Alandis, N.M.; Sharmin, E.; Ahmad, N.; Husain, F.M.; Khan, A. Mechanically Strong, Hydrophobic, Antimicrobial, and Corrosion Protective Polyesteramide Nanocomposite Coatings from Leucaena Leucocephala Oil: A Sustainable Resource. *ACS Omega* **2020**, *5*, 30383–30394. [CrossRef] [PubMed]
17. Naebe, M.; Wang, J.; Amini, A.; Khayyam, H.; Hameed, N.; Li, L.H.; Chen, Y.; Fox, B. Mechanical Property and Structure of Covalent Functionalised Graphene/Epoxy Nanocomposites. *Sci. Rep.* **2014**, *4*, 4375. [CrossRef]
18. Martinez-Felipe, A.; Cook, A.G.; Abberley, J.P.; Walker, R.; Storey, J.M.D.; Imrie, C.T. An FT-IR Spectroscopic Study of the Role of Hydrogen Bonding in the Formation of Liquid Crystallinity for Mixtures Containing Bipyridines and 4-Pentoxybenzoic Acid. *RSC Adv.* **2016**, *6*, 108164–108179. [CrossRef]
19. Huang, X.; Nakagawa, S.; Houjou, H.; Yoshie, N. Insights into the Role of Hydrogen Bonds on the Mechanical Properties of Polymer Networks. *Macromolecules* **2021**, *54*, 4070–4080. [CrossRef]
20. Li, W.; Ma, J.; Wu, S.; Zhang, J.; Cheng, J. The Effect of Hydrogen Bond on the Thermal and Mechanical Properties of Furan Epoxy Resins: Molecular Dynamics Simulation Study. *Polym. Test.* **2021**, *101*, 107275. [CrossRef]
21. Al-Otaibi, W.; Alandis, N.M.; Alam, M. Leucaena Leucocephala Oil-Based Poly Malate-Amide Nanocomposite Coating Material for Anticorrosive Applications. *e-Polymers* **2023**, *23*, 20230036. [CrossRef]
22. Alam, M.; Alandis, N.M.; Zafar, F.; Ghosal, A.; Ahmed, M. Linseed Oil Derived Terpolymer/Silica Nanocomposite Materials for Anticorrosive Coatings. *Polym. Eng. Sci.* **2021**, *61*, 2243–2256. [CrossRef]
23. Ramezanzadeh, B.; Ahmadi, A.; Mahdavian, M. Enhancement of the Corrosion Protection Performance and Cathodic Delamination Resistance of Epoxy Coating through Treatment of Steel Substrate by a Novel Nanometric Sol-Gel Based Silane Composite Film Filled with Functionalized Graphene Oxide Nanosheets. *Corros. Sci.* **2016**, *109*, 182–205. [CrossRef]
24. Zhang, C.; Zhang, R.Z.; Ma, Y.Q.; Guan, W.B.; Wu, X.L.; Liu, X.; Li, H.; Du, Y.L.; Pan, C.P. Preparation of Cellulose/Graphene Composite and Its Applications for Triazine Pesticides Adsorption from Water. *ACS Sustain. Chem. Eng.* **2015**, *3*, 396–405. [CrossRef]
25. Ghosh, T.; Karak, N. Mechanically Robust Hydrophobic Interpenetrating Polymer Network-Based Nanocomposite of Hyper-branched Polyurethane and Polystyrene as an Effective Anticorrosive Coating. *New J. Chem.* **2020**, *44*, 5980–5994. [CrossRef]
26. Liang, G.; Yao, F.; Qi, Y.; Gong, R.; Li, R.; Liu, B.; Zhao, Y.; Lian, C.; Li, L.; Dong, X.; et al. Improvement of Mechanical Properties and Solvent Resistance of Polyurethane Coating by Chemical Grafting of Graphene Oxide. *Polymers* **2023**, *15*, 882. [CrossRef]
27. Jiao, X.; Qiu, Y.; Zhang, L.; Zhang, X. Comparison of the Characteristic Properties of Reduced Graphene Oxides Synthesized from Natural Graphites with Different Graphitization Degrees. *RSC Adv.* **2017**, *7*, 52337–52344. [CrossRef]
28. Ashok Kumar, S.S.; Bashir, S.; Ramesh, K.; Ramesh, S. A Comprehensive Review: Super Hydrophobic Graphene Nanocomposite Coatings for Underwater and Wet Applications to Enhance Corrosion Resistance. *FlatChem* **2022**, *31*, 100326. [CrossRef]
29. Selim, M.S.; El-Safty, S.A.; Shenashen, M.A.; El-Sockary, M.A.; Elenien, O.M.A.; EL-Saeed, A.M. Robust Alkyd/Exfoliated Graphene Oxide Nanocomposite as a Surface Coating. *Prog. Org. Coat.* **2019**, *126*, 106–118. [CrossRef]
30. Jiang, W.; Sun, C.; Zhang, Y.; Xie, Z.; Zhou, J.; Kang, J.; Cao, Y.; Xiang, M. Preparation of Well-Dispersed Graphene Oxide-Silica Nanohybrids/Poly(Lactic Acid) Composites by Melt Mixing. *Polym. Test.* **2023**, *118*, 107912. [CrossRef]

1. Liu, W.; Wang, Q.; Hao, J.; Zou, G.; Zhang, P.; Wang, G.; Ai, Z.; Chen, H.; Ma, H.; Song, D. Corrosion Resistance and Corrosion Interface Characteristics of Cr-Alloyed Rebar Based on Accelerated Corrosion Testing with Impressed Current. *J. Mater. Res. Technol.* **2023**, *22*, 2996–3009. [CrossRef]
2. Nunes, M.S.; Bandeira, R.M.; Figueiredo, F.C.; dos Santos Junior, J.R.; de Matos, J.M.E. Corrosion Protection of Stainless Steel by a New and Low-cost Organic Coating Obtained from Cashew Nutshell Liquid. *J. Appl. Polym. Sci.* **2023**, *140*, e53420. [CrossRef]

Disclaimer/Publisher's Note: The statements, opinions and data contained in all publications are solely those of the individual author(s) and contributor(s) and not of MDPI and/or the editor(s). MDPI and/or the editor(s) disclaim responsibility for any injury to people or property resulting from any ideas, methods, instructions or products referred to in the content.

Article

Incorporation of Graphene Nanoplatelets into Fiber-Reinforced Polymer Composites in the Presence of Highly Branched Waterborne Polyurethanes

Ayşe Durmuş-Sayar [1,2], Murat Tansan [1], Tuğçe Çinko-Çoban [1], Dilay Serttan [1], Bekir Dizman [1,2], Mehmet Yildiz [1,2] and Serkan Ünal [1,2,*]

[1] Integrated Manufacturing Technologies Research and Application Center & Composite Technologies Centre of Excellence, Sabanci University, Teknopark Istanbul, Pendik, Istanbul 34906, Turkey; aysedurmus@sabanciuniv.edu (A.D.-S.); bekirdizman@sabanciuniv.edu (B.D.); meyildiz@sabanciuniv.edu (M.Y.)
[2] Faculty of Engineering and Natural Sciences, Sabanci University, Tuzla, Istanbul 34956, Turkey
* Correspondence: serkan.unal@sabanciuniv.edu

Citation: Durmuş-Sayar, A.; Tansan, M.; Çinko-Çoban, T.; Serttan, D.; Dizman, B.; Yildiz, M.; Ünal, S. Incorporation of Graphene Nanoplatelets into Fiber-Reinforced Polymer Composites in the Presence of Highly Branched Waterborne Polyurethanes. *Polymers* **2024**, *16*, 828. https://doi.org/10.3390/polym16060828

Academic Editors: Zhuohong Yang and Chang-An Xu

Received: 26 January 2024
Revised: 5 March 2024
Accepted: 12 March 2024
Published: 16 March 2024

Copyright: © 2024 by the authors. Licensee MDPI, Basel, Switzerland. This article is an open access article distributed under the terms and conditions of the Creative Commons Attribution (CC BY) license (https://creativecommons.org/licenses/by/4.0/).

Abstract: Enhancing interfacial interactions in fiber-reinforced polymer composites (FRPCs) is crucial for improving their mechanical properties. This can be achieved through the incorporation of nanomaterials or chemically functional agents into FRPCs. This study reports the tailoring of the fiber–matrix interface in FRPCs using non-functionalized graphene nanoplatelets (GNPs) in combination with a waterborne, highly branched, multi-functional polyurethane dispersion (HBPUD). A unique ultrasonic spray deposition technique was utilized to deposit aqueous mixtures of GNP/HBPUDs onto the surfaces of carbon fiber fabrics, which were used to prepare epoxy-prepreg sheets and corresponding FRPC laminates. The influence of the polyurethane (PU) and GNP content and their ratio at the fiber–matrix interface on the tensile properties of resulting high-performance composites was systematically investigated using stress–strain analysis of the produced FRPC plates and SEM analysis of their fractured surfaces. A synergistic stiffening and toughening effect was observed when as low as 20 to 30 mg of GNPs was deposited per square meter of each side of the carbon fiber fabrics in the presence of the multi-functional PU layer. This resulted in a significant improvement in the tensile strength from 908 to 1022 MPa, while maintaining or slightly improving the initial Young's modulus from approximately 63 to 66 MPa.

Keywords: highly branched functional waterborne polyurethane; carbon fiber-reinforced composites; graphene nanoplatelets; fiber sizing; interfacial properties; ultrasonic spray coating

1. Introduction

Carbon fiber-reinforced polymer composites (CFRPCs) continue to replace traditional materials due to their distinctive features such as high strength, stiffness, and long service life for lightweight structural composites [1–3]. CFRPCs are increasingly used in aerospace, automotive, electronics, and other applications. Interfacial properties between the fiber and polymer matrix are an important factor determining the performance of CFRPCs in both thermosetting and thermoplastic resin-based systems [4]. Typically, while the surface of the virgin fiber is non-polar, the polymer matrix in CFRPCs tends to have a polar character. This inherent difference in polarity necessitates the enhancement of the naturally weak interfacial interaction between the fiber and the matrix to meet the performance requirements expected from composite materials. The effectiveness of load transfer is often ascribed to the interaction between the fiber and the matrix. If this interaction is too weak, stress transfer becomes limited in composite structures, leading to a compromise in performance. Consequently, poor interfacial adhesion diminishes the magnitude of load transfer between the matrix and fibers. Conversely, when the interfacial interaction intensifies significantly, cracks tend to propagate diagonally in the matrix, breaking the fibers. Striking a balance in interfacial adhesion is crucial as overly weak or overly strong interactions can adversely impact the load transfer phenomenon and, consequently, the overall structural integrity of the composite material. Achieving an optimal level of interfacial interaction is imperative for maximizing the performance and mechanical properties of CFRPCs. As a result, the stress concentration tends to be higher around these breakages [5–7]. The load-carrying capacity of composite materials hinges predominantly on the nature of the fiber–matrix bonding, encompassing both chemical and frictional interactions [8,9]. Extensive studies in the literature have elucidated diverse methodologies for the surface modification of carbon fibers, including wet chemical or electrochemical treatments, polymer coating, and plasma treatment [10–17]. These techniques aim to introduce various functional groups onto the carbon fiber surface, fostering robust adhesion between the fiber and the matrix. Through such modifications, researchers strive to optimize the interface, ensuring enhanced compatibility and, consequently, bolstering the composite material's overall mechanical performance.

Carbon fibers inherently exhibit brittleness and low elongation, resulting in challenges such as yarn breakage and fluffiness during the manufacturing of CFRPCs. This inherent fragility necessitates surface treatment interventions. An effective sizing component becomes crucial, not only for enhancing chemical interactions between the fiber and the matrix to elevate interfacial adhesion properties but also for improving fiber bundling and overall performance characteristics [18]. Various functional groups, such as alcohol, carbonyl, and carboxylic acid, can be strategically incorporated onto the fiber surface through diverse sizing methods [19]. Furthermore, sizing facilitates the modification of the surface free energy of carbon fibers, thereby refining the interfacial features of composites. This, in turn, contributes to heightened mechanical characteristics in comparison with composites produced with untreated carbon fibers.

In contemporary fiber sizing applications, waterborne polyurethane dispersions (PUDs) have garnered significant attention owing to their exceptional coating behavior and multi-functionality. PUDs also stand out for their environmental friendliness, non-toxicity, low viscosity, and remarkable adhesion capabilities with diverse polymeric matrices in composites [20–22]. An especially attractive feature is their ability to establish robust adhesion without requiring pretreatment of the fibers. This not only simplifies the sizing process but also aligns with environmentally conscious practices, making PUDs a compelling choice in fiber sizing for their versatile and eco-friendly attributes. PUDs emerge as highly suitable sizing agents also for carbon fibers in CFRPCs [23]. This suitability is attributed to their inherent polarity, marked by an ability to form effective bonds with the carbon fiber surface. Furthermore, the high elasticity and ductility of polyurethanes present an advantageous combination, offering the flexibility to tailor these properties based on the specific hard and soft segment structures within the backbone. This tailoring capability enables one to match the requirements of the carbon fiber reinforcement, contributing to enhanced compatibility and overall performance in CFRPCs [24,25]. Zhang et al. [26] previously documented that the treatment of carbon fibers with PUDs resulted in an elevation of surface energy. This increase was attributed to the introduction of nitrogen (N) atoms on the fiber surface through the treatment with PUDs. Consequently, the carbon fibers exhibited heightened wettability when combined with epoxy resin in CFRPCs. Such enhanced wettability is a key factor in promoting a more effective and intimate bonding between the carbon fibers and the epoxy resin matrix, contributing to improved overall performance in the resulting composite materials. Li et al. [27] conducted a study wherein waterborne PUDs based on a tartaric acid polyol were synthesized specifically for carbon fiber sizing. CFRPCs incorporating carbon fibers sized with PUDs demonstrated a remarkable improvement, exhibiting a 14.3% increase in tensile strength, a 24.4% increase in flexural strength, and an impressive 119.6% increase in impact strength when compared with CFRPCs from pristine carbon fibers. Fazeli et al. [28] conducted a study on the surface treatment of recycled carbon fibers using a combination of PUDs and silane compounds, exploring its impact on the mechanical properties of ensuing composites. The application of a flexible coating comprising PUDs crosslinked with silane coupling agents onto the recycled carbon fiber surface yielded noteworthy enhancements in impact, tensile, and flexural strengths in epoxy-based CFRPCs.

In addition to the use of linear PUDs, there is a growing interest in incorporating hyperbranched polymers into fiber-reinforced polymer composites (FRPCs) as a means to enhance interfacial properties. Hyperbranched polymers exhibit spherical dendritic structures with cavities and numerous terminal functional groups. These distinctive features allow hyperbranched polymers to contribute to both mechanical interlocking and chemical bonding between the polymer matrix and carbon fibers. The incorporation of hyperbranched polymers represents a versatile strategy for optimizing the interface in composite materials based on carbon or glass fibers, which are typically supplied with commercial sizings with limited information on their nature and functionality. Thus, the incorporation of HBPs may broaden the range of materials and methodologies available for advancing FRPC technology [29,30].

The incorporation of carbon nanomaterials into FRPCs, specifically on the fiber surface, has been reported as a feasible approach to enhance the mechanical properties of such composite structures [31]. For example, Zhang et al. [32] produced composites with dispersed graphene oxide (GO) layers directly integrated onto the surface of individual carbon fibers as part of the fiber sizing process. The incorporation of 5 wt% GO sheets in this manner led to significant improvements in the interfacial and tensile properties of the resulting CFRPCs, highlighting the potential of integrating GO into fiber sizing as an effective strategy. Xiong et al. [33] introduced a novel strategy for enhancing the interface and mechanical properties of CFRPCs by grafting GO onto carbon fibers with HBPs using thiol-ene click chemistry and a vinyl-terminated hyperbranched polyester. The tensile and flexural strengths of corresponding CFRPCs increased by 47.6% and 65.8%, respectively.

CFRPCs obtained from carbon fibers modified with graphene nanoplatelets (GNPs) have been shown to exhibit enhanced mechanical and thermal properties [34]. A solution comprising GNPs in acetone, along with a small amount of resin/hardener, was formulated as a spraying solution for modifying dry fabrics suitable for the vacuum-assisted resin transfer infusion (VARI) process. Through this method, GNP-reinforced FRPCs were successfully fabricated, with GNPs uniformly distributed in the interlaminar regions. Analyses revealed effective immobilization of GNPs on the surfaces of carbon fibers post-spray coating. Moreover, significant enhancements were observed in the mechanical properties and thermal conductivity of the resulting epoxy-based CFRPCs. Specifically, the incorporation of 0.3 wt% GNPs led to the highest levels of flexural strength and interlaminar shear strength. Other studies have explored different matrices beyond epoxy in conjunction with GNPs and GOs to enhance the mechanical properties of FRPCs. Li et al. [35] successfully enhanced the interfacial properties of CF/copoly(phthalazinone ether sulfone)s (PPBESs)-based composites by incorporating multi-scale hybrid carbon fiber/GO (CF/GO) reinforcements. An optimized GO loading of 0.5% with a homogeneous distribution of GO by coating the hybrid fiber surface led to significant improvements in the PPBES composite's interlaminar shear strength, reaching 91.5 MPa, and flexural strength, reaching 1886 MPa. These enhancements represented increases of 16.0% and 24.1%, respectively, compared with the non-reinforced counterpart. Furthermore, a reduction in the interface debonding in CF/GO (0.5%) composites suggested superior interface adhesion due to the incorporation of GO into the interface. Choi et al. [36] investigated the influence of nanomaterials and fiber interface angles on the mode I fracture toughness of woven CFRPCs. Three types of carbon nanomaterials—COOH-functionalized short multi-walled carbon nanotubes (S-MWCNT-COOH), MWCNTs, and GNPs—were investigated. Specimens were fabricated using the hand lay-up method, comprising 12 woven carbon fiber fabrics with or without 1 wt% nanomaterials. The incorporation of nanomaterials led to a mode I fracture toughness exceeding that of pure CFRP. Notably, the utilization of GNPs demonstrated superior effectiveness in enhancing the fracture toughness compared with other nanomaterials. Costa et al. [37] reported the improvement of the tensile strength of a high-density polyethylene-based FRPC with natural fibers by incorporating GNPs into the matrix. An increase of over 20% in the Young's modulus was achieved compared with the high-density polyethylene composite alone, reaching 1.63 ± 0.15 GPa.

Although the vast majority of the literature studies offer unique strategies for enhancing interfacial interactions between the matrix and the fiber surface in FRPCs by the incorporation of various nanoparticles into this interface for improved mechanical properties, it becomes increasingly difficult to demonstrate such enhancements and improvements, particularly in the tensile properties of high-performance FRPCs with tensile strength (>900 MPa) and Young's modulus (>60 GPa) values that are notably high to start with. On the other hand, the individual use of PUDs as post-sizing agents and GNPs as reinforcing agents has been separately demonstrated to effectively tailor the interface of FRPCs in previous studies. However, the combined incorporation of carbon nanomaterials in the presence of chemically functional PUDs into the fiber–matrix interface of FRPCs remains an area worthy of investigation for potential synergistic effects in enhancing the mechanical properties of high-performance FRPCs.

In this manuscript, the fiber–matrix interface of a high-performance CFRPC structure was tailored with GNPs in the presence of a waterborne, multi-functional polyurethane in an effort to improve the tensile properties of corresponding CFRPCs. GNPs were incorporated into the fiber–matrix interface of CFRPCs in the presence of a waterborne, highly branched, multi-functional polyurethane dispersion (HBPUD). The in-house synthesized HBPUD, possessing both amine and silane terminal groups in the backbone, was designed to act as both a dispersing agent for GNPs in aqueous media during the spray deposition and a reactive sizing agent for covalent bridging between the carbon fiber, epoxy matrix, and GNPs at the interface of the corresponding composite structures upon curing. For this purpose, aqueous dispersions containing various ratios of HBPUDs and GNPs were introduced onto carbon fiber fabric surfaces via a novel ultrasonic spray deposition technique to ensure a fine distribution of GNP particles and homogeneous surface coverage, followed by the fabrication of prepreg laminates using hot melt epoxy films to obtain CFRPC plates by stacking them and curing in an autoclave. In a comprehensive examination, the effects of the presence of a multi-functional polyurethane layer at the interface, the ratio of the polyurethane to GNPs, and the overall GNP content per unit area of the carbon fiber fabric on the tensile strength and Young's modulus of the corresponding CFRPC plates were systematically investigated.

2. Materials and Methods

2.1. Materials

Hexamethylene diisocyanate (HDI), polyol of ethylene glycol/adipic acid/butane diol (M_n = 2000 g/mole, Desmophen 1652), was purchased from Covestro (previously Bayer MaterialScience AG, Leverkusen, Germany). Diethylenetriamine (DETA) and acetone (99.5%) were purchased from Aldrich Chemical Corporation (Milwaukee, WI, USA). The 3-isocyanatopropyltriethoxysilane (IPTES) was kindly donated by Momentive Performance Materials (Niskayuna, NY, USA). Sodium 2-[(2-aminoethyl) amino] ethanesulphonate (Vestamin A-95, AEAS) was kindly donated by Evonik Industries (Essen, Germany). Graphene nanoplatelets (purity: 99.9%) (GNPs) with 3 nm of thickness, 1.5 µ diameter, and 800 m^2/g specific surface area were purchased from Nanografi (Ankara, Turkey) [38]. Twill weave carbon fiber fabric woven with 400 gsm DowAksa (Istanbul, Turkey) 12K yarns (TW400) and uncured epoxy resin films with 145 gsm were provided by KordSA (Pendik, Turkey). The 3M Scotch-Weld AF163-2K adhesive was used for bonding specimens for tensile tests.

2.2. Synthesis of Functional HBPUDs

Waterborne HBPUD samples with (i) only amino- (HBPUD-0) and (ii) both amino- and silane-functional terminal groups (HBPUD-50) were designed and synthesized using the oligomeric $A_2 + B_3$ approach based on the chemical compositions given in Table 1 as previously reported [39]. The synthesis route (Figure 1) contained four steps: (a) preparation of the A_2 oligomer, (b) branching by the reaction of the A_2 oligomer with the B_3 monomer, (c) functionalization, and (d) dispersion and distillation. A precalculated amount of polyester polyol (Table 1) was charged into the dried 3 L four-necked round-bottom flask equipped with an overhead stirrer, a reflux condenser, and a thermocouple that was connected to a heating mantle to control the reaction temperature. The polyol was dewatered by applying a vacuum (~2 mbar) for 15 min, at 75–85 °C. Upon the removal of the vacuum, the temperature was set to 60 °C, HDI was slowly added into the reaction flask, and the reaction was stirred for 3 h at 80 °C. The NCO content of the reaction mixture during the prepolymer process was monitored by the back-titration method [40], and the completion of the reaction was verified when the measured NCO content reached the theoretical NCO value. Upon the completion of the prepolymer reaction, the reaction temperature was set to 55 °C, and acetone was added to obtain the NCO-terminated polyurethane prepolymer with a concentration of 35–40 wt% in acetone at 48–50 °C. Following the prepolymer synthesis, 15% aqueous solution of a precalculated amount of the ionic monomer AEAS was fed dropwise into the reaction mixture at 48 °C to form the NCO-terminated, anionic A_2 oligomer. The freshly prepared anionic A_2 oligomer was then immediately transferred into an addition funnel, which was then slowly added into the preweighed DETA solution in acetone and water in a 5 L, four-necked round-bottom flask equipped with a reflux condenser, a mechanical stirrer, and a thermocouple. Upon the completion of the branching step by the slow addition of the ionic A2 oligomer into the B3 monomer solution, amino-functional, branched polyurethane was obtained in acetone. Next, distilled water was slowly added into the reaction mixture while vigorously stirring to disperse polyurethane chains in water while cooling the mixture to 42 °C. Finally, acetone was removed from the reaction mixture by vacuum distillation, and the complete removal of the acetone was ensured at 42 °C, 50 mbar. The final product, amino-functional HBPUD (denoted as HBPUD-0) with a solid content of 33 wt%, was collected by filtering through a 50-micron filtration medium.

Table 1. Chemical compositions of synthesized HBPUD samples.

Component	Sample Name	
	HBPUD-0	HBPUD-50
Polyester polyol (g)	911.60	911.60
Diisocyanate (g)	144.62	144.65
AEAS (g)	109.74	109.74
DETA (g)	21.33	21.33
IPTES (g)	0	37.32

In order to attain silane functionality and obtain HBPUDs with both amino- and silane-terminal groups, in a separate synthesis reaction, after the synthesis of amino-functional polyurethane in acetone as described above, a precalculated amount of IPTES compound was added dropwise into the reaction mixture immediately after the branching step at 48 °C to achieve a 50:50 ratio of amino–silane terminal groups, denoted as HBPUD-50. Figure 2 shows the structure and terminal groups of HBPUD-50.

FT-IR spectroscopy was used to monitor the functionalization step, by ensuring the absence of any NCO stretching vibration band (~2260 cm^{-1}) from IPTES as shown in Figure 3. Upon the completion of the functionalization step, the dispersion and acetone distillation steps were applied as described above for the synthesis of amino-functional HBPUD (HBPUD-0). The chemical compositions of the synthesized HBPUD samples are given in Table 1.

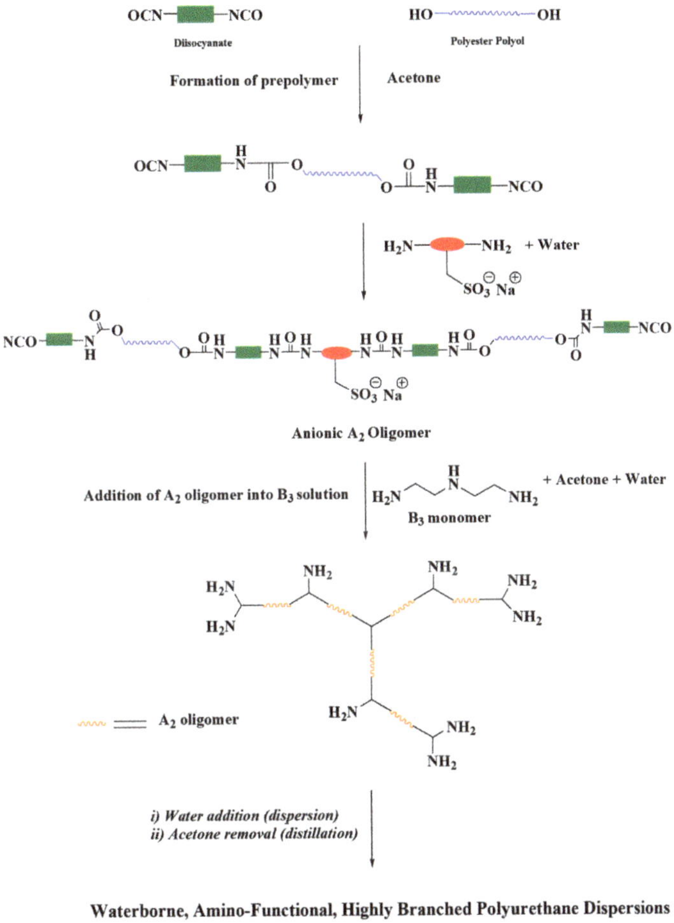

Figure 1. Synthesis of waterborne, amino-functional, highly branched PUDs via $A_2 + B_3$ approach (Green: HDI, Blue: polyol, Red: AEAS).

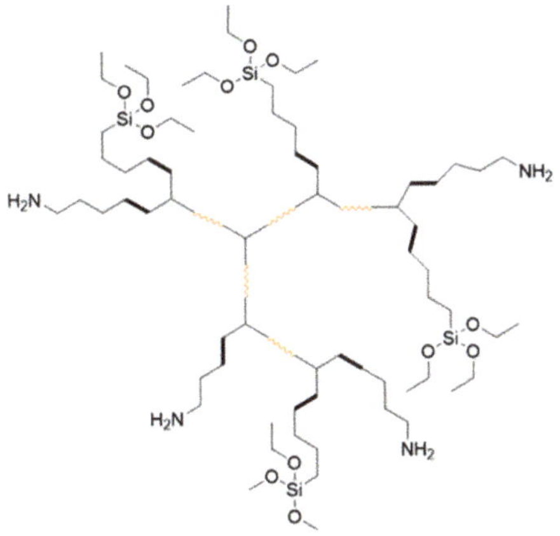

Figure 2. Representative structure of HBPUD-50.

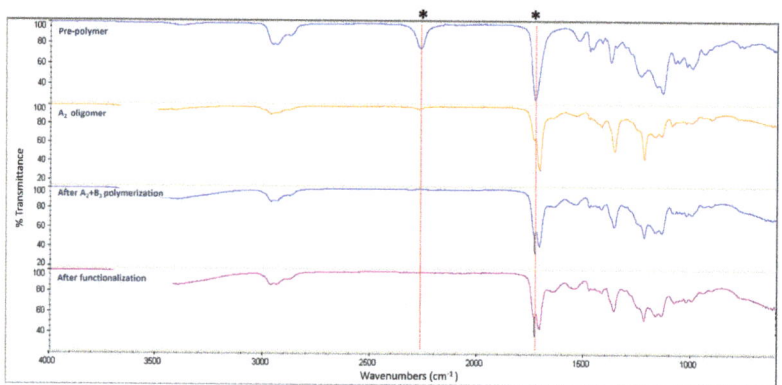

Figure 3. HBPUD-50 synthesis followed using FT-IR spectroscopy (*: bands corresponding to ~2260 and ~1725 cm^{-1} for NCO and C=O stretching vibrations, respectively).

2.3. Characterization of HBPUDs and PU Films

HBPUD syntheses were monitored, and final solid PU films were analyzed using a Nicolet IS10 Fourier Transform Infrared (FTIR) Spectrometer (Waltham, MA, USA) equipped with an ATR system with a 4 cm^{-1} resolution over 120 scans and ASTM D2572-97 back-titration method.

The particle sizes and distributions of HBPUDs were determined using a ZetaSizer, Malvern Instruments (Malvern, UK), provided with laser diffraction and polarized light of three wavelength detectors. Approximately 0.1 mL of HBPUD was diluted with distilled water to an adequate concentration in the cell and measured at room temperature. The refractive indices of PU and water were 1.50 and 1.30, respectively. The HBPUD-50/GNP mixtures were analyzed using a Partica LA-960V2 (Horiba, Kyoto, Japan) wet circulation system equipped with a dispersant filling pump, liquid level sensor, circulation pump, 30 W in-line ultrasonic probe, and relief valve. The particle size analysis was conducted by gauging the angular deviation of light scattered by particles when traversing a laser beam. The machine employed the principles of Mie scattering and Fraunhofer diffraction to ascertain precise measurements [41].

Tensile stress–strain tests of dried polyurethane films were performed on a universal testing machine (Zwick Roell Z100 UTM, Ulm, Germany), with a load cell of 200 N, a crosshead speed of 25 mm/min, and a grip-to-grip separation of 22 mm, and dog bone-shaped samples were prepared and tested according to a standard test method [42]. Three to five specimens were measured, and their average stress–strain values with standard deviations were reported.

Gel content (%) tests were carried out using Soxhlet extraction with toluene. For the gel content measurements, the dried polyurethane film (G1) and the thimble (G) were precisely weighed, and the polyurethane film was put into the thimble and extracted with toluene for 24 h. The thimble containing the film after the extraction was weighed again (G2) after drying. To calculate the gel content (%) of each polyurethane film, Equation (1) given below was used.

$$\text{Gel content (\%)} = [(G2 - G)/G1] \times 100 \quad (1)$$

Scanning electron microscopy (SEM) was employed using a Leo Supra 35VP FEG-SEM (Miami Beach, FL, USA) to examine the surface morphology of the dried HBPUD film samples, which were coated with gold and palladium for enhanced conductivity and imaging quality.

2.4. Preparation of Aqueous HBPUD/GNP Mixtures and Their Deposition onto Carbon Fiber Fabric Surfaces

The preparation of HBPUD-50/GNP mixtures in water and the deposition of this mixture onto the carbon fiber surface is schematically given in Figure 4. First, 0.25 g of GNPs was dispersed in 500 g of deionized water using probe ultrasonication (SONICA Q700 equipment, Niles, IL, USA) for 15 min under 70% amplitude with 5 s pulse on and 5 s pulse off. Then a predetermined amount of HBPUD-50 was mixed into this dispersion by mechanical stirring such that two different HBPUD-50/GNP mixtures were obtained with 0.05 wt% GNP concentration and solid PU:GNP weight ratios of 0.33:1 and 1:1. In addition, an aqueous dispersion of pure GNPs with 0.05 wt% and a pure HBPUD-50 dispersion with 0.05 wt% solid PU were prepared, which corresponded to 0:1 and 1:0 solid PU:GNP ratios, respectively.

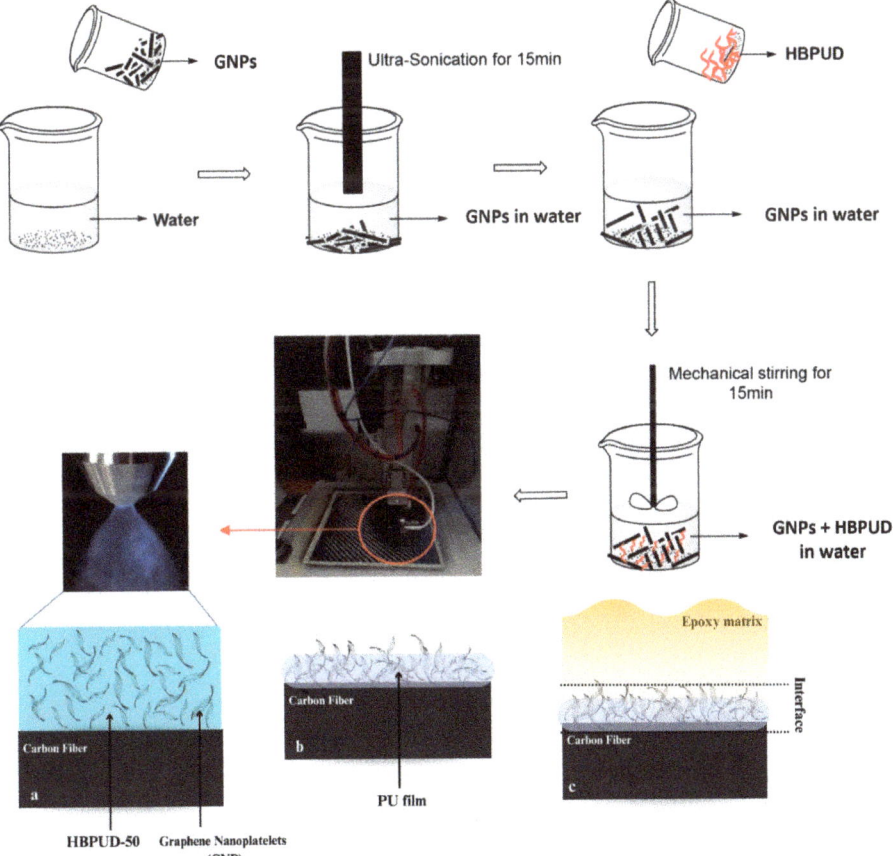

Figure 4. Preparation of HBPUD-50/GNP dispersions, their ultrasonic spray deposition onto the carbon fiber fabric surface, and the formation of the epoxy–carbon fiber interface containing PU and GNPs: (**a**) carbon fiber surface upon ultrasonic spray deposition, (**b**) carbon fiber surface upon drying, and (**c**) epoxy–carbon fiber interface in the final CFRPC.

Freshly prepared aqueous HBPUD-50/GNP mixtures were sprayed onto each side of 350 mm × 350 mm TW400 carbon fiber fabrics using a SONO-TEK Inc. (Milton, NY, USA), Flexi Coat ultrasonic spray coater with a 48 kHz, impact-type ultrasonic spray shaping nozzle. The spray was guided onto the substrate using jet air deflection aided by compressed air gas. At the end of the spray deposition, fabrics were dried on a heated plate at 50 °C for 1 day. The amount of GNPs deposited per unit surface area of carbon fiber fabric was controlled by the amount of HBPUD-50/GNP sample to be sprayed, such that 10, 20, and 30 mg of GNPs was deposited per m^2 of each side of the carbon fiber fabric (denoted as 10, 20, and 30 mgsm) from each HBPUD-50/GNP sample, corresponding to 15 different carbon fiber fabrics in total, coated with HBPUD-50/GNP aqueous mixtures.

Samples of HBPUD-50/GNP mixtures with solid PU:GNP weight ratios of 0.33:1 and 1:1 dried in an oven overnight at 80 °C and pure GNPs (PU:GNP ratio of 0:1) were analyzed using a Nicolet iS50 FTIR Spectrometer (Waltham, MA, USA) in transmission mode by preparing KBr pellets containing approximately 0.05 wt% of each dried sample.

2.5. Fabrication of Prepregs and Manufacturing of CFRPC Plates

An in-house method was developed and used for the fabrication of prepreg materials from ultrasonic spray-coated carbon fiber fabrics and hot melt epoxy resin films utilizing a hot-press technique with optimized parameters. The detailed fabrication process and setup are illustrated in Figure 5, while specifications of the desired prepreg materials are detailed in Table 2, for which the production process was precisely tailored.

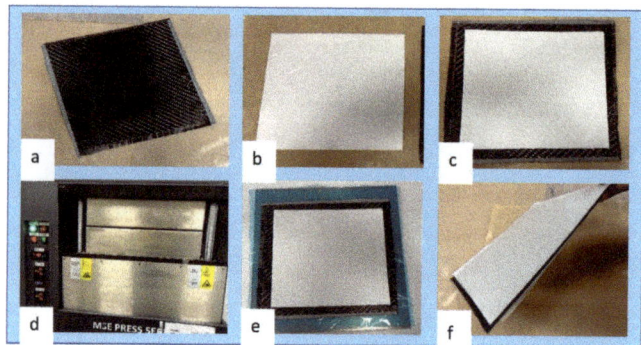

Figure 5. In-house prepreg laminate fabrication: (**a**) TW400 carbon fiber fabric (330 × 330 mm), (**b**) 145 gsm epoxy resin film (320 × 320 mm), (**c**) TW400 fabric sandwiched in between epoxy resin film, (**d**) resin impregnation in hot press, (**e**) prepreg (initial form out of hot press), and (**f**) prepreg after trimming the edges (300 × 300 mm, ready to lay up).

Table 2. Specifications of desired prepreg materials.

Test Standard	Test Result Type	Desired Range	Average Result	Unit
[43]	Prepreg areal weight (PAW)	690 ± 27	708	g/m²
[43]	Fiber areal weight (FAW)	400 ± 12	411	g/m²
[44]	Glass transition temperature (T_g)	−3.5 ± 3.5	−1.4	°C
[44]	Curing enthalpy (ΔH)	100 ± 50	81	(J/g)
[45]	Gel time	15 ± 5	15	min

TW400 fabric layers that were either as is or spray coated with each aqueous HBPUD-50/GNP mixture were cut into the precise dimensions of 330 mm × 330 mm (Figure 5a), and a roll of 145 gsm epoxy resin film was accurately cut into preforms with dimensions of 320 mm × 320 mm (Figure 5b) to ensure complete coverage without resin overflow. Concurrently, two layers of epoxy resin film were strategically placed on both surfaces of the TW400 carbon fiber fabric (Figure 5c) to promote an even resin distribution upon the application of pressure and heat. Each carbon fiber fabric sandwiched between epoxy resin films was placed in the hot press (Figure 5d) at 60 °C, under 0.1–0.2 ton-force pressure for

30–40 s ensuring the transformation of the sandwich structure into prepreg (Figure 5e). Last, prepregs out of the hot press were precisely trimmed to dimensions of 300 mm × 300 mm (Figure 5f), rendering them suitable for use in the manufacturing of CFRPC test plates.

The carbon/epoxy prepregs prepared in house, as depicted in Figure 5f, were utilized in the manufacturing of CFRPC laminates from five layers of prepregs with stacking sequences of $[0]_{5s}$ for tensile tests. Subsequently, stacked prepreg laminates were bagged for the autoclave manufacturing process with Teflon-coated glass fabrics (Fiberflon 828-25) on the autoclave trays, facilitating the plates' removal post-curing and peel ply material application to both the top and bottom surfaces of all laminates. A vacuum blanket (Airtech N10) (Airtech, Huntington Beach, CA, USA) was then carefully placed over the laminates, and the assembly was sealed using leak-proof tape (AT 200Y) (Airtech) around the tray edges. Two vacuum valves were positioned diagonally on each tray, and the setup was enclosed with a vacuum bag (Airtech WL7400).

The autoclave curing cycle was conducted at 120 ± 3 °C, under a vacuum of -0.2 ± 0.05 bar and a positive pressure of 7 ± 0.2 bar, sustained for 1 h. The cooling phase ensued at a controlled rate of 3 ± 0.5 °C/min. Upon curing, the plates underwent thickness verification via the non-destructive A-Scan inspection method. Plates that passed this inspection that had thickness values of 2.30 ± 0.15 mm (fabricated from five layers of prepreg sheets with individual cured ply thicknesses of 0.46 ± 0.03 mm) were then subjected to the tab bonding process first, followed by the coupon cutting process. The tensile specimens were then accurately sectioned using a water jet milling system equipped with KUKA KR-16 Ultra F Robot (Kuka, Ausburg, Germany).

A reference CFRPC sample denoted as CFRPC-Ref was manufactured using TW400 carbon fiber fabric as is, while CFRPC samples manufactured using prepregs from each spray-coated TW400 fabric were named as CFRPC-x-y:z, where x denoted the GNP deposition amount per meter square of TW400 fabric (mgsm), and y:z denoted the solid PU:GNP ratio in each spray deposition.

2.6. Characterization and Testing of CFRPC Plates

The surface morphologies of spray-coated carbon fiber samples and fractured CFRP sample surfaces were analyzed using a Leo SUPRA 35VP FEG-SEM. The images were taken at varying accelerating voltages between 2 kV and 5 kV using secondary electron imaging.

The tensile properties of manufactured CFRPC plates were measured according to a test standard [46]. Test samples were prepared from each CFRPC plate with dimensions of 250 mm (length) × 15 mm (width). Tension test plates were tabbed with $[+45°/-45°]_{4s}$ glass fiber-reinforced epoxy prepregs using 3M Scotch-Weld AF163-2K adhesive film (3M, St. Paul, MN, USA) and cured in the vacuum oven at 105 °C for 2 h. Tensile tests of the CFRPC plates were performed using INSTRON 5982 100 kN Universal Testing Systems (Norwood, MA, USA). A clip-on biaxial extensometer was initially attached to each tensile test specimen, which was removed before 0.5% strain during each test.

3. Results and Discussion

An anionic, isocyanate-terminated prepolymer was synthesized as an A_2 oligomer, which was polymerized with DETA as the B_3 monomer in dilute acetone solution to obtain highly branched, amino-functional polyurethane as shown in Figure 1. The polymerization was carried out in acetone medium, and the resulting polyurethane was dispersed in water to obtain waterborne, amino-functional HBPUDs as reported previously [39]. In this study, the reaction of amino-terminal groups with IPTES prior to the dispersion step enabled the partial conversion of amine terminal groups to silane groups as shown in Figure 2. The presence of both amine and silane terminal groups on the highly branched polyurethane backbone was envisioned to enhance interactions between the carbon fiber, GNPs, and the polymeric matrix both covalently and non-covalently when incorporated into the interface of CFRPCs. In this context, while silane terminal groups of the polyurethane were expected to react with residual hydroxyl groups present on the fiber and GNP surfaces, amine terminal units on the same polyurethane backbone were expected to react with the epoxy resin during the curing stage of the prepreg laminates. For this purpose, the HBPUD-50 sample was synthesized according to the composition given in Table 3, with an amine–silane terminal group molar ratio of 50:50. This sample was successfully obtained with a solid content of 33 wt% and an average particle size value of 84 nm (Table 3), which was stable over prolonged shelf storage.

Table 3. Physical and mechanical properties of HBPU-0 and HBPU-50 (*: HBPUD-0 sample did not form a self-standing film).

	Sample Name	
	HBPUD-0	HBPUD-50
Particle size of dispersion (nm)	80	84
Film properties		
Tensile strength at break (MPa)	*	6.1 ± 1.1
Elongation at break (%)	*	301.2 ± 15.6
Young's modulus (MPa)	*	3.2 ± 0.6
Gel content of film (wt%)	0	84

The presence and the effect of silane terminal groups in the HBPUD-50 sample (Figure 2) were first evaluated in the pure polyurethane film as they were expected to lead to the self-crosslinking of the corresponding film upon casting and drying. SEM images of the surfaces of solid polyurethane films from the HBPUD-0 and HBPUD-50 samples are presented in Figure 6. Both samples formed continuous films. The amino-functional polyurethane had a smooth surface, yet it did not form a self-standing film with a mechanical integrity. The polyurethane film from the silane functional HBPUD-50 sample was self-standing, and its SEM images revealed a rougher surface with micro-voids, possibly due to the hydrolysis and self-condensation of silane terminal groups.

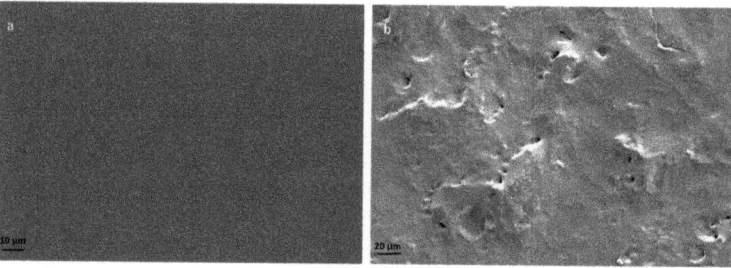

Figure 6. SEM images of the surfaces of solid polyurethane films from (**a**) HBPUD-0 and (**b**) HBPUD-50 samples.

In order to assess the effects of silane terminal groups on the physical properties of the resulting polyurethane films, the gel content and tensile properties of standalone polyurethane films from HBPUD-0 and HBPUD-50 were compared as summarized in Table 3. While the HBPUD-0 sample resulted in a fully soluble film in toluene with 0 wt% gel content, the HBPUD-50 sample had >80 wt% gel content, which was attributed to the hydrolysis, self-condensation, and crosslinking of silane terminal groups (Figure S1) in the highly branched polyurethane backbone from the HBPUD-50 sample. While the HBPUD-0 sample did not form a self-standing film with mechanical integrity, the crosslinking mechanism resulted in self-standing polyurethane films from HBPUD-50 with the tensile stress–strain behavior shown in Figure 7 and tensile properties given in Table 3. Last, the presence of Si-O-Si groups in the polyurethane film from the HBPUD-50 sample was also verified by the peaks observed around 1200, 1050, and 750 cm^{-1} in the FT-IR spectrum of the film as shown in Figure 8.

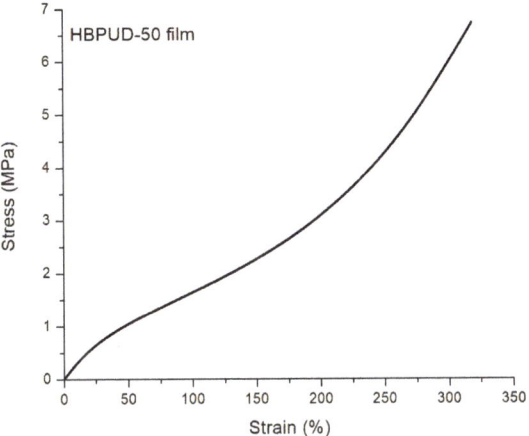

Figure 7. Tensile stress–strain curve of polyurethane film from HBPUD-50 sample.

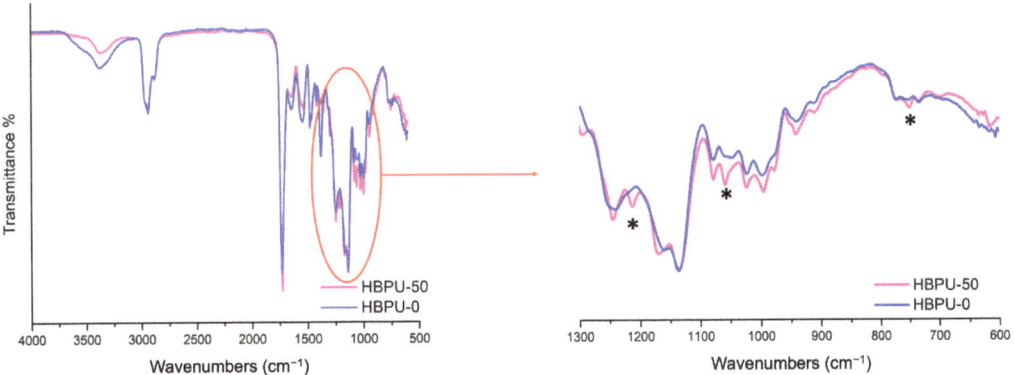

Figure 8. FT-IR spectra of dried PU films from HBPUD-0 and HBPUD-50 samples (*: bands corresponding to Si-O-Si at ~1200, 1050 and ~750 cm^{-1}).

Upon the synthesis and characterization of the HBPUD-50 sample, aqueous mixtures of GNPs and HBPUD-50 with different weight ratios were prepared for their deposition onto the carbon fiber fabric surface by ultrasonic spray deposition as depicted in Figure 4. The GNPs used in this study were formed of individual platelets with an average particle diameter of approximately 1.5 µm and a thickness less than 5 nm as previously analyzed using transmission electron microscopy (TEM) in the literature [47]. Considering the fact that these platelets were expected to form agglomerations rapidly in water, HBPUD was expected to enhance the stability and dispersibility of GNPs in water, which was a critical factor during their ultrasonic spray deposition onto the carbon fiber surface. As demonstrated in Figure S2, freshly prepared HBPUD-50/GNP mixtures with different weight ratios had relatively broad, uniform particle size distributions in water. Such broad distributions demonstrate the fact that GNPs are agglomerated in water, and ultrasonic spray deposition could play a key role in depositing them onto carbon fiber surfaces in smaller forms. When these mixtures were allowed to sit on the shelf for 24 h and shaken gently, an HBPUD-50/GNP mixture with a 1:1 solid PU:GNP weight ratio was observed to retain its original particle size distribution. The other two mixtures with less

(0.33:1 ratio of solid PU:GNP) and no HBPUD-50 (0:1 ratio of solid PU:GNP) showed new, larger particle size shoulders, indicating that an adequate amount of HBPUD may assist in obtaining homogeneous, stable GNP dispersions in water. Dried samples of HBPUD-50/GNP mixtures were analyzed using FT-IR spectroscopy as shown in Figure S3. Pure GNPs (HBPUD-50/GNP 0:1) were characterized by a broad peak around 3400 cm^{-1} due to hydroxyl groups around the edges of the GNP sheets, small C–H stretching peaks below 3000 cm^{-1} presumably due to imperfections in the graphitic structure, and the main peak around 1615 cm^{-1} due to C=C bond stretching. On the other hand, pure polyurethane film was characterized by a strong C=O bond stretching peak around 1730 cm^{-1}, along with C–H stretching peaks below 3000 cm^{-1} and a small peak arising from amine groups around 3300 cm^{-1}. The FT-IR spectra of the HBPUD-50/GNP mixtures with 0.33:1 and 1:1 ratios of solid PU:GNP verified the presence of both polyurethane and GNPs in these mixtures by the presence of C=C bonds arising from GNPs and both C=O and C–H bonds increasing parallel with the polyurethane content.

After the preparation of the HBPUD-50/GNP mixtures with different solid PU:GNP ratios, they were introduced onto carbon fiber fabric surfaces using a novel ultrasonic spray deposition method as demonstrated in Figure 4. The ultrasonic shaping nozzle of the spray equipment was expected to break up the agglomerates of GNPs in the aqueous medium immediately prior to the deposition of GNPs and enable uniform distribution of them on the coated carbon fiber surface. In this study, HBPUD-50/GNP mixtures with 0:1, 0.33:1, 1:1, and 1:0 ratios of solid PU:GNP were sprayed onto each side of 350 mm × 350 mm carbon fiber fabrics by ultrasonic spray deposition. While the samples with 0:1 and 1:0 weight ratios of PU:GNP corresponded to the deposition of pure GNPs and pure PU, respectively, samples with 0.33:1 and 1:1 weight ratios allowed the investigation of the presence of both PU and GNPs with two different ratios at the fiber–matrix interface. Each HBPUD-50/GNP mixture, as well as the pure HBPUD-50 (1:0 ratio) and GNP dispersion (0:1 ratio), was spray deposited in specific amounts to achieve depositions of 10, 20, and 30 mg of GNPs per m^2 (mgsm) of each side of the carbon fiber fabric. The amount of the pure HBPUD-50 sample was adjusted to deposit a solid PU amount the same as that of the 1:1 solid PU:GNP sample for each mgsm deposition series. It should be noted that the depositions of 10, 20, and 30 mgsm GNPs corresponded to approximately 0.003, 0.006, and 0.009 wt% GNPs in the overall composite structure, respectively, when calculated based on the average areal weight of each prepreg sheet given in Table 2. Upon the spray deposition of HBPUD-50/GNP mixtures with different ratios in each deposition series, each sprayed carbon fiber fabric sample underwent an overnight drying process, during which water was removed and a nanocomposite film layer was formed by facilitating the self-crosslinking or reaction of silane terminal groups with GNP and carbon fiber surfaces. SEM images of the uncoated carbon fiber surface and the ones coated with 20 mgsm GNP from pure GNP dispersion and from HBPUD-50/GNP mixtures with 1:1 PU:GNP ratios are shown in Figure 9. The successful deposition of pure GNPs onto the originally smooth carbon fiber surfaces is visible in Figure 9b, showing a significant change in the surface morphology of fibers with the aid of the ultrasonic spray deposition. Visually, the surface of fibers coated with the HBPUD-50/GNP sample (with a 1:1 ratio of solid PU:GNP) appears similar to that of the pure GNP-coated one (Figure 9c), while a better attachment of GNP particles onto the fiber surface is expected due to the presence of a polyurethane layer, although it is not visible in the SEM images. A chemical bond is expected to develop between silane terminal groups of polyurethane and GNP or carbon fiber surfaces while retaining amine terminal groups, resulting in the establishment of an intricate interface between the fiber and the matrix to be introduced.

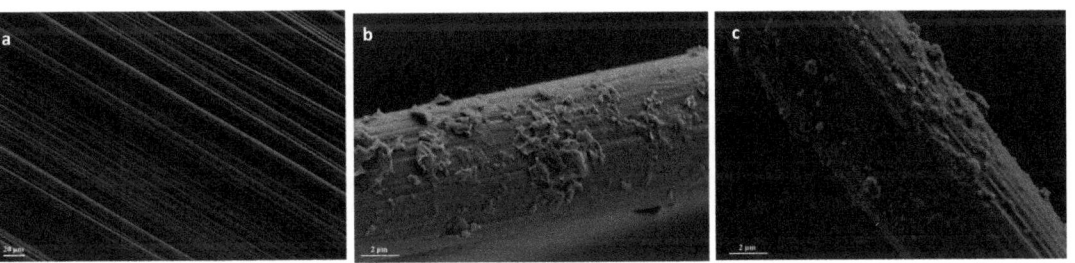

Figure 9. SEM images of (**a**) pristine carbon fiber, (**b**) fiber coated with 20 mgsm GNPs only, and (**c**) fiber coated with HBPUD/GNP-1:1 with 20 mgsm GNP deposition by ultrasonic spray deposition.

Following the ultrasonic spray deposition of each HBPUD-50/GNP, as well as pure HBPUD-50 and GNP dispersions onto carbon fiber fabrics with 10, 20, and 30 mgsm GNP depositions from each dispersion, prepreg laminates were fabricated by sandwiching each carbon fiber fabric in between epoxy resin films with standard gsm values. The optimized parameters used in the in-house process depicted in Figure 5 ensured the robust adhesion of the resin to the fiber without resin overflow, while maintaining the fiber fabric's integrity without causing damage, ensuring the transformation of the sandwich structure into prepreg form. The carbon/epoxy prepregs prepared in house were utilized in the manufacturing of CFRPC test plates by stacking laminates in $[0]_{5s}$ orientation, followed by autoclave curing in vacuum bags. Fifteen different CFRPC test plates, each formed of five prepreg layers, with varying GNP amounts or PU:GNP ratios, were manufactured, along with a reference CFRPC manufactured from TW400 carbon fiber fabric without any HBPUD-50 and/or GNP deposition.

The fabrication and testing of the CFRPC series with 10, 20, and 30 mgsm GNP deposition on the carbon fiber surface allowed a systematic investigation of the influence of varying the content of GNPs and/or solid PU at the fiber–matrix interface on the tensile properties of CFRPCs. In Figure 10a, representative stress–strain curves of the CFPRC-10 series are shown, while Figure 10b displays the variation of the average tensile strength at break and the Young's modulus values of the CFRPC samples with a GNP content of 10 mgsm and varying PU:GNP ratios at the interface. The deposition of 10 mgsm GNPs in the CFRPC-10-0:1 sample resulted in a slight increase in the tensile strength and modulus; however, a relatively large standard deviation especially in the modulus value indicated that GNPs alone may not have been homogeneously distributed at the interface. On the other hand, by the incorporation of 10 mgsm PU only from HBPUD-50, the CFRPC-10-1:0 sample showed an approximately 12% increase in the tensile strength reaching 1014.6 MPa, albeit with a slight decrease in the Young's modulus value. Furthermore, the CFRPC-10-0.33:1 with both PU and GNPs showed similar tensile properties to those of the CFRPC-10-1:0 sample, whereas an increased amount of PU in the HBPUD-50/GNP mixture corresponding to the CFRPC-10-1:1 sample resulted in a slight decrease in the tensile strength value reaching 982.2 MPa, still remaining above the reference CFRPC. In conclusion, although a clear trend was not observed as a function of the PU:GNP ratio, the incorporation of PU only or GNPs in the presence of PU resulted in increased tensile strength values with no change in the Young's modulus.

Stress–strain curves of the CFRPC-20 series with 20 mgsm GNPs deposited alone or in the presence of HBPUD-50 are plotted in Figure 11a, and the variation in tensile properties as a function of PU:GNP ratios is given in Figure 11b. The increased amount of incorporated GNPs from 10 mgsm to 20 mgsm resulted in a significantly increased Young's modulus but reduced tensile strength compared with the reference CFRPC sample. This suggested that a certain amount of GNPs at the interface without any attachment purely contributed to an increase in the modulus values. On the other hand, the incorporation of 20 mgsm PU only from HBPUD-50 in CFRPC-20-1:0 resulted in a significantly increased tensile strength value reaching above 1000 MPa and a slightly decreased Young's modulus value compared with both the CFRPC-Ref and CFRPC-20-0:1 samples. Interestingly, the incorporation of a combination of PU and GNPs at weight ratios of 0.33:1 and 1:1 resulted in a synergistic effect. In the CFRPC sample having a combination of PU and GNPs at a weight ratio of 0.33:1 (CFRPC-20-0.33:1), the tensile strength value was moderately increased above 950 MPa, while the Young's modulus value remained similar to that of the 20 mgsm pure GNPs incorporated CFRPC sample (CFRPC-20-0:1). In the case of the CFRPC-20-1:1 sample with increased PU content in combination with GNPs, the tensile strength value further increased compared with the CFRPC-20-0.33:1 sample, reaching the tensile strength value of the CFRPC-20-1:0 sample with pure PU, with a Young's modulus value in between those of the CFRPC-pure and CFRPC-20-0:1 samples.

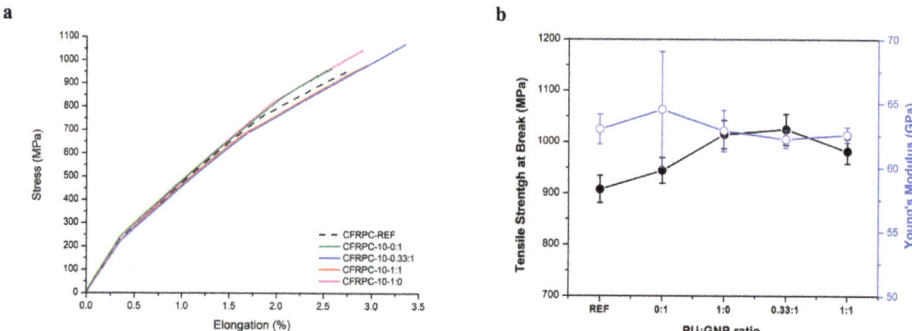

Figure 10. (a) Tensile stress–strain curves of CFRPC-10 test plate series and (b) variation in tensile properties of CFRPC-10 series as a function of PU:GNP ratio.

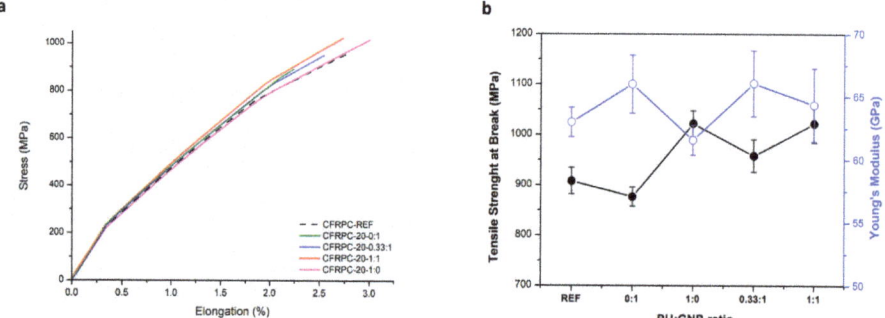

Figure 11. (a) Tensile stress–strain curves of CFRPC-20 series and (b) variation in tensile properties of CFRPC-20 series as a function of PU:GNP ratio.

Figure 12 shows the tensile properties of CFRPC samples with 30 mgsm GNPs at the interface alone or in combination with amine and silane functional polyurethane. The increased content of pure GNPs at the fiber–matrix interface resulted in a drastic decrease in not only the tensile strength but also the Young's modulus value, contrary to the 20 mgsm pure GNPs incorporated CFRPC sample. On the other hand, 30 mgsm incorporation of only PU at the interface showed a slight improvement in the tensile strength without any changes in the Young's modulus compared with the reference CFRPC. The incorporation of a combination of PU and GNPs at different weight ratios resulted in a visible trend of improved tensile strength and Young's modulus behavior such that while the CFRPC-30-0.33:1 sample was similar to the CFRPC-30-1:0 sample with only PU at the interface, the CFRPC-30-1:1 sample containing equivalent weights of solid PU and GNPs stood out among all samples with a significantly improved average tensile strength value above 1000 MPa and a moderately increased Young's modulus value around 65 MPa. It should be noted that our preliminary studies on increasing the incorporated GNP and/or PU content beyond 30 mgsm did not show any significant changes in the mechanical properties of the corresponding CFRPC laminates. Yet, the incorporation of high amounts of GNPs or other nanomaterials onto fiber fabric surfaces by ultrasonic spray deposition can be a promising approach in improving the thermal or electrical conductivity of FRPCs.

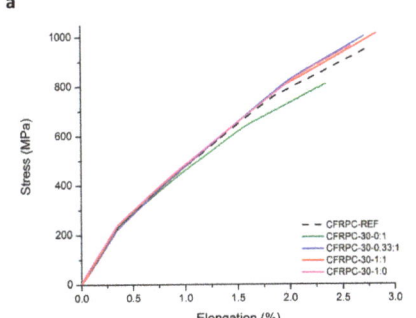 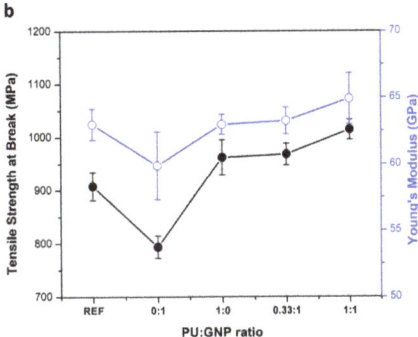

Figure 12. (a) Tensile stress–strain curves of CFRPC-30 series and (b) variation in tensile properties of CFRPC-30 series as a function of PU:GNP ratio.

A comprehensive analysis of the tensile behavior of all samples clearly indicated that the incorporation of relatively rigid GNPs alone into the interface without any attachment or chemical interactions solely improved the stiffness of corresponding samples up to 20 mgsm GNP incorporation, above which all tensile properties significantly decreased presumably due to an agglomeration effect of GNPs. On the other hand, the incorporation of a chemically functional PU layer alone into the fiber–matrix interface resulted in the improvement of mechanical properties through the enhancement of interfacial interactions, which was reflected as a significant increase in the tensile properties and clearly evidenced in the stress–strain curves of corresponding samples. In the case of the combined use of GNPs and a functional PU, a stiffening effect with the aid of GNPs and enhancement of interfacial interactions with the aid of a multi-functional PU layer through chemical bonding and interactions resulted in the improvement of tensile strength while maintaining or improving the initial Young's modulus with the optimum content of PU and GNPs, such as in the CFRPC-20 series.

The presented enhancement of interfacial interactions with the use of GNPs and multi-functional PU has been further assessed by SEM analysis of selected CFRPC samples after fracture. As illustrated in Figure 13, the reference CFRPC sample's failure occurred predominantly through progressive interfacial debonding and fiber pullout, leading to arbitrary fiber breakage at multiple levels along the fiber direction and voids in the matrix. In contrast, when one of the best performing CFRPC samples' (CFRPC-20-1:1) fractured surface was analyzed, the interface between the fiber and matrix remained almost intact after the failure, showing fewer fiber pullouts and more uniform fiber breakage, which provided evidence of strong interfacial bonding and contribution to improved tensile properties.

Here, we demonstrated a novel approach to enhance the interfacial interactions and improve the tensile properties of fiber-reinforced polymer composites (FRPCs) by combining graphene nanoplatelets (GNPs) and a multi-functional polyurethane at the fiber–matrix interface using ultrasonic spray deposition. This method resulted in significant improvements in the tensile properties of FRPCs with as little as 20 to 30 mg of GNPs and PU/m^2 of carbon fiber fabric, corresponding to approximately 0.006 to 0.009 wt% of each component in the overall composite structure. Notably, our study achieved these improvements with much lower amounts of carbon nanomaterials compared with previous studies. For instance, a prior study with a similar approach and composition of composite structure reported a notable increase in the tensile strength of CFRPCs from approximately 700 MPa to 850 MPa with the interfacial incorporation of 0.3 wt% GNPs, which is over 30 times higher than the GNP content used in our study [34]. It is important to point out that the waterborne, multi-functional polyurethane described in our study shows promise as a chemical compatibilizer and sizing agent, potentially enhancing interfacial interactions between dissimilar surfaces in composite materials synergistically when combined with various nanoparticles.

CFRPC-REF   **CFRPC-20-1:1**

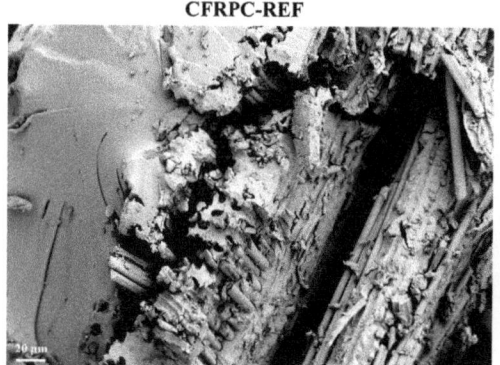 

Figure 13. SEM images of a cross-section of the fractured surface of CFRPC-REF and CFRPC-20-1:1 samples.

4. Conclusions

The incorporation of GNPs in the presence of a multi-functional polyurethane via their deposition onto the carbon fabric surface from an aqueous dispersion using an ultrasonic spraying deposition technique was systematically investigated. The synthesis and characterization of HBPUDs possessing only amine or both amine and silane terminal groups were carried out, aiming to understand the role of silane terminal groups on the polyurethane backbone. The presence of silane terminal groups on the polyurethane backbone led to a very high gel content and decent mechanical properties in resulting films from the HBPUD-50 sample, whereas the solid PU film from the HBPUD-0 sample with only amine terminal groups did not form a self-standing film. The subsequent preparation of GNPs with HBPUD-50 at different ratios assisted in keeping a homogeneous dispersion for the ultrasonic spray deposition of these mixtures onto carbon fiber fabrics to incorporate both GNPs and multi-functional polyurethane chains at the fiber–matrix interface of CFRPCs by the preparation of their prepreg laminates, stacking, and autoclave curing. A systematic study on the relative content of GNPs and the PU:GNP weight ratio at the fiber–matrix interface showed that a synergistic effect of both stiffening and enhancement of interfacial interactions was achieved, resulting in the improvement of the tensile strength values from approximately 908 MPa up to 1022 MPa and Young's modulus values from 63 MPa up to 66 MPa. This study underscored the importance of carefully tuning the GNP content and PU:GNP ratio in tailoring these tensile properties in high-performance CFRPCs.

Supplementary Materials: The following supporting information can be downloaded at https://www.mdpi.com/article/10.3390/polym16060828/s1: Figure S1: Self-crosslinking mechanism of HBPUD-50 upon film formation; Figure S2: Particle size analyses of HBPUD-50/GNP dispersion mixtures (a) freshly prepared and (b) gently shaken after one day of storage (all measurements were performed at 0.05 wt% concentration); Figure S3: FT-IR spectra of dried HBPUD-50/GNP samples.

Author Contributions: Conceptualization, A.D.-S., S.Ü., B.D. and M.Y.; methodology, A.D.-S., S.Ü., B.D. and M.Y.; validation, A.D.-S.; formal analysis, A.D.-S.; investigation, A.D.-S., M.T., T.Ç.-Ç. and D.S.; data curation, A.D.-S.; writing—original draft preparation, A.D.-S.; writing—review and editing, S.Ü.; visualization, A.D.-S.; supervision, S.Ü.; project administration, D.S., M.Y. and S.Ü. All authors have read and agreed to the published version of the manuscript.

Funding: This research received no funding.

Institutional Review Board Statement: Not applicable.

Data Availability Statement: Data are contained within the article and Supplementary Materials.

Acknowledgments: Authors would like to thank KordSA for the donation of epoxy resin film and carbon fiber fabric samples.

Conflicts of Interest: The authors declare no conflicts of interest.

References

1. Cho, J.; Daniel, I.M.; Dikin, D. Effects of block copolymer dispersant and nanotube length on reinforcement of carbon/epoxy composites. *Compos. Part A Appl. Sci. Manuf.* **2008**, *39*, 1844–1850. [CrossRef]
2. Tang, Y.; Liu, L.; Li, W.; Shen, B.; Hu, W. Interface characteristics and mechanical properties of short carbon fibers/Al composites with different coatings. *Appl. Surf. Sci.* **2009**, *255*, 4393–4400. [CrossRef]
3. Zhang, R.; Huang, Y.; Liu, L.; Tang, Y.; Su, D.; Xu, L. Effect of the molecular weight of sizing agent on the surface of carbon fibres and interface of its composites. *Appl. Surf. Sci.* **2011**, *257*, 1840–1844. [CrossRef]
4. Yao, S.-S.; Jin, F.-L.; Rhee, K.Y.; Hui, D.; Park, S.-J. Recent advances in carbon-fiber-reinforced thermoplastic composites: A review. *Compos. Part B Eng.* **2018**, *142*, 241–250. [CrossRef]
5. Ochiai, S.; Osamura, K. Influences of matrix ductility, interfacial bonding strength, and fiber volume fraction on tensile strength of unidirectional metal matrix composite. *Metall. Trans. A* **1990**, *21*, 971–977. [CrossRef]
6. Mullin, J. *Influence of Fiber Property Variation on Composite Failure Mechanisms*; ASTM International: West Conshohocken, PA, USA, 1973.
7. Gatti, A.; Mullin, J.; Berry, J. The role of bond strength in the fracture of advanced filament reinforced composites. In *Composite Materials: Testing and Design*; ASTM International: West Conshohocken, PA, USA, 1969.
8. Tze, W.T.; Gardner, D.J.; Tripp, C.P.; O'Neill, S.C. Cellulose fiber/polymer adhesion: Effects of fiber/matrix interfacial chemistry on the micromechanics of the interphase. *J. Adhes. Sci. Technol.* **2006**, *20*, 1649–1668. [CrossRef]
9. Jin, Z.; Han, Z.; Chang, C.; Sun, S.; Fu, H. Review of methods for enhancing interlaminar mechanical properties of fiber-reinforced thermoplastic composites: Interfacial modification, nano-filling and forming technology. *Compos. Sci. Technol.* **2022**, *228*, 109660. [CrossRef]
10. Park, S.-J.; Seo, M.-K.; Lee, Y.-S. Surface characteristics of fluorine-modified PAN-based carbon fibers. *Carbon* **2003**, *41*, 723–730. [CrossRef]
11. Bachinger, A.; Rössler, J.; Asp, L.E. Electrocoating of carbon fibres at ambient conditions. *Compos. Part B Eng.* **2016**, *91*, 94–102. [CrossRef]
12. Shin, H.K.; Park, M.; Kim, H.-Y.; Park, S.-J. An overview of new oxidation methods for polyacrylonitrile-based carbon fibers. *Carbon Lett.* **2015**, *16*, 11–18. [CrossRef]
13. Park, S.-J.; Jang, Y.-S.; Shim, J.-W.; Ryu, S.-K. Studies on pore structures and surface functional groups of pitch-based activated carbon fibers. *J. Colloid Interface Sci.* **2003**, *260*, 259–264. [CrossRef]
14. Nagura, A.; Okamoto, K.; Itoh, K.; Imai, Y.; Shimamoto, D.; Hotta, Y. The Ni-plated carbon fiber as a tracer for observation of the fiber orientation in the carbon fiber reinforced plastic with X-ray CT. *Compos. Part B Eng.* **2015**, *76*, 38–43. [CrossRef]
15. Lee, H.S.; Kim, S.-y.; Noh, Y.J.; Kim, S.Y. Design of microwave plasma and enhanced mechanical properties of thermoplastic composites reinforced with microwave plasma-treated carbon fiber fabric. *Compos. Part B Eng.* **2014**, *60*, 621–626. [CrossRef]
16. Park, S.-J.; Jang, Y.-S.; Kawasaki, J. Studies on nanoscaled Ni-P plating of carbon fiber surfaces in a composite system. *Carbon Lett.* **2002**, *3*, 77–79.
17. Wu, G.; Ma, L.; Liu, L.; Wang, Y.; Xie, F.; Zhong, Z.; Zhao, M.; Jiang, B.; Huang, Y. Interfacially reinforced methylphenylsilicone resin composites by chemically grafting multiwall carbon nanotubes onto carbon fibers. *Compos. Part B Eng.* **2015**, *82*, 50–58. [CrossRef]
18. Brito-Santana, H.; de Medeiros, R.; Rodriguez-Ramos, R.; Tita, V. Different interface models for calculating the effective properties in piezoelectric composite materials with imperfect fiber–matrix adhesion. *Compos. Struct.* **2016**, *151*, 70–80. [CrossRef]
19. Gao, Y.; Guo, Z.; Song, Z.; Yao, H. Spiral interface: A reinforcing mechanism for laminated composite materials learned from nature. *J. Mech. Phys. Solids* **2017**, *109*, 252–263. [CrossRef]
20. Zhao, Y.; Liu, F.; Lu, J.; Shui, X.; Dong, Y.; Zhu, Y.; Fu, Y. Si-Al hybrid effect of waterborne polyurethane hybrid sizing agent for carbon fiber/PA6 composites. *Fibers Polym.* **2017**, *18*, 1586–1593. [CrossRef]

21. Zhang, T.; Zhao, Y.; Li, H.; Zhang, B. Effect of polyurethane sizing on carbon fibers surface and interfacial adhesion of fiber/polyamide 6 composites. *J. Appl. Polym. Sci.* **2018**, *135*, 46111. [CrossRef]
22. Guo, X.; Lu, Y.; Sun, Y.; Wang, J.; Li, H.; Yang, C. Effect of sizing on interfacial adhesion property of glass fiber-reinforced polyurethane composites. *J. Reinf. Plast. Compos.* **2018**, *37*, 321–330. [CrossRef]
23. Ma, Y.; Yokozeki, T.; Ueda, M.; Sugahara, T.; Yang, Y.; Hamada, H. Effect of polyurethane dispersion as surface treatment for carbon fabrics on mechanical properties of carbon/Nylon composites. *Compos. Sci. Technol.* **2017**, *151*, 268–281. [CrossRef]
24. Andideh, M.; Esfandeh, M. Effect of surface modification of electrochemically oxidized carbon fibers by grafting hydroxyl and amine functionalized hyperbranched polyurethanes on interlaminar shear strength of epoxy composites. *Carbon* **2017**, *123*, 233–242. [CrossRef]
25. Song, X.; Yu, M.; Niu, H.; Li, Y.; Chen, C.; Zhou, C.; Liu, L.; Wu, G. Poly (methyl dihydroxybenzoate) modified waterborne polyurethane sizing coatings with chemical and hydrogen-bonded complex cross-linking structures for improving the surface wettability and mechanical properties of carbon fiber. *Prog. Org. Coat.* **2024**, *187*, 108112. [CrossRef]
26. Dai, Z.; Zhang, B.; Shi, F.; Li, M.; Zhang, Z.; Gu, Y. Chemical interaction between carbon fibers and surface sizing. *J. Appl. Polym. Sci.* **2012**, *124*, 2127–2132. [CrossRef]
27. Dai, S.; Li, P.; Li, X.; Ning, C.; Kong, L.; Fang, L.; Liu, Y.; Liu, L.; Ao, Y. Waterborne polyurethanes from self-catalytic tartaric acid-based polyols for environmentally-friendly sizing agents. *Compos. Commun.* **2021**, *27*, 100849. [CrossRef]
28. Fazeli, M.; Liu, X.; Rudd, C. The effect of waterborne polyurethane coating on the mechanical properties of epoxy-based composite containing recycled carbon fibres. *Surf. Interfaces* **2022**, *29*, 101684. [CrossRef]
29. Popeney, C.S.; Lukowiak, M.C.; Böttcher, C.; Schade, B.; Welker, P.; Mangoldt, D.; Gunkel, G.; Guan, Z.; Haag, R. Tandem coordination, ring-opening, hyperbranched polymerization for the synthesis of water-soluble core–shell unimolecular transporters. *ACS Macro Lett.* **2012**, *1*, 564–567. [CrossRef] [PubMed]
30. Yang, Y.; Xie, X.; Wu, J.; Yang, Z.; Wang, X.; Mai, Y.W. Multiwalled carbon nanotubes functionalized by hyperbranched poly (urea-urethane) s by a one-pot polycondensation. *Macromol. Rapid Commun.* **2006**, *27*, 1695–1701. [CrossRef]
31. Sharma, H.; Kumar, A.; Rana, S.; Guadagno, L. An overview on carbon fiber-reinforced epoxy composites: Effect of graphene oxide incorporation on composites performance. *Polymers* **2022**, *14*, 1548. [CrossRef] [PubMed]
32. Zhang, X.; Fan, X.; Yan, C.; Li, H.; Zhu, Y.; Li, X.; Yu, L. Interfacial microstructure and properties of carbon fiber composites modified with graphene oxide. *ACS Appl. Mater. Interfaces* **2012**, *4*, 1543–1552. [CrossRef]
33. Xiong, M.; Xiong, L.; Xiong, K.; Liu, F. A new strategy for improvement of interface and mechanical properties of carbon fiber/epoxy composites by grafting graphene oxide onto carbon fiber with hyperbranched polymers via thiol-ene click chemistry. *Polym. Compos.* **2023**, *44*, 5490–5498. [CrossRef]
34. Wang, F.; Cai, X. Improvement of mechanical properties and thermal conductivity of carbon fiber laminated composites through depositing graphene nanoplatelets on fibers. *J. Mater. Sci.* **2019**, *54*, 3847–3862. [CrossRef]
35. Li, N.; Yang, X.; Bao, F.; Pan, Y.; Wang, C.; Chen, B.; Zong, L.; Liu, C.; Wang, J.; Jian, X. Improved mechanical properties s of copoly (phthalazinone ether sulphone) s composites reinforced by multiscale carbon fibre/graphene oxide reinforcements: A step closer to industrial production. *Polymers* **2019**, *11*, 237. [CrossRef]
36. Truong, G.T.; Tran, H.V.; Choi, K.-K. Investigation on mode i fracture toughness of woven carbon fiber-reinforced polymer composites incorporating nanomaterials. *Polymers* **2020**, *12*, 2512. [CrossRef] [PubMed]
37. Costa, U.O.; Garcia Filho, F.d.C.; Río, T.G.-d.; Rodrigues, J.G.P.; Simonassi, N.T.; Monteiro, S.N.; Nascimento, L.F.C. Mechanical Properties Optimization of Hybrid Aramid and Jute Fabrics-Reinforced Graphene Nanoplatelets in Functionalized HDPE Matrix Nanocomposites. *Polymers* **2023**, *15*, 2460. [CrossRef]
38. Graphene Nanoplatelet, Purity: 99.9%, Size: 3 nm, S.A: 800 m2/g, Dia: 1.5 µm. Available online: https://nanografi.com/graphene/graphene-nanoplatelet-purity-99-9-size-3-nm-s-a-800-m2-g-dia-1-5-m/?ysclid=lsvo0bqej0975861049 (accessed on 18 January 2024).
39. Durmuş Sayar, A. Synthesis and Characterization of Branched, Functional Polyurethane Dispersions as a Technology Platform for Self-Crosslinking Textile Coatings. Master's Thesis, Sabanci University, Istanbul, Turkey, 2017.
40. *ASTM D2572-97*; Standard Test Method for Isocyanate Groups in Urethane Materials or Prepolymers. ASTM International: West Conshohocken, PA, USA, 2010.
41. *ISO 13320:2020*; Particle Size Analysis—Laser Diffraction Methods. International Organization for Standardization: Geneva, Switzerland, 2020.
42. *ASTM D1708-10*; Standard Test Method for Tensile Properties of Plastics by Use of Microtensile Specimens. ASTM International: West Conshohocken, PA, USA, 2013.
43. *ASTM C613-14*; Standard Test Method for Constituent Content of Composite Prepreg by Soxhlet Extraction. ASTM International: West Conshohocken, PA, USA, 2014.
44. *DIN EN 6041:2018*; Aerospace Series—Non-Metallic Materials—Test Method—Analysis of Non-Metallic Materials (uncured) by Differential Scanning Calorimetry (DSC). European Standards: Brussels, Belgium, 2018.
45. *DIN EN 6043:1996*; Aerospace Series—Thermosetting Resin Systems—Test Method; Determination of Gel Time and Viscosity. European Standards: Brussels, Belgium, 1996.

6. ASTM 3039/3039M-17; Standard Test Method for Tensile Properties of Polymer Matrix Composite Materials. ASTM International: West Conshohocken, PA, USA, 2000.
7. Martin, K. Performance analysis of a thermosiphon charged with deionized water/ethylene glycol mixture based graphene nano platelet nanofluid. *Konya J. Eng. Sci.* **2022**, *10*, 679–691. [CrossRef]

Disclaimer/Publisher's Note: The statements, opinions and data contained in all publications are solely those of the individual author(s) and contributor(s) and not of MDPI and/or the editor(s). MDPI and/or the editor(s) disclaim responsibility for any injury to people or property resulting from any ideas, methods, instructions or products referred to in the content.

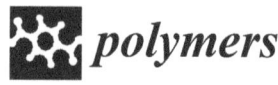

Article

Flexural Response of Concrete Specimen Retrofitted with PU Grout Material: Experimental and Numerical Modeling

Sadi Ibrahim Haruna [1,2,*], Yasser E. Ibrahim [1,*], Zhu Han [2] and Abdulwarith Ibrahim Bibi Farouk [2]

1. Engineering Management Department, College of Engineering, Prince Sultan University, Riyadh 11586, Saudi Arabia
2. School of Civil Engineering, Tianjin University, Tianjin 300072, China; hanzhu2000@tju.edu.cn (Z.H.); aifg1986@tju.edu.cn (A.I.B.F.)
* Correspondence: sharuna@psu.edu.sa (S.I.H.); ymansour@psu.edu.sa (Y.E.I.)

Citation: Haruna, S.I.; Ibrahim, Y.E.; Han, Z.; Farouk, A.I.B. Flexural Response of Concrete Specimen Retrofitted with PU Grout Material: Experimental and Numerical Modeling. *Polymers* 2023, 15, 4114. https://doi.org/10.3390/polym 15204114

Academic Editors: Zhuohong Yang and Chang-An Xu

Received: 20 September 2023
Revised: 10 October 2023
Accepted: 13 October 2023
Published: 17 October 2023

Copyright: © 2023 by the authors. Licensee MDPI, Basel, Switzerland. This article is an open access article distributed under the terms and conditions of the Creative Commons Attribution (CC BY) license (https://creativecommons.org/licenses/by/4.0/).

Abstract: Polyurethane (PU) composite is increasingly used as a repair material for civil engineering infrastructure, including runway, road pavement, and buildings. Evaluation of polyurethane grouting (PUG) material is critical to achieve a desirable maintenance effect. This study aims to evaluate the flexural behavior of normal concrete repaired with polyurethane grout (NC-PUG) under a three-point bending test. A finite element (FE) model was developed to simulate the flexural response of the NC-PUG specimens. The equivalent principle response of the NC-PUG was analyzed through a three-dimensional finite element model (3D FEM). The NC and PUG properties were simulated using stress–strain relations acquired from compressive and tensile tests. The overlaid PUG material was prepared by mixing PU and quartz sand and overlayed on the either top or bottom surface of the concrete beam. Two different overlaid thicknesses were adopted, including 5 mm and 10 mm. The composite NC-PUG specimens were formed by casting a PUG material using different overlaid thicknesses and configurations. The reference specimen showed the highest average ultimate flexural stress of 5.56 MPa ± 2.57% at a 95% confidence interval with a corresponding midspan deflection of 0.49 mm ± 13.60%. However, due to the strengthened effect of the PUG layer, the deflection of the composite specimen was significantly improved. The concrete specimens retrofitted at the top surface demonstrated a typical linear pattern from the initial loading stage until the complete failure of the specimen. Moreover, the concrete specimens retrofitted at the bottom surface exhibit two deformation regions before the complete failure. The FE analysis showed good agreement between the numerical model and the experimental test result. The numerical model accurately predicted the flexural strength of the NC-PUG beam, slightly underestimating Ke by 4% and overestimating the ultimate flexural stress by 3%.

Keywords: concrete; polyurethane; finite element analysis; polyurethane grout material; flexural strength

1. Introduction

Repair and protective techniques are applied to civil engineering infrastructure, including reinforced concrete structures, roads, and runways, by exposing damaged sections and changing them with cement-based composites [1–5]. Due to the cracking of the replaced materials and further penetration of degradation substances into concrete structures, the functions of the repaired section may deteriorate again [6,7]. Water can be an essential carrier for aggressive, penetrating substances [6]. Thus, efficient means to restrict water penetration into repaired areas and underlying concrete structures is critical for preserving their high resilience and extended lifetimes. Alternative materials of polymeric resins, including methyl methacrylate (MMA), epoxy resins, furan resins, polyurethane resins, urea formaldehyde, and unsaturated polyester resins, are available to maintain concrete structure effectively. Moreover, latex (polymer suspension in water), powder, and resin (liquid form) are all common types of polymer modifiers. Similarly, there are polymer modifiers, which include styrene butadiene rubber (SBR) emulsion, ethylene vinyl acetate (EVA), polyacrylate (PAE) emulsion, epoxy resin (EP), polyvinyl alcohol vinyl acetate, ethylene vinyl acetate, and acrylic acid [8–12]. PU is a hard polymer with good wear-resistance characteristics [13]. The PU-cement-based composite was reportedly employed in retrofitting structures after a seismic event because of its high bending and low compressive strength reduction [14]. Polyurethane gout materials have been used in several repair projects, including highway crack treatment, high-speed railway track slab raising, emergency reinforcement of water conservation projects, and repair of rigid pavement [15–18], as depicted in Figure 1. Similarly, closed-cell one-component hydrophobic polyurethane foam can be used to stabilize expansive soil [19], whereas polyurethane grouting materials are employed for repair of road and runway facilities [20].

Figure 1. Engineering application of polyurethane-based polymer composites.

The numerical analytical method has been utilized to investigate the performance of concrete materials under static loadings, which is considered isotropic and homogenous for numerical simulation [21–24]. The numerical analysis was carried out using the finite element approach, demonstrating that the numerical analysis could be utilized as an evaluation tool for analyzing the different performance of the polyurethane-concrete composite [25–28]. Hala et al. [25] conducted an experimental study and numerical analysis on the ballistic resistance of high-performance fiber-reinforced concrete panels coated with polyurethane materials. The numerical models adequately predict the ballistic strength of the panels under independent ballistics tests. Sing et al. [29] evaluated the compression, tension, and flexural properties of four epoxy grouts and developed a finite element model to simulate composite repaired pipes. The result showed good agreement between the FE models and the experimental test result with a margin error of less than 10%. It was discovered that by modifying the infill parameters in the finite element model to simulate the usage of different infill materials for the repair, a 4–8% increase in burst pressure can be produced. Shigang et al. [30] performed numerical simulations of the polyurethane polymer concrete specimens in compression under different strain rates utilizing explicit numerical methods based on LS-DYNA codes. The failure factors of polymer concrete at the mesoscale level were numerically analyzed. The result indicated that the novel dynamic properties of the material attributed to the damage and failure mode of the interface and elastic/plastic properties of the polyurethane polymer composites. Chen et al. [31] conducted a long-term study and numerical simulation of PU foam insulation on concrete dams under extreme cold conditions. The FE analysis on the composite profiled sheet deck formed by applying polyurethane and polyvinyl chloride tubes was studied by [32]. Manjun et al. [28] established the FE model to simulate the shear failure process of polyurethane–bentonite composite specimens under variable angle shear test; the results indicated that the FE model result is consistent with the experimental result. Somarathna et al. [33] studied the dynamic mechanical properties of concrete retrofitted with polyurethane coating material subjected to quasi-static and dynamic loads simulated via a three-point bending test. The failure mechanism between the PU grout and concrete under the influence of moisture was investigated using digital image correlation [34]. Huang et al. [35] proposed a calculation technique to determine the deformation of precast concrete frame assembled with artificial controllable plastic hinges, and performed seismic analysis. The result showed that artificial controllable plastic hinges effectively reduced the base shear of the frame structure. The seismic performance of corrosion-damaged reinforced concrete columns strengthened with a bonded steel plate (BSP) and a high-performance ferrocement laminate (HPFL) was evaluated [36]. Zhang et al. [37] developed a numerical model and reliability-based analysis of the flexural strength of concrete beams reinforced with hybrid basalt fiber-reinforced polymer and steel rebars.

Studying the performance properties of composite concrete retrofitted with polyurethane grout under FE simulation requires considerable attention, as most previous studies paid attention to the experimental investigation, even though experimental studies were an appropriate means of understanding the structural response. Experimental studies are expensive, time-consuming, and unviable, especially for comprehensive or parametric studies. Thus, when accurately calibrated and validated, the FE analysis technique is an alternative way of investigating the structural responses. This study intends to investigate the effectiveness and capacity of the PU material prepared by combining bio-based polyurethane (castor oil) and quartz sand as a coating material for a concrete beam subjected to a three-point bending test and a developed FE simulation of the concrete–polyurethane grout (NC-PUG) under a flexural load. The equivalent principle response of the NC-PUG was analyzed through a three-dimensional finite element model (3D FEM). The NC and PUG properties were simulated using stress–strain relations acquired from compressive and tensile tests. The concrete damage plasticity model (CDPM) available in commercially available FE software ABAQUS 2021 [38] is utilized to model the response of normal concrete and polyurethane grout material.

2. Materials and Methods

2.1. Materials

Grade 42.5R ordinary Portland cement was used to produce a concrete mixture, and its chemical compositions are summarized in Table 1. The fine aggregate in this study was a river sand with 2650 kg/m^3 density 2.63 fineness modulus. The coarse aggregate is the crushed natural stone with 10 mm aggregate size and 2.67 fineness modulus. Figure 2 presents the distribution of the aggregate sizes used. The quartz sand has particles the size of 0.5–1.0 mm and density of 1430 kg/m^3. The desired workability of the concrete was maintained by adding polycarboxylate-based superplasticizer at 0.15% of the weight of cement to obtain required workability.

Table 1. Cement chemical composition.

Material	Oxides									
	SiO_2	Al_2O_3	Fe_2O_3	CaO	MgO	K_2O	Na_2O	SO_3	TiO_2	LOI
Cement	23.27	4.41	2.45	62.85	1.42	0.48	0.21	2.57	0.08	1.82

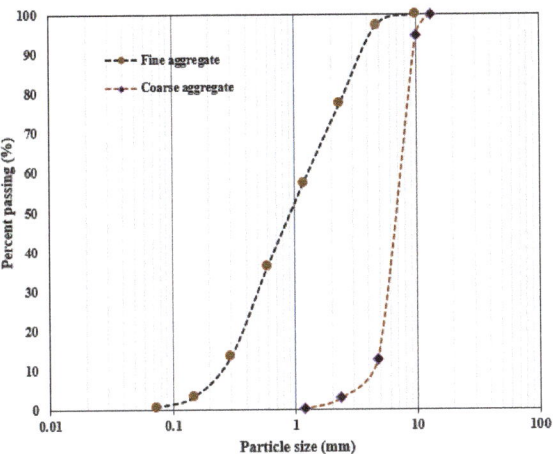

Figure 2. Grading curve of the aggregate materials.

The PU binder was synthesized through a polymerization reaction between polyaryl polymethylene, isocyanate (PAPI), and castor oil, mixed at the mix ratio of 6:1, and placed in a container. A homogenous solution was produced after 2 min of rigorous mixing with hand mixer set at high speed [17,39,40]. Table 2 shows the physical and performance indexes of the PU binder.

Table 2. Physical and performance parameter of PU binder.

PU Materials	Viscosity (CPS)	Appearance	Curing Age (h)		Tension Property (MPa)
			Initial	Final	
Castor oil	35,000	Grey/white sticky	-	-	-
PAPI	250	Brown transparent	-	-	-
PU binder	-	-	3.5	72	5.5

2.2. Specimen Preparation

Table 3 shows the mix proportion for the preparation of concrete mixture and PU grouting materials. The NC mix was poured into beam molds with defined dimensions of 100 × 100 × 400 mm^3. To obtain a satisfactory level of compaction, the cast specimens were put on the vibrating table. They were then left at room temperature for around 24 h before being removed from the mold. Furthermore, all the specimens were cured for 28 days prior to the test.

Table 3. NC and PUG design mix.

Mix ID	Cement (kg/m^3)	Sand (kg/m^3)	Coarse aggregate (kg/m^3)	Water (kg/m^3)
NC	425	718	966	170
	PU/Quartz sand (weight ratio)	PU binder (200 g)		
		Castor oil (g)	PAPI (g)	Solvent (g)
PU grout	1:0.5	167	33	8.4

After 28 days, all samples were air dried for seven days to make sure that all the surface moisture was completely dried before PU grouting was overlaid. The prepared PU grouting materials were synthesized by mixing quartz sand and PU binder using a mixing ratio of 1:0.5 in relation to weight. A homogenous mixture of quartz sand and PU binder was obtained by rigorous mixing of the two components using a hand mixer at high speed. Table 3 presents the preparation process of the PUG. The synthetic route toward producing the PUG binder and its microphase structure is shown in Figure 3a,b. Therefore, the PUG was cast either at the top and bottom or both surfaces of the concrete beam at 5 mm and 10 mm thicknesses, as indicated in Table 4. The NC-PUG composite configuration used to conduct flexural tests is shown in Table 4. The graphical representation process of preparing the composite beam and strengthened with PU grouting material is depicted in Figure 4.

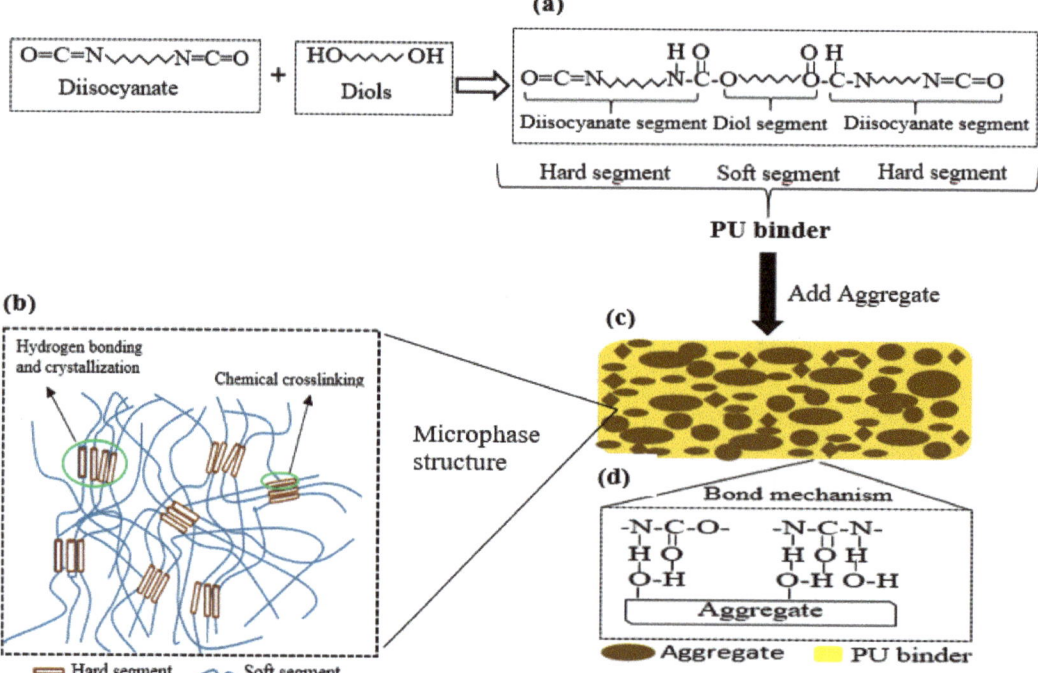

Figure 3. Systemic illustration of (**a**) production of PU binder, (**b**) PU molecular structure, (**c**) PUG geometric model, and (**d**) bond mechanism between PU binder and quartz sand.

Table 4. The NC-PUG configuration.

Sample ID	Sample Type	Sample Designation	PU Grout Layer Thickness (mm)	
			Top Surface	Bottom Surface
NC-PUG0		-	-	-
NC-PUGT5		T	5	-
NC-PUGB5		B	-	5
NC-PUGTB5		T&B	5	5
NC-PUGT10		T	10	-

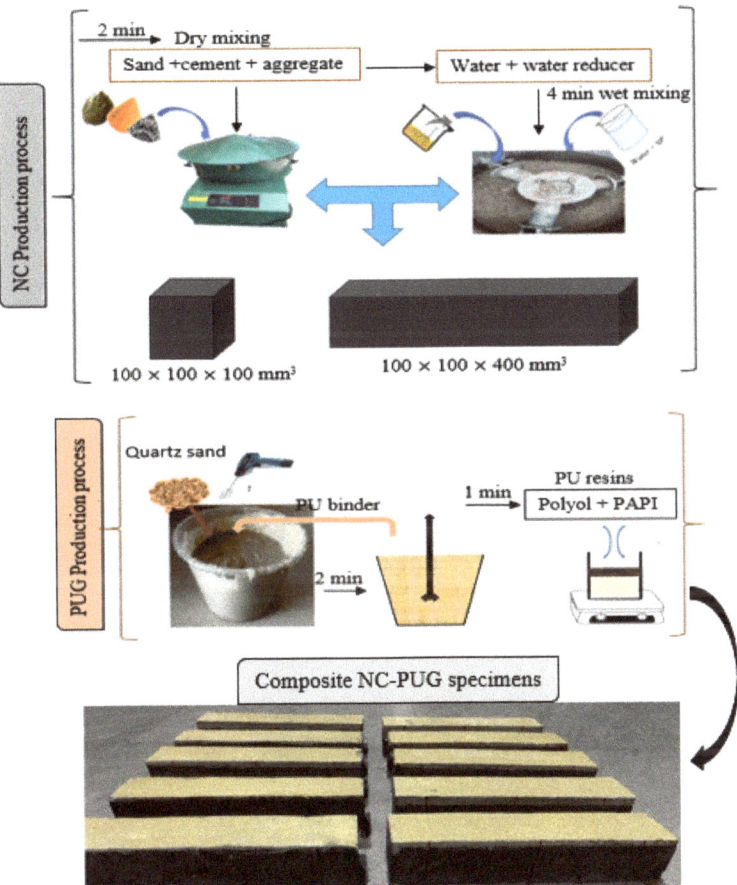

Figure 4. The graphical representation of the process of preparing NC-PUG composite.

2.3. Test Methodology
2.3.1. NC Compressive Test

The NC cube compression test was conducted following 50081-2002 [41] using 100 cube specimens. The 20-ton loading capacity (WDW 200E) universal testing machine was used to test the NC compressive strength. Three samples were tested after 28 days of curation, and the average was considered as the strength of normal concrete.

2.3.2. The NC-PUG Flexural Test

Figure 5 presents the setup for the flexural strength test. The NC-PUG flexural behavior was tested according to the Chinese national standard GB/T 50081-2002 [41]. Using a 300 mm clear span loading and a 0.85 size reduction coefficient, a three-point bending method was used to calculate the flexural stress [41]. The UTM was set at 0.05 mm/min displacement-controlled loading and the specimens were loaded until complete specimen failure. The NC-PUG flexural strength was determined based on Equation (1). Moreover, three LVDTs were attached to the test specimen to monitor the deflection at midspan and two end supports, as illustrated in Figure 5. To add the load to the test sample, a load cell was coupled to the apparatus. The datasets for time, load, and displacement were simultaneously captured with a static data collecting system.

$$f_y = \frac{Pl}{bh^2} \times 0.85 \quad (1)$$

where f_y is the NC-PUG and the flexural strength (MPa); P represents the maximum applied load (KN); l represents the distance of the two supports (mm); and h and b represent the height and width of the beam section (mm), respectively.

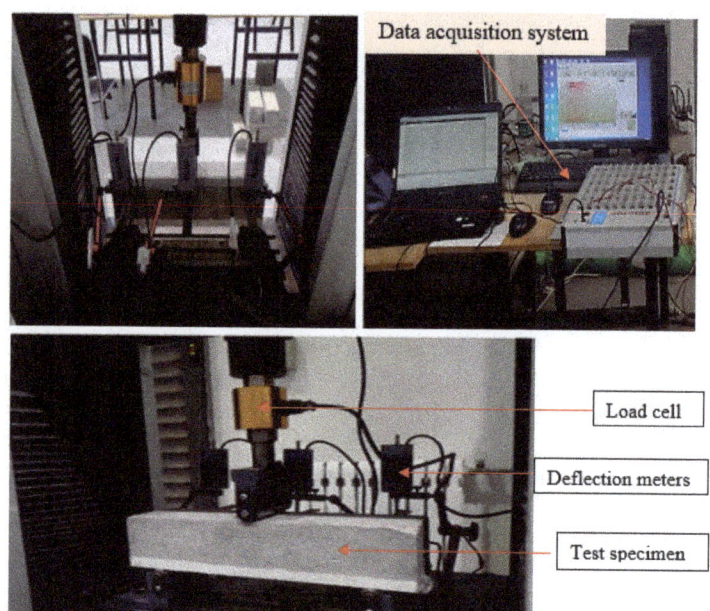

Figure 5. Experimental setup for flexural test.

3. Result and Discussion

3.1. Flexural Response of NC-PUG

Table 5 summarizes the response of the NC repaired with the PU grout material under flexural load. The average of the three samples tested under each testing condition was computed, and the flexural and deflection response was considered based on the 95% confidence level. As shown in Table 5, the control sample (NC-PUG0) reveals the highest average ultimate flexural stress of 5.56 MPa ± 2.57% against the concrete specimen retrofitted with the PUG grouting material; furthermore, a minimum mid-span deflection of 0.49 mm ± 13.60% was recorded for the control sample. However, the ultimate flexural stress of the specimens retrofitted with the PU grouting materials demonstrated reduced flexural strength with increasing deflection, as presented in Table 6. The flexural response is seen in Figure 6, showing the decreasing and increasing pattern of the NC-PUG composite due to the retrofitting effect of the PU grout material and casting configuration. The specimen repaired with a 5 mm thick PUG overlaid at the bottom surface (NC-PUGB5) and top-bottom surface (NC-PUGTB5) showed nearly the same ultimate flexural stress of 4.30 MPa ± 1.77% and 4.35 MPa ± 3.62%, respectively, which are lower than that of the reference specimen by 22.66% and 21.76%, respectively. The mid-span deflection of NC-PUGB5 is 2.19 mm ± 5.26%, and that of NC-PUGB5 is 1.40 mm ± 1.66%. The specimen retrofitted with 5 mm and 10 mm overlaid the PU grout material at the top surface exhibits the ultimate flexural stress of 3.39 MPa ± 6.80% and 3.63 MPa ± 3.72%, which are lower than the flexural strength of the specimen retrofitted at the bottom surface. This behavior is attributed to the specimen attaining a maximum carrying load and then failing. The concrete section no longer bears the applied load. At this stage, the overlaid PU grout sustained the applied load to some specific point before the ultimate failure of the NC-PUG composite, indicating the viscoelastic properties of polyurethane, which tend to make concrete more ductile and less brittle. Following the experimental research utilizing polyurethane demonstrates that PU is a very strain rate-sensitive elastomer, with a significant change in performance from rubbery to leathery in response to increased strain rates [42–45].

Table 5. The flexural response of concrete retrofitted with PU grout material.

Specimen ID	Code	Flexural Strength (MPa)	L-Deflection (mm)	Max Deflection (mm)	R-Deflection (mm)
Reference	1	5.478	0.42	0.57	0.42
	2	5.746	0.36	0.43	0.34
	3	5.478	0.35	0.47	0.38
	Confidential level (0.95)	5.56 ± 2.57%	0.38 ± 9.29%	0.49 ± 13.60%	0.38 ± 9.73%
NC-PUGB5	1	4.371	2.07	2.46	2.71
	2	4.32	2.2	2.60	2.43
	3	4.21	2.32	2.56	2.30
	Confidential level (0.95)	4.30 ± 1.77%	2.19 ± 5.26%	2.54 ± 2.62%	2.48 ± 7.81%
NC-PUGT5	1	3.662	2.32	2.46	2.42
	2	3.174	1.73	1.81	1.71
	3	3.330	2.24	2.22	2.21
	Confidential level (0.95)	3.39 ± 6.80%	2.09 ± 14.10%	2.16 ± 14.04%	2.11 ± 15.95%
NC-PUGT10	1	3.723	1.35	1.34	1.24
	2	3.702	0.97	1.1	1.04
	3	3.46	1.14	1.36	1.21
	Confidential level (0.95)	3.63 ± 3.72%	1.15 ± 15.25%	1.27 ± 10.55%	1.16 ± 8.57%
NC-PUGTB5	1	4.440	1.38	1.48	1.27
	2	4.155	1.43	1.2	1.28
	3	4.67	1.40	1.45	1.39
	Confidential level (0.95)	4.35 ± 3.62%	1.40 ± 1.66%	1.38 ± 10.32%	1.31 ± 4.68%

Table 6. Mechanical characteristics of NC and PUG.

Material	Compressive Strength (MPa)	Elastic Modulus (GPa)	Tensile Strength (MPa)	Density (kg/m^3)
NC (C50)	48.67	32.29	4.76	2400
PUG	19.89	36.67	14.29	2400

Additionally, deflection due to the applied load was recorded at the two end supports, as shown in Figure 7. As shown in Figure 7, the deflection at the end support showed increasing and decreasing behavior according to the overlaid thickness and configuration. Reference specimens exhibit the lowest average deflection of 0.49 mm ± 13.60% and 0.38 mm ± 9.73% at the left and right support, respectively. This deflection is drastically increased due to the PU grouting effect, as exhibited in NC-PUGB5, which records an average deflection of 2.54 ± 2.62% and 2.48 ± 7.81% at the left and right support, respectively. The support deflection tends to decrease with a change in the PUG overlaid thickness and casting position. The support deflection of composite specimens with a 5 mm thick PUG cast at the top surface is reduced by 14.6% compared to the composite specimen cast with a 5 mm PUG overlaid thickness at the bottom surface. The support deflection decreased further in the specimen cast with a 10 mm thick PUG overlaid at the top surface. Hence, the record lowest recorded support deflection of 1.27 mm ± 10.55% and 1.16 mm ± 8.57%, respectively, was obtained. An improvement in the support deflection was observed in the NC-PUGBT5 specimen when compared to the NC-PUGT10 composite specimen. This result indicated that the deflection behavior of concrete can be improved with PUG grouting materials, and was more pronounced when the PUG overlaid was cast at the bottom surface. A related study by Somarathna et al. [33] reported that concrete specimens retrofitted externally reveal a higher strain during ultimate failure due to the elastomeric coating on the impact face under quasi-static testing.

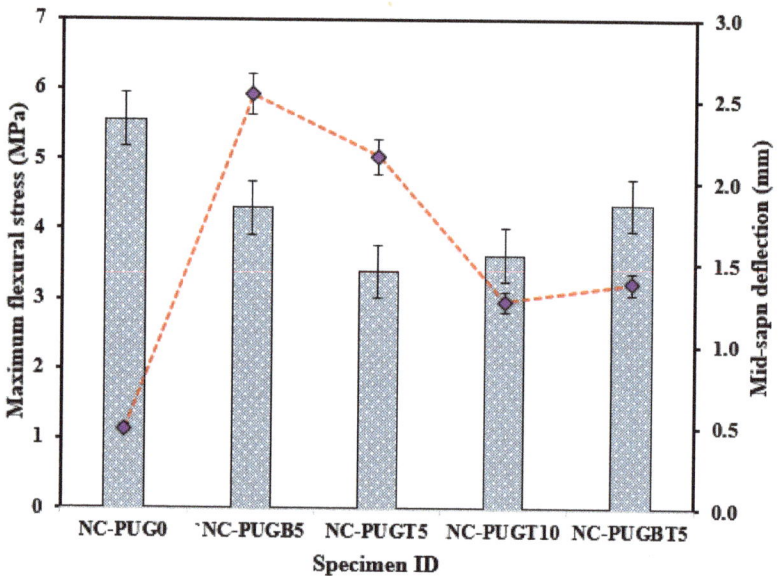

Figure 6. The flexural response of NC-PUG concrete.

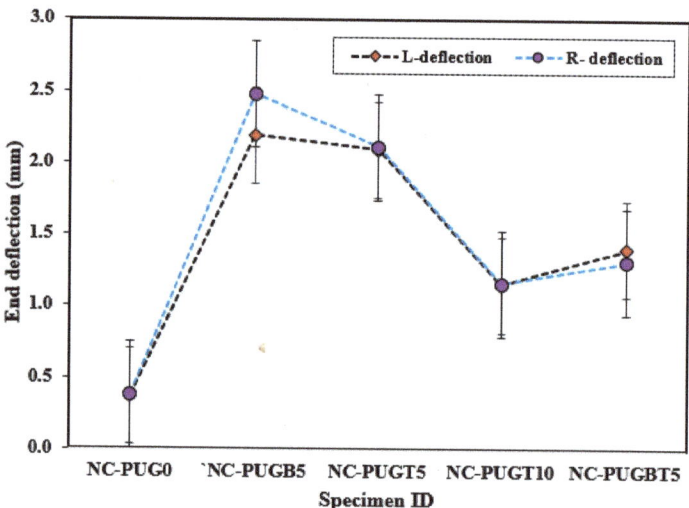

Figure 7. Deflection behavior at the two ends of the specimens.

3.2. Finite Element Modeling of NC-PUG Beam

The finite element modeling (FEM) and analysis (FEA) of the NC-PUG beam were carried out using the FE software ABAQUS 2021 [46]. FEM can help evaluate the properties of the concrete-to-polyurethane grout composite (NC-PUG). The details of the FEM, including the geometry, meshing, and boundary conditions, are presented in this section. This section presents the FE model development to evaluate the flexural responses of the NC-PUG beam.

3.3. Boundary Conditions, Loading Analysis, and Interaction

The specimens were fixed from the bottom steel support (Figure 8a). Displacement-controlled type of loading was applied to the simulation system. The NC-PUG contact was regarded for surface-to-surface relations. The firm contact without penetration was employed, and the shear characteristic was defined using the "penalty" function. The friction coefficient 0.4 between the NC substrate and PUG layer interface was considered. The NS surface was designed as the master surface, while the PUG surface was set as the slave surface.

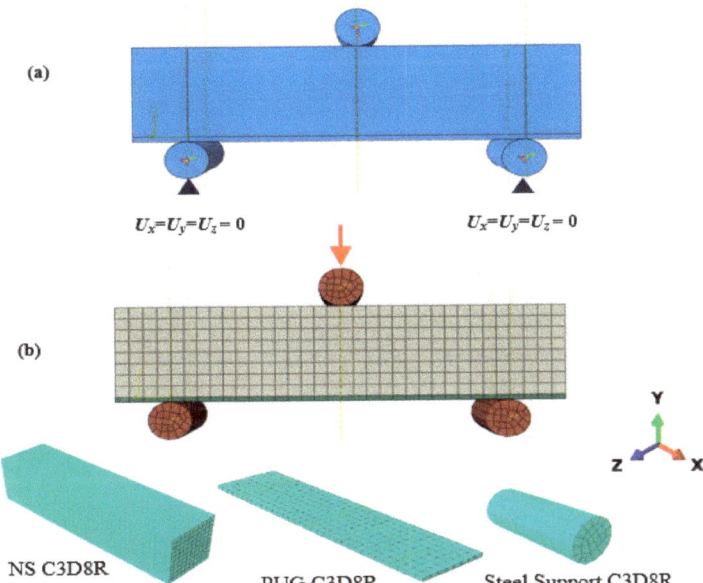

Figure 8. FE model of the NC-PUG beam. (a) Boundary condition; (b) meshing.

3.4. Element Type and Mesh Size

Figure 8 presents a typical FE model of the NC, PUG, and NC-PUG composite beams. The NC and PUG were designed as three-dimensional eight-node composites with a reduced integration point (C3D8R) that was adopted for NC and PUG. The contact surface between NC and PUG used common mesh seeds to ensure accuracy. A sensitivity analysis was conducted with various mesh sizes, and the finest mesh with a size of 10 mm was selected according to the mesh convergence learning process.

3.5. Contact Modeling

The interaction between the NC and PUG grouting material was defined using surface-to-surface contact obtained in ABAQUS. The contact pair in the FE models consists of NC and PUG. The contact surface associated with the NC was used for the master surface, and the PUG contact surface was selected as the slave surface. Similarly, the contact surface associated with the NC-PUG composite was used as the master surface. A friction coefficient of 0.4 was used between the two components [47]. The normal contact behavior was assumed to be hard contact, allowing for the transmission of the surface's pressure and separation when the pressure was zero or negative. The NC was designated as the host element.

3.6. Material Constitutive Model

The ABAQUS concrete damage plasticity (CDP) model simulated the NC and PUG's material behavior. The CDP is a uniaxial compression and uniaxial tension plasticity model that describes the inelastic and damage behavior of concrete. The yielding criteria defined by Lee et al. [48] were adopted in this study. The concrete was characterized using the CDP model. The material properties of grade C50 concrete and PUG grouting material are summarized in Table 6.

3.7. Material Constitutive Model of NC

The CDP model consisted of concrete compression and tensile damage; Figure 9 presents the stress–strain curves of the NC. The compressive stress–strain curve is classified into three parts. The first part is the elastic up to $0.4 f_{mm}$. The second section is the ascending parabolic part starting from $0.4 f_{mm}$ to f_{mm}, which is calculated from Equation (2)—according to EC2—and the third part linearly descends from f_{mm} to $0.85 f_{mm}$.

$$\sigma_c = f_{cm} \left[\frac{kn - n^2}{1 + (k-2)n} \right]$$
$$n = \frac{\varepsilon_c}{\varepsilon_{co}},\ k = E_c \varepsilon_{co} / f_{mm} \quad (2)$$

where ε_c represents the compressive strain, σ_c represents the NSC compressive stress, f_{mm} represents the ultimate compressive stress, ε_{co} represents the strain corresponding to f_{mm} which is equal to 0.002, and the ultimate strain ε_{cu} is equal to 0.0035 [49]. E_c is the modulus of elasticity. Figure 1b shows the tensile stress–strain curve, in which the stress rises proportionately to the strain before cracking. The tensile strength (f_t) is calculated using Equation (3) [50], and the equivalent strain (ε_{ck}) is defined as f_t / E_c.

$$f_t = 0.395 f_{cu}^{0.55} \quad (3)$$

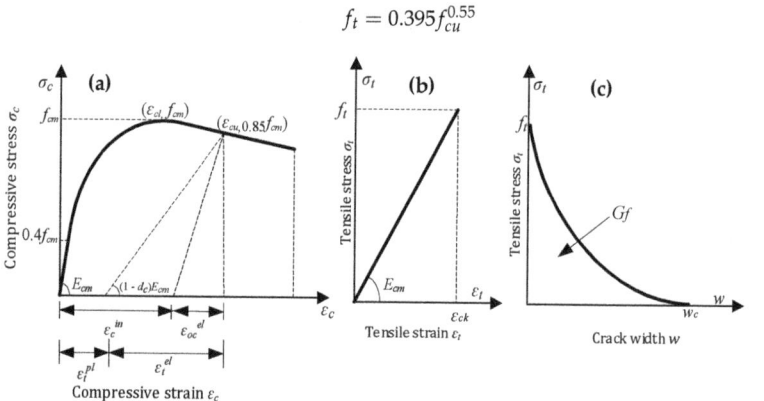

Figure 9. Concrete-damaged plasticity model of NC in (**a**) stress–strain in compression, (**b**) stress–strain in tension, and (**c**) stress–crack width.

The relationship between tensile stress and cracking width is calculated using the fracture energy cracking model to define the tensile behavior of the NSC after cracking. The fracture was obtained from CEP-FIP [51] using Equation (4).

$$G_f = G_{f10} \left(\frac{f_c}{10} \right)^{0.7} \quad (4)$$

where parameter G_f represents the fracture energy in Nmm/mm², and the diameter of the coarse aggregate is approximately 16 mm. Hence, $G_{f10} = 0.003$ Nmm/mm². The concrete compressive and damage coefficient, i.e., d_c and d_t, represents the damage behavior of the concrete. Thus, $d_c = 1 - f_{mm}/\sigma_c$ and $d_t = 1 - f_t/\sigma_t$.

3.8. Model Validations

Figures 10–14 compare the flexural load–deflection curves between the experiments and the numerical model for each specimen condition. The FE analysis showed good agreement between the numerical model and the experimental test result. The initial stiffness (Ke) of the load–deflection curves between the numerical and experiment were compared for each specimen. Initial stiffness is explained as the ratio of 45% of the maximum load (P0.45) to the corresponding deflection (Δ0.45) as described by the ACI 318M-05 [52] and given in Equation (5).

$$Ke = \frac{P_{0.45}}{\Delta_{0.45}} \quad (5)$$

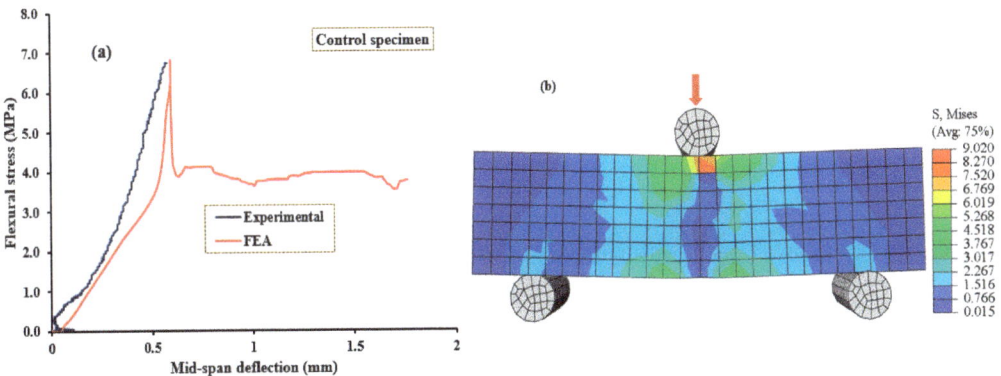

Figure 10. Validation of CDPM under flexure. (**a**) Numerical and experimental stress–strain models; (**b**) FEM and stress distribution for NC-PUG0.

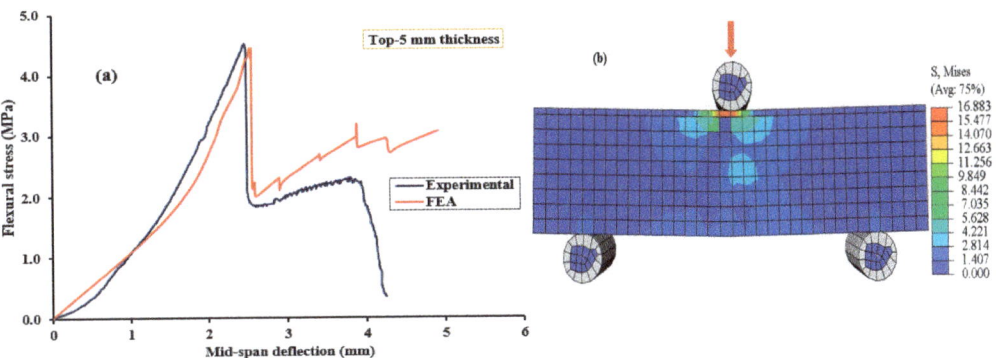

Figure 11. Validation of CDPM under flexure. (**a**) Numerical and experimental stress–strain models; (**b**) FEM and stress distribution for NC-PUGT5.

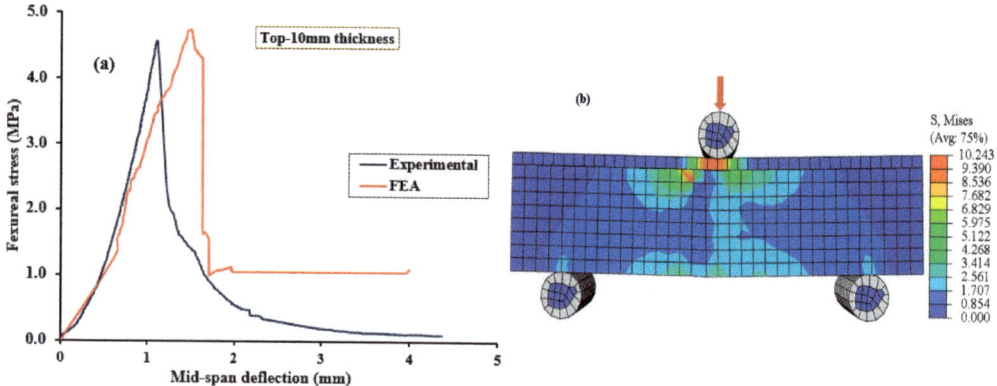

Figure 12. Validation of CDPM under flexure. (**a**) Numerical and experimental stress–strain models; (**b**) FEM and stress distribution for NC-PUGT10.

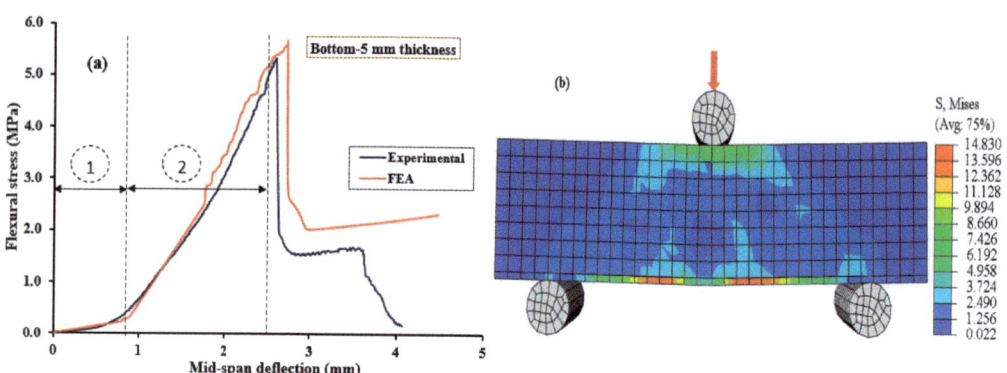

Figure 13. Validation of CDPM under flexure. (**a**) Numerical and experimental stress–strain models; (**b**) FEM and stress distribution for NC-PUGB5.

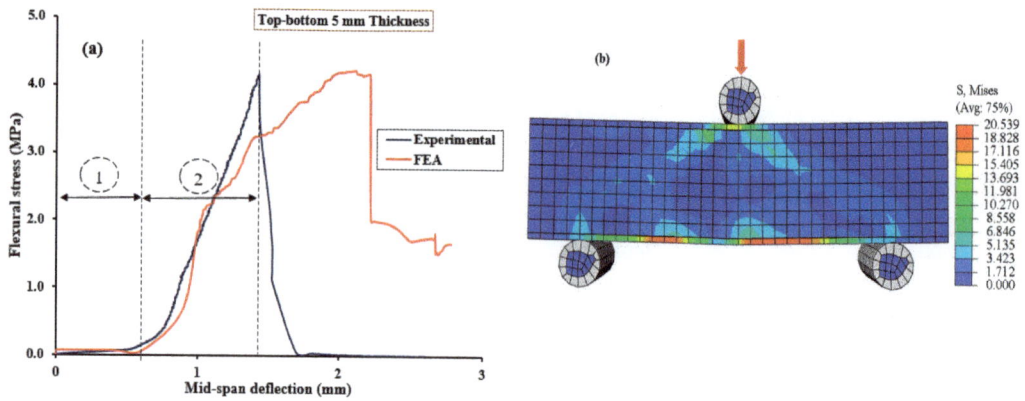

Figure 14. Validation of CDPM under flexure. (**a**) Numerical and experimental stress–strain models; (**b**) FEM and stress distribution for NC-PUGTB5.

The relationship between the flexural stress and mid-span deflection showed a linear pattern at the initial loading stage for the reference, and all the specimens were retrofitted at the top surface, as shown in Figures 10–12, regardless of the PUG grout overlaid thickness. Common behaviors were observed for both the FE model and experimental test results. Under these conditions, deflection continuously increases with the applied loads until the specimen reaches the ultimate load and then fails in flexure. Nearly equal ultimate flexural stress was observed between the FE simulation and experimental test result.

On the other hand, two deformation regions were observed in the NC-PUG composite specimens strengthened with the PUG overlaid cast at the bottom surface or top-bottom surface, as revealed in Figures 13 and 14, before complete failure. Under this configuration, polyurethane grout material was cast at the tension zone of the composite specimen, and polyurethane exhibited viscoelastic properties in response to the increased deflection without significance stress. This behavior tends to change the concrete's brittle nature to a ductile state; polyurethane is a very strain rate-sensitive elastomer, with a significant change in performance from rubbery to leathery in response to increased strain rates. As a result, two deformation regions occurred during testing. Region (i): the initial loading condition; the relationship between flexural stress and deflection exhibits a flat slop with a drastic increase in mid-span deflection under continuous loading. The magnitude of the applied load became noticeable when the deflection reached a significant level, which marked the second deformation region. Region (ii): under this region, the relationship between the flexural stress and mid-span deflection exhibits a sharp slope. Both the applied load and deflection are increased rapidly until the test specimen fails in flexure. The two deformation regions demonstrated by these test specimens are caused by the viscoelastic behavior of polyurethane subjected to tensile stress at the tension zone of the composite specimens.

Table 7 presents the observed initial stiffness (Ke) and ultimate flexural stress (Pu) with numerical values, and the experimental test result-to-predicted value ratio of Ke and Pu is depicted in Figure 15. Table 7 and Figure 15 show that the FE model had accurately estimated the flexural responses of the NC-PUG beam and slightly underestimated Ke by 4% and overestimated Pu by 3%. The standard deviation of the flexural strength of the test specimen-to-prediction ratios of Ke and Pu are 0.1 and 0.02, respectively. The results indicated that the developed FE model predicts the flexural behavior and elastic stiffness of the NC-PUG specimens more accurately. It indicates that 80% of the predictions are within the range of ±10% of the prediction, as shown in Figure 15.

Table 7. Comparison of experimental results with FE model predictions.

No.	Specimens	Ke kN/mm	K_N kN/mm	P_u (kN)	P_N (kN)	Ke/K_N	P_u/P_N
1	NC	22.880	18.511	18.260	18.427	1.230	0.991
2	NC-PUGB5	3.661	3.963	14.400	15.365	0.924	0.937
3	NC-PUGT5	7.914	7.585	12.340	12.788	1.048	0.964
4	NC-PUGT10	3.541	3.275	12.210	12.034	1.081	1.014
5	NC-PUGBTB5	6.038	6.245	13.850	14.100	0.966	0.982
	Standard deviation	7.221	5.519	2.194	2.251	0.106	0.026
	Mean	8.807	7.916	14.212	14.543	1.049	0.977
	Cov (%)	82	69.7	15.4	15.5	10.1	2.6

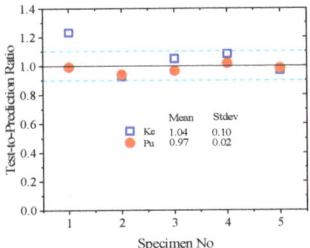

Figure 15. Scatter of test-to-prediction ratios for K_e and P_u.

4. Conclusions

This study investigated the flexural behavior of the NC specimens retrofitted with a polyurethane grouting material of different thicknesses and configurations. The concrete beam specimens were subjected to a three-point bending test. The composite NC-PUG specimens were formed by casting the polyurethane grouting material at either top and/or bottom or both top-bottom surfaces. Moreover, finite element models were developed to simulate the flexural response of the NC-PUG specimens. The following conclusion can be drawn to summarize the findings.

The reference specimen (NC-PUG0) showed the highest average ultimate flexural stress of 5.56 MPa $\pm$ 2.57% at a 95% confidence interval with a corresponding mid-span deflection of 0.49 mm $\pm$ 13.60%. However, due to the strengthened effect of the polyurethane grout, the deflection of the composite specimen was significantly improved.

1. The configuration and/or position of the PU grout material cast influenced the relationship between the flexural stress and mid-span deflection. Specimens retrofitted at the bottom surface exhibit two deformation regions.
2. The effect of the PU grouting material changes the brittle nature of concrete to a more ductile state due to the viscoelastic behavior of polyurethane. This behavior is more effective on the specimen retrofitted at the bottom surface.
3. The FE analysis showed good agreement between the numerical model and the experimental test result. The numerical model accurately predicted the flexural strength of the NC-PUG beam, slightly underestimating K_e by 4% and overestimating P_u by 3%.

Author Contributions: S.I.H. and A.I.B.F.: conceptualization, methodology, investigation, data curation, writing—original draft, writing—review and editing, and visualization; Z.H. and Y.E.I.: conceptualization, investigation, supervision, resources, project administration, and funding acquisition. All authors have read and agreed to the published version of the manuscript.

Funding: This research was funded by the Natural Science Foundation of China (No. 51708314) and the APC was funded by the Structures and Materials Laboratory (S&M Lab) of the College of Engineering, Prince Sultan University, Riyadh, Saudi Arabia.

Institutional Review Board Statement: Not applicable.

Data Availability Statement: The data presented in this study are available on request from the corresponding author.

Acknowledgments: The authors greatly acknowledge the financial support of this research by the Natural Science Foundation of China (No. 51708314), and the Structures and Materials Laboratory (S&M Lab) of the College of Engineering, Prince Sultan University, Riyadh, Saudi Arabia, for funding the article process fees.

Conflicts of Interest: The authors declare no conflict of interest.

References

1. Aabid, A.; Ibrahim, Y.E.; Hrairi, M.; Ali, J.S.M. Optimization of Structural Damage Repair with Single and Double-Sided Composite Patches through the Finite Element Analysis and Taguchi Method. *Materials* **2023**, *16*, 1581. [CrossRef] [PubMed]
2. Aabid, A. Optimization of Reinforcing Patch Effects on Cracked Plates Using Analytical Modeling and Taguchi Design. *Materials* **2023**, *16*, 4348. [CrossRef] [PubMed]
3. Al-shawafi, A.; Zhu, H.; Haruna, S.I.; Bo, Z.; Laqsum, S.A.; Borito, S.M. Impact resistance of ultra-high-performance concrete retrofitted with polyurethane grout material: Experimental investigation and statistical analysis. *Structures* **2023**, *55*, 185–200. [CrossRef]
4. Park, D.; Park, S.; Seo, Y.; Noguchi, T. Water absorption and constraint stress analysis of polymer-modified cement mortar used as a patch repair material. *Constr. Build. Mater.* **2012**, *28*, 819–830. [CrossRef]
5. Assaad, J.J. Development and use of polymer-modified cement for adhesive and repair applications. *Constr. Build. Mater.* **2018**, *168*, 139–148. [CrossRef]
6. Zaccardi, Y.A.V.; Alderete, N.M.; De Belie, N. Improved model for capillary absorption in cementitious materials: Progress over the fourth root of time. *Cem. Concr. Res.* **2017**, *100*, 153–165. [CrossRef]

1. Alderete, N.M.; Zaccardi, Y.A.V.; De Belie, N. Mechanism of long-term capillary water uptake in cementitious materials. *Cem. Concr. Compos.* **2020**, *106*, 103448. [CrossRef]
2. Ohama, Y. Recent progress in concrete-polymer composites. *Adv. Cem. Based Mater.* **1997**, *5*, 31–40. [CrossRef]
3. Giustozzi, F. Polymer-modified pervious concrete for durable and sustainable transportation infrastructures. *Constr. Build. Mater.* **2016**, *111*, 502–512. [CrossRef]
4. Shao, J.; Zhu, H.; Zuo, X.; Lei, W.; Mirgan, S.; Liang, J.; Duan, F. Effect of waste rubber particles on the mechanical performance and deformation properties of epoxy concrete for repair. *Constr. Build. Mater.* **2020**, *241*, 118008. [CrossRef]
5. Afzal, A.; Kausar, A.; Siddiq, M. Role of polymeric composite in civil engineering applications: A review. *Polym. Technol. Mater.* **2020**, *59*, 1023–1040. [CrossRef]
6. Pendhari, S.S.; Kant, T.; Desai, Y.M. Application of polymer composites in civil construction: A general review. *Compos. Struct.* **2008**, *84*, 114–124. [CrossRef]
7. Liu, G.; Otsuka, H.; Mizuta, Y.; Shimitsu, A. A foundational study on static mechanical characteristics of the super lightweight and high strength material using fly-ash. *J. Soc. Mater. Sci. Jpn.* **2006**, *55*, 738. [CrossRef]
8. Hussain, H.K.; Zhang, L.Z.; Liu, G.W. An experimental study on strengthening reinforced concrete T-beams using new material poly-urethane-cement (PUC). *Constr. Build. Mater.* **2013**, *40*, 104–117. [CrossRef]
9. Li, B.; Wang, F.; Fang, H.; Yang, K.; Zhang, X.; Ji, Y. Experimental and numerical study on polymer grouting pretreatment technology in void and corroded concrete pipes. *Tunn. Undergr. Space Technol.* **2021**, *113*, 103842. [CrossRef]
10. Fang, H.; Li, B.; Wang, F.; Wang, Y.; Cui, C. The mechanical behaviour of drainage pipeline under traffic load before and after polymer grouting trenchless repairing. *Tunn. Undergr. Space Technol.* **2018**, *74*, 185–194. [CrossRef]
11. Haruna, S.I.; Zhu, H.; Jiang, W.; Shao, J. Evaluation of impact resistance properties of polyurethane-based polymer concrete for the repair of runway subjected to repeated drop-weight impact test. *Constr. Build. Mater.* **2021**, *309*, 125152. [CrossRef]
12. Jiang, W.; Zhu, H.; Haruna, S.I.; Shao, J.; Yu, Y.; Wu, K. Mechanical properties and freeze–thaw resistance of polyurethane-based polymer mortar with crumb rubber powder. *Constr. Build. Mater.* **2022**, *352*, 129040. [CrossRef]
13. Al-Atroush, M.E.; Shabbir, O.; Almeshari, B.; Waly, M.; Sebaey, T.A. A Novel Application of the Hydrophobic Polyurethane Foam: Expansive Soil Stabilization. *Polymers* **2021**, *13*, 1335. [CrossRef]
14. Liu, K.; Tong, J.; Huang, M.; Wang, F.; Pang, H. Model and experimental studies on the effects of load characteristics and polyurethane densities on fatigue damage of rigid polyurethane grouting materials. *Constr. Build. Mater.* **2022**, *347*, 128595. [CrossRef]
15. Ibrahim, Y.E.; Nabil, M. Assessment of structural response of an existing structure under blast load using finite element analysis. *Alex. Eng. J.* **2019**, *58*, 1327–1338. [CrossRef]
16. Shahbeyk, S.; Hosseini, M.; Yaghoobi, M. Mesoscale finite element prediction of concrete failure. *Comput. Mater. Sci.* **2011**, *55*, 1973–1990. [CrossRef]
17. Słowik, M.; Nowicki, T. The analysis of diagonal crack propagation in concrete beams. *Comput. Mater. Sci.* **2012**, *52*, 261–267. [CrossRef]
18. Guo, L.-P.; Carpinteri, A.; Roncella, R.; Spagnoli, A.; Sun, W.; Vantadori, S. Fatigue damage of high performance concrete through a 2D mesoscopic lattice model. *Comput. Mater. Sci.* **2009**, *44*, 1098–1106. [CrossRef]
19. Hála, P.; Perrot, A.; Vacková, B.; Kheml, P.; Sovják, R. Experimental and numerical study on ballistic resistance of polyurethane-coated thin HPFRC plate. *Mater. Today Proc.* **2023**. [CrossRef]
20. Niu, S.; Wang, Z.; Wang, J.; Wang, D.; Sun, X. Experimental investigation on flexural performance of polyester polyurethane concrete steel bridge deck composite structure. *Case Stud. Constr. Mater.* **2023**, *18*, e01915. [CrossRef]
21. Goushis, R.; Mini, K.M. Finite element analysis of polymeric and cementitious materials to secure cracks in concrete. *Mater. Today Proc.* **2022**, *49*, 1599–1606. [CrossRef]
22. Li, M.; Xue, B.; Fang, H.; Zhang, S.; Wang, F. Variable angle shear test and finite element simulation of polyurethane—Bentonite composite structure. *Structures* **2023**, *48*, 1722–1729. [CrossRef]
23. Sing, L.K.; Yahaya, N.; Valipour, A.; Zardasti, L.; Azraai, S.N.A.; Noor, N.M. Mechanical properties characterization and finite element analysis of epoxy grouts in repairing damaged pipeline. *J. Press. Vessel Technol.* **2018**, *140*, 61701. [CrossRef]
24. Ai, S.; Tang, L.; Mao, Y.; Liu, Y.; Fang, D. Numerical analysis on failure behaviour of polyurethane polymer concrete at high strain rates in compression. *Comput. Mater. Sci.* **2013**, *69*, 389–395. [CrossRef]
25. Chen, B.; He, M.; Huang, Z.; Wu, Z. Long-tern field test and numerical simulation of foamed polyurethane insulation on concrete dam in severely cold region. *Constr. Build. Mater.* **2019**, *212*, 618–634. [CrossRef]
26. Moslemi, A.; Neya, B.N.; Davoodi, M.-R.; Dehestani, M. Proposing and finite element analysis of a new composite profiled sheet deck—Applying PU and PVC and stability considerations. *Structures* **2021**, *34*, 3040–3054. [CrossRef]
27. Somarathna, H.M.C.C.; Raman, S.N.; Mohotti, D.; Mutalib, A.A.; Badri, K.H. Behaviour of concrete specimens retrofitted with bio-based polyurethane coatings under dynamic loads. *Constr. Build. Mater.* **2021**, *270*, 121860. [CrossRef]
28. Li, X.; Wang, M.; Zheng, D.; Fang, H.; Wang, F.; Wan, J. Study on the failure mechanism between polyurethane grouting material and concrete considering the effect of moisture by digital image correlation. *J. Build. Eng.* **2023**, *67*, 105948. [CrossRef]
29. Huang, H.; Li, M.; Yuan, Y.; Bai, H. Theoretical analysis on the lateral drift of precast concrete frame with replaceable artificial controllable plastic hinges. *J. Build. Eng.* **2022**, *62*, 105386. [CrossRef]

36. Huang, H.; Huang, M.; Zhang, W.; Pospisil, S.; Wu, T. Experimental investigation on rehabilitation of corroded RC columns with BSP and HPFL under combined loadings. *J. Struct. Eng.* **2020**, *146*, 4020157. [CrossRef]
37. Zhang, W.; Liu, X.; Huang, Y.; Tong, M.-N. Reliability-based analysis of the flexural strength of concrete beams reinforced with hybrid BFRP and steel rebars. *Arch. Civ. Mech. Eng.* **2022**, *22*, 171. [CrossRef]
38. ABAQUS. *ABAQUS Standard User's Manual, Version 6.14*; Dassault Systèmes: Providence, RI, USA, 2014.
39. Jiang, W.; Zhu, H.; Haruna, S.I.; Zhao, B.; Shao, J.; Yu, Y. Effect of crumb rubber powder on mechanical properties and pore structure of polyurethane-based polymer mortar for repair. *Constr. Build. Mater.* **2021**, *309*, 125169. [CrossRef]
40. Laqsum, S.A.; Zhu, H.; Bo, Z.; Haruna, S.I.; Al-shawafi, A.; Borito, S.M. Static properties and impact resistance performance of U-shaped PU-modified concrete under repeated drop-weight impact load. *Arch. Civ. Mech. Eng.* **2023**, *23*, 227. [CrossRef]
41. *GB/T50081-2002*; Standard for Test Method of Mechanical Properties on Ordinary Concrete. China Academy of Building Research: Beijing, China, 2012.
42. Somarathna, H.M.C.C.; Raman, S.N.; Mohotti, D.; Mutalib, A.A.; Badri, K.H. Rate dependent tensile behavior of polyurethane under varying strain rates. *Constr. Build. Mater.* **2020**, *254*, 119203. [CrossRef]
43. Somarathna, H.; Raman, S.N.; Mohotti, D.; Mutalib, A.A.; Badri, K.H. Hyper-viscoelastic constitutive models for predicting the material behavior of polyurethane under varying strain rates and uniaxial tensile loading. *Constr. Build. Mater.* **2020**, *236*, 117417. [CrossRef]
44. Haruna, S.I.; Zhu, H.; Ibrahim, Y.E.; Shao, J.; Adamu, M.; Farouk, A.I.B. Experimental and Statistical Analysis of U-Shaped Polyurethane-Based Polymer Concrete under Static and Impact Loads as a Repair Material. *Buildings* **2022**, *26*, 1986. [CrossRef]
45. Al-kahtani, M.S.M.; Zhu, H.; Haruna, S.I.; Shao, J. Evaluation of Mechanical Properties of Polyurethane-Based Polymer Rubber Concrete Modified Ground Glass Fiber Using Response Surface Methodology. *Arab. J. Sci. Eng.* **2022**, *48*, 4695–4710. [CrossRef]
46. Dassault, A. *6.14-Abaqus Analysis User's Manual*; Dassault Systèmes Simulia Corp.: Providence, RI, USA, 2014.
47. Farouk, A.I.B.; Zhu, J.; Yuhui, G. Finite element analysis of the shear performance of box-groove interface of ultra-high-performance concrete (UHPC)-normal strength concrete (NSC) composite girder. *Innov. Infrastruct. Solut.* **2022**, *7*, 212. [CrossRef]
48. Lee, J.; Fenves, G.L. Plastic-Damage Model for Cyclic Loading of Concrete Structures. *J. Eng. Mech.* **1998**, *124*, 892–900. [CrossRef]
49. *EN 1992-1-1*; Eurocode 2: Design of Concrete Structures—Part 1-1: General Rules and Rules for Buildings. British Standard Institution: London, UK, 2004.
50. *GB50010-2010*; Code for Design of Concrete Structures. National Standard of China: Beijing, China, 2010.
51. Béton, C.E.-I.D. *CEB-FIP Model Code 1990: Design Code*; Thomas Telford Publishing: London, UK, 1993.
52. *ACI 318M-05*; Building Code Requirements for Structural Concrete and Commentary. American Concrete Institute: Farmington Hills, MI, USA, 2005.

Disclaimer/Publisher's Note: The statements, opinions and data contained in all publications are solely those of the individual author(s) and contributor(s) and not of MDPI and/or the editor(s). MDPI and/or the editor(s) disclaim responsibility for any injury to people or property resulting from any ideas, methods, instructions or products referred to in the content.

Article

Effects of Grafting Degree on the Formation of Waterborne Polyurethane-Acrylate Film with Hard Core–Soft Shell Structure

Yong Rok Kwon [1,2], Seok Kyu Moon [1,3], Hae Chan Kim [1,2], Jung Soo Kim [1], Miyeon Kwon [1] and Dong Hyun Kim [1,*]

1. Materials & Component Convergence R&D Department, Korea Institute of Industrial Technology (KITECH), 143, Hanggaul-ro, Sangnok-gu, Ansan-si 15588, Republic of Korea; yongrok@kitech.re.kr (Y.R.K.); anstjrrb@kitech.re.kr (S.K.M.); coolskawk@kitech.re.kr (H.C.K.); kimjungsoo11@kitech.re.kr (J.S.K.); mykwon@kitech.re.kr (M.K.)
2. Department of Material Chemical Engineering, Hanyang University, 55, Hanggaul-ro, Sangnok-gu, Ansan-si 15588, Republic of Korea
3. School of Integrative Engineering, Chung-Ang University, 84, Heukseok-ro, Dongjak-gu, Seoul 06974, Republic of Korea
* Correspondence: dhkim@kitech.re.kr

Abstract: Waterborne polyurethane-acrylate (WPUA) grafted with polyurethane was prepared to improve the film-forming ability of hard-type acrylic latex. To balance the film-formation ability and hardness, the WPUA latex was designed with a hard core (polyacrylate) and soft shell (polyurethane). The grafting ratio was controlled through varying the content of 2-hydroxyethyl methacrylate (HEMA) used to cap the ends of the polyurethane prepolymer. The morphologies of the latex particles, film surface, and fracture surface of the film were characterized through transmission electron microscopy, atomic force microscopy, and scanning electron microscopy, respectively. An increase in the grafting ratio resulted in the enhanced miscibility of polyurethane and polyacrylate but reduced adhesion between particles and increased minimum film formation temperature. In addition, grafting was essential to obtain transparent WPUA films. Excessive grafting induced defects such as micropores within the film, leading to the decreased hardness and adhesive strength of the film. The optimal HEMA content for the preparation of a WPUA coating with excellent film-forming ability and high hardness in ambient conditions was noted to be 50%. The final WPUA film was prepared without coalescence agents that generate volatile organic compounds.

Keywords: waterborne polyurethane-acrylate; 2-hydroxyethyl methacrylate; minimum film formation temperature; grafting; nano-indentation

1. Introduction

Acrylic latexes are widely used in coating formulations to reduce volatile organic compounds (VOCs) that impact the environment and human health [1]. However, solvent-based binders are still typically used in certain coating applications with high performance requirements, such as enhanced hardness, gloss, and scratch resistance [2]. These characteristics of waterborne coatings are inferior to those of solvent-based coatings due to their suboptimal film-forming ability. In waterborne systems, the latex polymer is dispersed in the form of nano- or micro-sized individual particles in the water phase. To achieve a continuous film during film formation, smooth coalescence must be ensured between these particles and chain diffusion [3]. Hard latexes can be used to achieve strong mechanical properties. However, their high transition glass temperature (T_g) limits intergranular coalescence [4]. Owing to these aspects, it remains challenging to balance film formation and mechanical properties. Slow-drying solvents or low-molecular-weight plasticizers have been incorporated in waterborne coatings to facilitate polymer diffusion in the rigid latex [5,6]. However, such additives result in increased VOC emissions and adversely affect hardness development, chemical resistance, and durability [7].

Various methods based on T_g heterogeneity, such as oligomer integration, the use of hard/soft latex blends, and the introduction of core–shell latexes, have been developed to simultaneously achieve low VOC emissions, high hardness, and low minimum film formation temperature (MFFT) [8–10]. The introduction of an oligomeric plasticizer that was prepared in situ using a chain transfer agent helped decrease the MFFT of the acrylic latex and VOC emissions by less than half [8]. In the case of hard/soft latex blends, the hard phase enhances the film hardness, and the soft phase promotes film formation [9]. Using a hard core–soft shell latex, a more uniform distribution than that of a blend can be achieved through incorporating the different phases into one structure. The improved mechanical and film-forming properties of acrylic core–shell latex depend on various parameters, such as the shell T_g, shell specific gravity, and particle size [10].

An alternative approach is to introduce polyurethane (PU), known for its excellent flexibility and hydrogen bonding ability, into acrylic latex [11]. Wang et al. used a PU dispersion as a polymer coagulant to aid film formation and demonstrated its potential as an alternative to conventional coalescence aids [12]. However, the thermodynamic immiscibility between acrylic polymers and PU may lead to significant phase separation [13]. To overcome this problem, vinyl-terminated PU can be grafted onto acrylic chains. Grafted waterborne polyurethane-acrylate (WPUA) latexes have been noted to be superior to conventional blend latexes owing to improved compatibility [14]. However, excessive grafting can reduce the mobility of the polymer chains, thereby inhibiting interparticle coalescence.

In most of the existing studies, the effect of grafting on WPUA latex has been interpreted in terms of the phase separation and crosslinking density, and the film development process has been rarely considered [15–19]. In particular, no study has been reported on the design of WPUA latex with a hard core–soft shell structure to prepare hard coatings without coalescence additives that generate VOCs. Latex formulations that emit low VOCs and form films under ambient conditions are highly attractive to consumers. Therefore, this study was aimed at developing a latex coating formulation with excellent mechanical properties and film-forming ability in ambient conditions through introducing PU with high coalescing ability. To this end, we attempted to clarify the relationship between the grafting degree and film-forming ability of WPUA latex with a hard core–soft shell structure.

2. Experiment

2.1. Materials and Methods

Polytetrahydrofuran (PTHF, M_n = 1000 g/mol, Sigma Aldrich, St. Louis, MI, USA) and dimethylol propionic acid (DMPA, Sigma Aldrich, St. Louis, MI, USA) were dried in a vacuum oven at 60 °C for 24 h before use. Isophorone diisocyanate (IPDI), 2-hydroxyethyl methacrylate (HEMA), methyl methacrylate (MMA), and n-butyl acrylate (BA) were purchased from Sigma Aldrich, St. Louis, MI, USA. Dibutyltin dilaurate (DBTDL, Sigma Aldrich, St. Louis, MI, USA) was used as the catalyst for the urethane reaction. Triethylamine (TEA), ethylene diamine (EDA), ammonium persulfate (APS), and 1-methyl-2-pyrrolidinone (NMP) were purchased from Alfa Aesar, Ward Hill, MA, USA, and used as received.

2.2. Preparation of WPUA Samples

Figure 1 shows the synthesis process of WPUA latex. PTHF and IPDI were added to a 500 mL four-necked reactor and heated to 70 °C. The mixture was stirred with a mechanical stirrer under a nitrogen atmosphere. DMPA dissolved in NMP was added dropwise to the mixture and then stirred at 80 °C for 30 min. Subsequently, DBTDL was added, and the reaction was sustained until the NCO content reached the theoretical value. The resulting NCO-terminated PU prepolymer was cooled to 60 °C, and the viscosity was lowered through adding MMA and BA. Next, HEMA was added and allowed to react at the same temperature for 3 h to synthesize the acrylic terminal PU. The system was cooled to

40 °C, and TEA was added to neutralize the –COOH groups of DMPA for 1 h. The reactor was then immersed in an ice bath, and a mixture of distilled water and EDA was slowly introduced into the dropping funnel under high-speed agitation (600–700 rpm). After the completion of water dispersion and chain extension, the reactor was heated to 80 °C, and APS was added to polymerize the acrylic monomers. Finally, a WPUA dispersion with a solid content of 30 wt% was obtained.

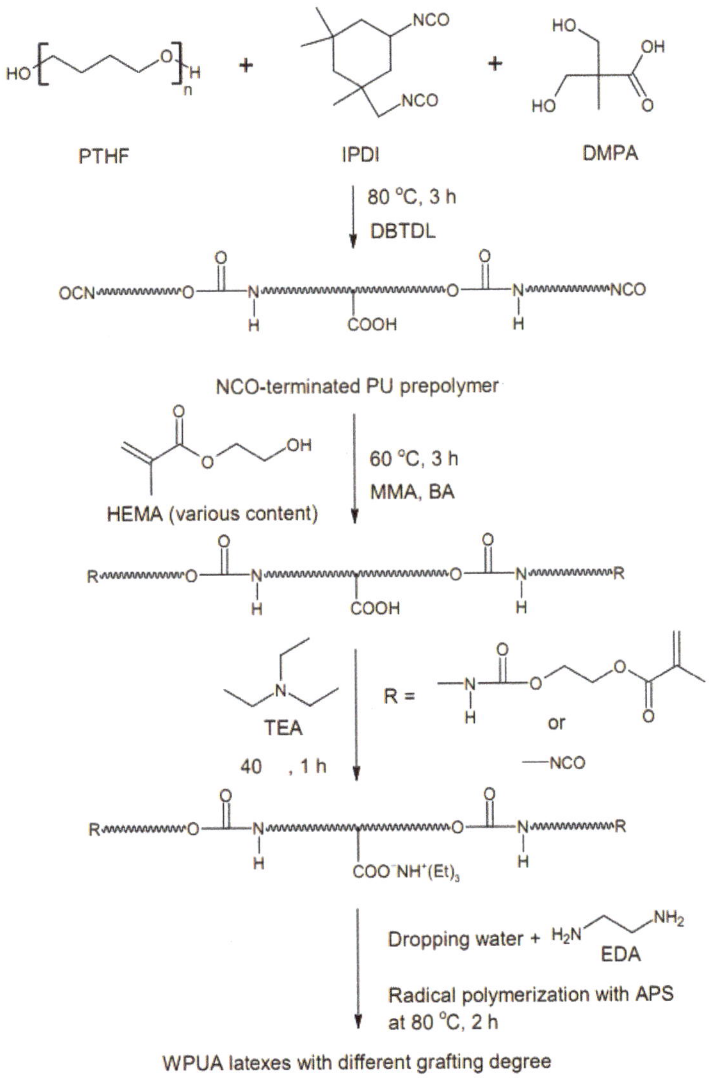

Figure 1. Synthesis process of WPUA latexes.

The HEMA content was calculated using the theoretical molar number of the NCO groups in the PU prepolymer. The synthesized samples were labeled as WPUA_$H_{0.00}$, WPUA_$H_{0.25}$, WPUA_$H_{0.50}$, WPUA_$H_{0.75}$, and WPUA_$H_{1.00}$, depending on the molar ratio of $NCO_{PU\ prepolymer}/OH_{HEMA}$ (Table 1).

Table 1. WPUA samples with different HEMA contents.

Sample	HEMA (mmol)	EDA (mmol)	OH_{HEMA}/NCO_{PUpre} (mol Ratio)
WPUA_$H_{0.00}$	0	30	0
WPUA_$H_{0.25}$	15	22.5	0.25
WPUA_$H_{0.50}$	30	15	0.50
WPUA_$H_{0.75}$	45	7.5	0.75
WPUA_$H_{1.00}$	60	0	1.00

PU prepolymer composition: PTHF = 0.02 mol, IPDI = 0.09 mol, DMPA = 0.04 mol, TEA = 0.04 mol; Acrylic composition: MMA = 61.2 g, BA = 6.8 g.

2.3. Preparation of WPUA Films

The WPUA film was prepared through casting the dispersion onto a Teflon mold or glass plate and drying for 48 h at 25 °C and 24 h in a vacuum oven at 40 °C.

2.4. Characterization

2.4.1. Particle Size Distribution and Zeta Potential

The particle size and zeta potential of the dispersions were measured using a particle size analyzer (SZ-100, HORIBA, Kyoto, Japan) after diluting the samples with distilled water. The measurements were conducted in triplicate at 25 °C, and the average values of the three trials were used as the final results.

2.4.2. Transmission Electron Microscopy (TEM)

The morphology of the WPUA particles was observed using a Tecnai F20 G2 (FEI, Hillsboro, OR, USA) instrument operating at an accelerating voltage of 100 kV. The latex was diluted to a solids content of 0.1 wt% and dyed with 3 wt% phosphotungstic acid solution [20]. The mixture was adjusted to pH 7.5 using aqueous ammonia to stabilize the latex particles. A drop of the sample solution was deposited onto a copper mesh and allowed to dry at room temperature.

2.4.3. Gel Fraction

To measure the gel content of the WPUA films, 1 cm × 1 cm samples were immersed in acetone for 48 h (initial weight, W_0). Subsequently, the samples were dried in a 60 °C vacuum oven for 24 h to eliminate any residual acetone (dried weight, W). The gel content (G) was calculated as

$$G = W/W_0 \times 100\ (\%) \qquad (1)$$

2.4.4. XPS

XPS measurements were performed using an X-ray photoelectron spectrometer (NEXSA, Thermo Fisher Scientific, Waltham, MA, USA) equipped with an Al Kα X-ray source (1486.6 eV) to determine the elemental composition of the latex surface. Latex samples for XPS were prepared in powder form via lyophilization for 72 h.

2.4.5. Fourier Transform Infrared Spectroscopy (FT-IR)

FT-IR spectroscopy (Cary 630, Agilent Technologies, Santa Clara, CA, USA) was performed to analyze the chemical structure of the WPUA. The spectra were recorded in the range of 650–4000 cm^{-1} at a resolution of 4 cm^{-1} with a scan count of 64. All measurements were obtained at room temperature in the attenuated total reflectance mode.

2.4.6. MFFT

The MFFT measurements were obtained in the temperature range of 0–90 °C, following the ASTM D2354 standard [21], using custom-made MFFT equipment.

2.4.7. Dynamic Mechanical Analysis (DMA)

The dynamic mechanical properties of the prepared films were characterized through a DMA (242C, Netzsch, Selb, Germany). Each film was heated from $-80\ °C$ to $120\ °C$ at a rate of $0.66\ K\ min^{-1}$ under a N_2 atmosphere, with a frequency of 1 Hz.

2.4.8. Atomic Force Microscopy (AFM)

The surface morphology of the WPUA film was investigated through tapping-mode AFM (XE-100, Park systems, Suwon, Republic of Korea). Topographic and phase images were observed in the 2 μm^2 range at a scan rate of 1 Hz using a silicon probe.

2.4.9. Scanning Electron Microscopy (SEM)

SEM (SU8000, Hitachi, Tokyo, Japan) was performed to observe the fracture surface morphology of the film. Observations were made at an accelerating voltage of 10 kV. Before the test, the film was frozen in liquid nitrogen and then fractured.

2.4.10. Nano-Indentation Test

To evaluate the surface hardness and elastic modulus of the WPUA film, a load–indentation depth curve was obtained using a nanoindenter (ZHN, Zwick Roell, Ulm, Germany). Tests for each sample were conducted 10 times, and the average value was obtained according to ASTM D1474 [22].

2.4.11. Cross-Cut Test

The adhesion of the WPUA coating to the substrate was evaluated using a cross-cutter, and the measurements were recorded according to the ISO 2409 standard [23].

2.4.12. Transparency Evaluation

Transmittance was recorded on a UV–vis spectrophotometer (Lambda 35, Perkin Elmer, Waltham, MA, USA) in the visible range. Transmittance values were averaged after five measurements.

3. Results

3.1. FT-IR Analysis

Figure 2 shows the FT-IR spectra for the WPUA prepolymers, attained through the dibutylamine back titration method at the theoretical NCO content of the prepolymer. The peaks for all samples were similar, except for the peak corresponding to –NCO at 2270 cm^{-1} [20]. The disappearance of the peak associated with the –OH group at 3300 cm^{-1} confirmed the complete reaction of IPDI, PTHF, DMPA, and HEMA [24]. A broad peak pertaining to urethane N–H was observed at 3320 cm^{-1} [25]. The peak intensity associated with –NCO at 2270 cm^{-1} reduced with increasing HEMA content. Peaks corresponding to the symmetric and asymmetric stretching vibrations of –CH_3 and –CH_2 were observed at 2855–2955 cm^{-1} [26]. The peak associated with PTHF C–O–C stretching vibrations appeared at 1110 cm^{-1}, whereas those for free C=O and hydrogen-bonded C=O appeared at approximately 1719 cm^{-1} [27]. The C=C peak for HEMA appeared at 1650 cm^{-1}, indicating that no polymerization occurred during the 3 h reaction period [28]. At the end of all of the reactions, the residual –NCO and C=C peaks disappeared in the FT-IR spectra for the prepared WPUA_H0.50 dispersion and film (Figure 2c).

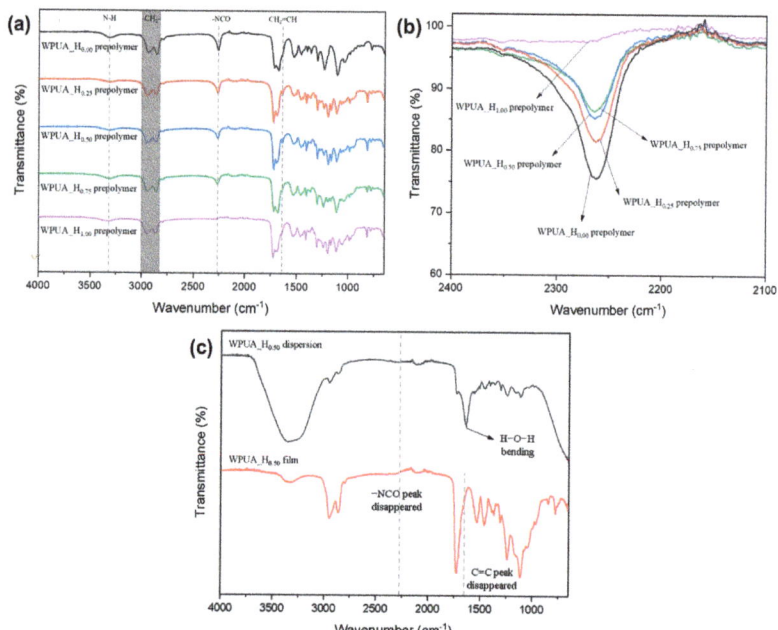

Figure 2. FT-IR spectra for the (**a**) overall frequency; (**b**) –NCO stretching region of the WPUA_$H_{0.00}$, WPUA_$H_{0.25}$, WPUA_$H_{0.50}$, WPUA_$H_{0.75}$, and WPUA_$H_{1.00}$ prepolymers; and (**c**) WPUA_$H_{0.50}$ dispersion and film.

3.2. Particle Size and Gel Fraction of WPUA Latex

The average particle size, zeta potential, and gel fraction of the WPUA latex samples with different degrees of grafting are summarized in Table 2. Increasing the HEMA content resulted in an increase in the average particle size of the WPUA latex and decrease in the absolute zeta potential. During the synthesis of WPUA, a higher HEMA content facilitated the generation of additional grafting points between the acrylic polymer (PA, core) and PU polymer (shell). Consequently, the mobility of PU polymers with hydrophilic functional groups was reduced, which limited the localization of COO⁻ ions to the particle surface. This phenomenon reduced the number of ionic groups exposed on the WPUA latex surface, weakening the electrostatic repulsion between latex particles. As a result, agglomeration between WPUA latex occurred and particle size increased [29].

Table 2. Gel fraction, particle size, and zeta potential for WPUA latex.

Sample	Gel Fraction (%)	Particle Size (nm)	Zeta Potential (mV)
WPUA_$H_{0.00}$	0	64.2	−72.4
WPUA_$H_{0.25}$	5.8	70.5	−52.3
WPUA_$H_{0.50}$	27.7	77.1	−42.9
WPUA_$H_{0.75}$	53.1	84.5	−38.4
WPUA_$H_{1.00}$	86.2	92.9	−35.2

The gel fraction measurement results indicated that the acrylic terminal PU facilitated the crosslinking of the WPUA latex particles. As the HEMA content increased, the gel fraction of the latex rapidly increased. Theoretically, the PU chain is chain-extended with EDA to have a diacrylate structure, which can act as a crosslinking agent. The highest gel fraction (86.2%) corresponded to the WPUA_$H_{1.00}$ sample, and highly crosslinked particles were formed.

3.3. Characterization of Core-Shell Structure

The particle morphology of WPUA_$H_{0.50}$ latex was visualized in negative-stain TEM images as shown in Figure 3. The diameter of the particles was approximately 82 nm, which is similar to the results measured using the particle size analyzer. The latex particles exhibited a distinct core–shell structure, where the dark domains in the outer layer were PU and the relatively bright domains in the core were PA. The PU moiety is more polar than the PA moiety [30]. Therefore, shell regions with higher electron cloud density were observed to be darker.

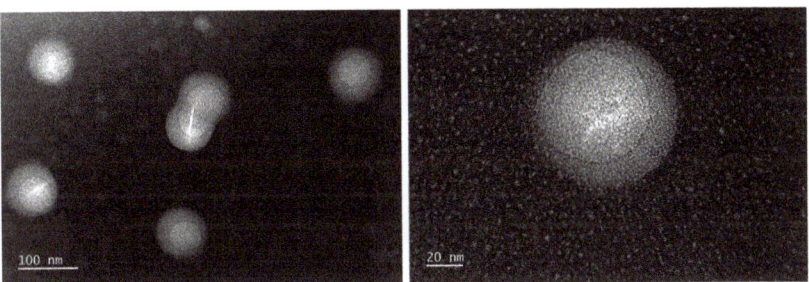

Figure 3. TEM images for the WPUA_$H_{0.50}$ latex.

The core–shell structure of WPUA was further characterized through comparing the surface elemental composition of WPUA and acryl-free WPU particles through XPS [31–33]. Here, WPU consists only of PU components in WPUA. WPUA_$H_{0.50}$ and WPU latex were lyophilized prior to analysis to maintain the shape and morphology of the particles. Figure 4 shows the XPS spectra of WPUA_$H_{0.50}$ and WPU particles. Although the WPUA_$H_{0.50}$ sample contained approximately 58% acrylic component, it was almost identical to the surface component composition of the WPU particles. In addition, the contents of nitrogen elements associated with urethane and urea groups in WPUA_$H_{0.50}$ (4.6%) and WPU (4.7%) were almost similar. These results proved that the hydrophobic acrylic component in WPUA is almost completely covered by the hydrophilic PU.

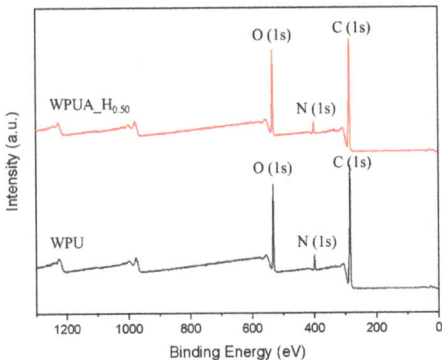

Figure 4. XPS spectra for WPUA_$H_{0.50}$ and WPU particles.

3.4. Viscoelastic and Thermal Characterization of WPUA Films

Figure 5a shows the changes in the storage modulus of the WPUA films with increasing temperature. The storage modulus of the WPUA films increased as the HEMA content in the vitreous region increased, attributable to the higher degree of grafting and crosslinking. The storage modulus of WPUA_$H_{0.00}$ at −70 °C was 2480 MPa, which increased to 5020 MPa at the same temperature. As the temperature increased, the molecular chain activity intensified, and the storage modulus gradually decreased. Notably, the storage moduli of the WPUA_H$_{0.00}$ and WPUA_H$_{0.25}$ rapidly decreased near 0 °C.

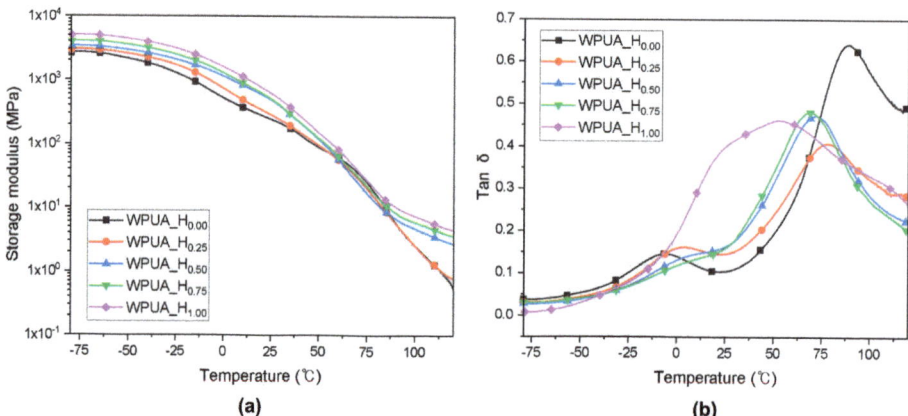

Figure 5. (a) Storage moduli and (b) tan δ vs. temperature curves for the WPUA_H$_{0.00}$, WPUA_H$_{0.25}$, WPUA_H$_{0.50}$, WPUA_H$_{0.75}$, and WPUA_H$_{1.00}$ films.

The crosslink density, the number of moles of elastically effective network chains per cubic centimeter of sample (v_e), can be determined through measuring the modulus in the rubbery region [34,35]. The network crosslink density of WPUA films with different HEMA contents was calculated from rubber elasticity theory using Equation (2) and is summarized in Table 3.

$$v_e = E'/3RT \qquad (2)$$

where v_e, E', R, and T are the crosslink density, elastic storage modulus, ideal gas constant, and temperature at which E' is obtained, respectively. The elastic storage modulus of the film was obtained at 393.15 K (>T_g). The calculated v_e of WPUA films increased with increasing HEMA content.

Table 3. T_g, v_e, and MFFT for WPUA films.

Sample	T_g, PU (°C)	T_g, PA (°C)	$v_e \times 10^3$ (mol/cm³)	MFFT (°C)
WPUA_H$_{0.00}$	−6.6	88.7	0.16	8.4
WPUA_H$_{0.25}$	2.9	77.2	0.23	11.6
WPUA_H$_{0.50}$	7.7	71.5	0.83	19.2
WPUA_H$_{0.75}$	8.0	68.5	1.09	29.7
WPUA_H$_{1.00}$		53.4	1.35	42.2

Figure 5b shows the relationship between the loss factor (tan δ) and temperature, which can reflect the thermal behavior and degree of phase separation of the hybrid polymer film. All samples, except for WPUA_H$_{1.00}$, exhibited two tan δ peaks, indicating the presence of two phases in the internal structure of the WPUA film [36]. The first phase transition, which occurred between −50 and 25 °C, was attributed to the chain activity of PU. The second phase transition, which occurred between 50 and 100 °C, was associated with the activity of the acrylic chain. The difference between the two peak temperatures decreased with increasing HEMA content, indicating improved miscibility between the PU and acrylic. WPUA_H$_{1.00}$ exhibited the best phase miscibility, with a broad single peak curve between 0 and 90 °C.

Table 3 summarizes the T_g and MFFT values of the WPUA samples obtained from DMA. The results indicate the clear dependence of the MFFT on the first T_g. In general, a lower T_g of the latex corresponds to superior interfacial adhesion and chain diffusion, resulting in a lower MFFT. The T_g of the core acrylic polymer, calculated through the Fox equation for the prepared WPUA latexes, was 86 °C, which highlights that film formation is challenging in ambient conditions. In contrast, the MFFT values of WPUA_$H_{0.00}$, WPUA_$H_{0.25}$, and WPUA_$H_{0.50}$ were lower than 25 °C, owing to the high adhesion capacity of the PU placed on the shell. PU is composed of soft segments, ionic groups, and urethane groups, and thus, it can be plasticized by water [37]. Consequently, the gaps between particles are filled, and chain entanglement is strengthened. Notably, the MFFT of WPUA_$H_{0.75}$ and WPUA_$H_{1.00}$ was significantly high, potentially because of the high degree of grafting and crosslinking. The covalent bonds between PU and acrylic polymer reduce the mobility of flexible PU chains, rendering it challenging to fill the voids between particles. Consequently, film cracking occurs, and a higher drying temperature is required to obtain a smooth film. Thus, although high-level grafting improves the compatibility of WPUA, it considerably limits its film-forming ability.

Figure 6 shows the TGA results for the WPUA samples. The initial weight loss values for the WPUA films with HEMA mol ratios of 0.25, 0.50, 0.75, and 1.00 were recorded at 240, 252, 262, and 264 °C, respectively, signifying the onset of the first decomposition temperature in the derivative thermogravimetry (DTG) graph. The temperatures corresponding to the first maximum degradation peaks were 310, 315, 315, 326, and 330 °C, associated with the decomposition of the hard segments of the PU [38]. The temperatures at which 50% weight loss was observed were 373, 381, 384, and 385 °C. The second maximum decomposition temperatures in the DTG analysis were 395, 396, 405, 408, and 409 °C, indicating the decomposition temperature of the soft segments and PA components [38]. Thus, as the HEMA content increased, the degradation temperature and maximum degradation temperature of the hard and soft segments increased by 20 °C and 14 °C, respectively. This phenomenon was attributable to the grafting and crosslinking of WPUA, which resulted in improved heat resistance [39].

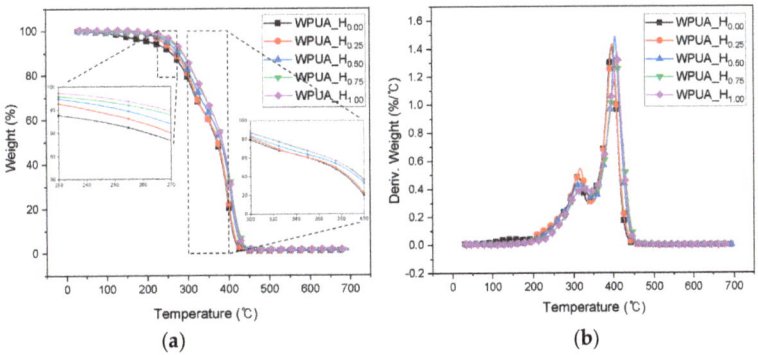

Figure 6. (**a**) TGA curves and (**b**) DTG graphs for the WPUA_$H_{0.00}$, WPUA_$H_{0.25}$, WPUA_$H_{0.50}$, WPUA_$H_{0.75}$, and WPUA_$H_{1.00}$ films.

3.5. Morphological Characteristics of WPUA Films

To evaluate the coalescence and film-forming ability of the WPUA latex, the surface of the dried film was imaged via AFM. Figure 7 shows the height images and surface profile roughness values (R_q) of the WPUA films with different HEMA contents. As predicted from the MFFT results, the increase in the HEMA content prevented the collapse of the particles and increased the surface roughness. The WPUA_$H_{0.25}$ and WPUA_$H_{0.50}$ samples, with a low degree of grafting, exhibited highly deformed surface morphologies. In contrast, most particles of the WPUA_$H_{0.75}$ and WPUA_$H_{1.00}$ samples, characterized by a high degree of grafting and crosslinking, did not collapse and maintained a spherical shape. Interestingly, smaller particles could be distinctly identified in the surface image of WPUA_$H_{0.00}$. A potential reason for this phenomenon is that the shell polymer, free from the core, easily migrated to fill the voids between the particles, and undeformed core particles appeared on the surface [3]. Consequently, the WPUA_H$_{0.00}$ film exhibited the lowest surface roughness.

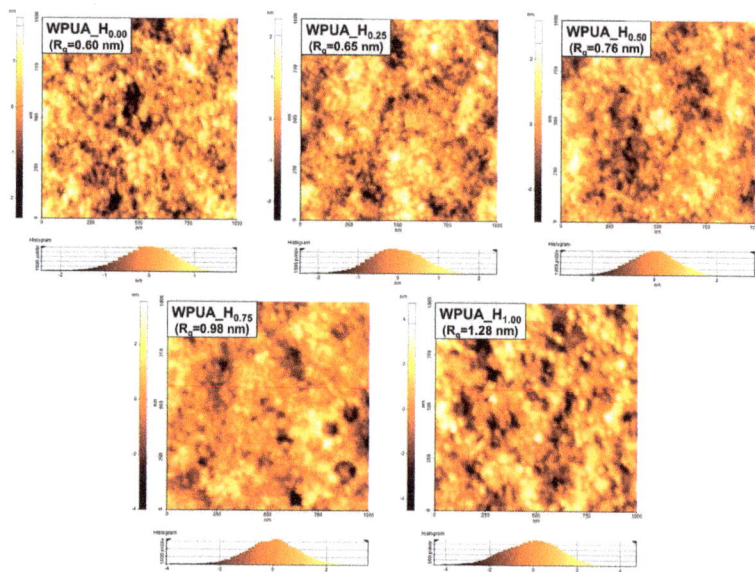

Figure 7. Tapping-mode-AFM height images for the WPUA_H$_{0.00}$, WPUA_H$_{0.25}$, WPUA_H$_{0.50}$, WPUA_H$_{0.75}$, and WPUA_H$_{1.00}$ films.

Figure 8 shows the fracture surface SEM images of the WPUA samples. The results were consistent with those derived from AFM phase images. As the grafting ratio of WPUA increased, distinct latex particles and rough fracture surfaces were observed. Specifically, the WPUA_H$_{1.00}$ film with the highest MFFT exhibited numerous micro-voids inside, attributable to its low particle-to-particle coalescence ability. Such discontinuous films may be characterized by several limitations, such as low barrier properties and mechanical strength.

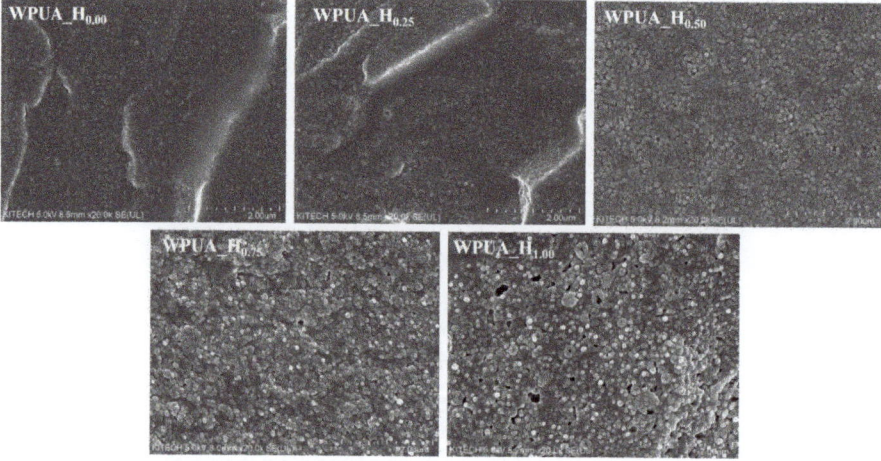

Figure 8. SEM images of the fracture surface for the WPUA_H$_{0.00}$, WPUA_H$_{0.25}$, WPUA_H$_{0.50}$, WPUA_H$_{0.75}$, and WPUA_H$_{1.00}$ films.

3.6. Mechanical and Optical Properties

The effect of the grafting ratio on the mechanical properties of WPUA coatings was investigated through nano-indentation and cross-cut tests. Figure 9 shows the load–indentation depth curve reflecting the mechanical surface behavior of WPUA coatings with different grafting ratios. Table 4 lists the average values of the indentation hardness (H_{IT}) and modulus of elasticity (E_{IT}). As the grafting ratio increased from 0 to 0.75, the indentation depth decreased at the same pressure. H_{IT} gradually increased from 67.50 MPa to 89.53 MPa, and E_{IT} increased from 2.87 GPa to 3.84 GPa. These trends were attributable to the strengthening effect induced by the high degree of grafting and crosslinking. Notably, the H_{IT} and E_{IT} of WPUA_$H_{1.00}$ were 67.5 MPa and 2.87 GPa, respectively. The voids in the WPUA_$H_{1.00}$ coating, as observed in the SEM images, resulted in decreased hardness and elastic modulus [40].

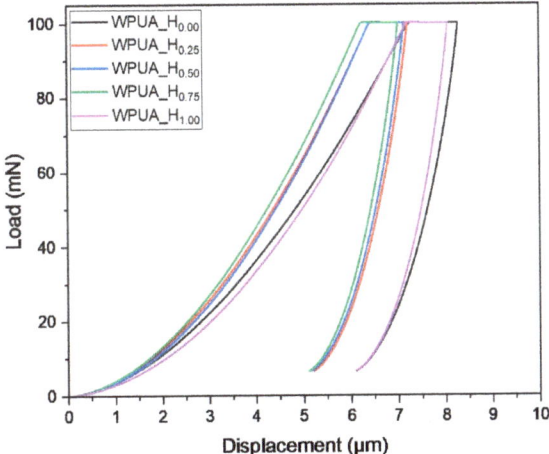

Figure 9. Typical load–indentation depth curves from nano-indentation tests on WPUA films.

Table 4. Hardness, elastic modulus, and light transmittance of WPUA films.

Sample	H_{IT}[a] (MPa)	E_{IT}[a] (GPa)	Adhesion [b]	Transmittance [c] (%)
WPUA_$H_{0.00}$	63.0 ± 2.2	3.1 ± 0.1	5B	74 ± 2
WPUA_$H_{0.25}$	84.9 ± 2.4	3.5 ± 0.1	5B	88 ± 2
WPUA_$H_{0.50}$	86.4 ± 3.2	3.5 ± 0.2	4B	90 ± 3
WPUA_$H_{0.75}$	89.5 ± 4.7	3.8 ± 0.3	3B	91 ± 3
WPUA_$H_{1.00}$	67.5 ± 5.1	2.9 ± 0.4	2B	87 ± 5

[a] Results obtained from nano-indentation test; [b] results obtained from cross-cut test; [c] light transmittance in the visible range.

According to the cross-cut test results, the adhesion of the WPUA coatings decreased as the grafting ratio increased (Table 4), likely because of the film-forming ability of WPUA. In general, a greater deformability of the particles is associated with superior adhesion with the substrate. In the case of the WPUA latex particles with a low degree of grafting, the soft-shell PU polymer is expected to move freely and completely cover the substrate. These characteristics are associated with enhanced film-forming ability and superior adhesion of the WPUA coating.

In general, the transparency of composite polymers is closely related to the particle size, crystallinity of the polymer matrix, phase separation, and refractive index differences [41,42]. The grafted and crosslinked WPUA films exhibited excellent light transmittance in the visible range (Table 4). However, the WPUA_$H_{0.00}$ film exhibited haze and recorded a 74% lower light transmittance. This is consistent with Desai's report that the transparency of PU/poly(MMA) latex films without HEMA was reduced [43]. This is due to the difference in the refractive index between the highly separated PU and acrylic domains, as in the results obtained with DMA [14]. In particular, WPUA with a hard core-soft shell structure may have intensified phase separation due to the heterogeneity of T_g. On the other hand, comparing WPUA_$H_{0.25}$ and WPUA_$H_{0.75}$, light transmittance increased even though the particle size increased from 70.5 to 84.5 nm. This may be because the improved compatibility had a greater impact on the transparency of the WPUA film than the particle size change in the 14 nm range. Therefore, improving the compatibility of the two phases is essential to obtain a transparent film. The reduced transparency of WPUA_$H_{1.00}$ is attributed to the formation of voids.

4. Conclusions

This study was aimed at preparing coatings with excellent film-forming ability and high hardness at room temperature without the use of coalescence agents. To this end, PU was introduced into acrylic latex as a coalescence aid, and the effect of the grafting ratio of the core (PA) and shell (PU) on the properties and film-forming ability of WPUA was systematically investigated. Increasing the grafting ratio resulted in the crosslinking of the WPUA latex particles and an increase in their size. The grafted WPUA coating exhibited improved hardness, modulus, thermal stability, and compatibility. However, excessive grafting reduced the chain mobility of the soft shell, which limited the collapse and coalescence of the latex particles during film formation. Therefore, the MFFT of WPUA increased, resulting in the formation of a discontinuous film with reduced hardness and adhesion. These results highlight that the grafting ratio must be optimized to ensure a balance between the film-forming ability and mechanical properties of WPUA latex. In follow-up studies, we plan to prepare various series of PUs and investigate their coalescence-supporting abilities in more detail.

Author Contributions: Conceptualization, D.H.K. and Y.R.K.; methodology, Y.R.K. and S.K.M.; software, M.K.; investigation, H.C.K. and S.K.M.; resources, J.S.K.; data curation, Y.R.K. and S.K.M.; writing—original draft preparation, Y.R.K. and S.K.M.; writing—review and editing, D.H.K. and J.S.K.; visualization, H.C.K.; supervision, D.H.K.; project administration, D.H.K.; funding acquisition, D.H.K. All authors have read and agreed to the published version of the manuscript.

Funding: This work was funded by the Technology Innovation Program, Parts and Materials Technology Development (20010566, Development of waterborne transparent coating vanish and related coating process based on eco-friendly VOC-free technology for transportation), funded by the Ministry of Trade, Industry and Energy (MOTIE, Republic of Korea).

Institutional Review Board Statement: Not applicable.

Data Availability Statement: The data are available upon request.

Conflicts of Interest: The authors declare no conflict of interest.

References

1. Jiménez-López, A.M.; Hincapié-Llanos, G.A. Identification of Factors Affecting the Reduction of VOC Emissions in the Paint Industry: Systematic Literature Review-SLR. *Prog. Org. Coat.* **2022**, *170*, 106945. [CrossRef]
2. Ifijen, I.H.; Maliki, M.; Odiachi, I.J.; Aghedo, O.N.; Ohiocheoya, E.B. Review on Solvents Based Alkyd Resins and Water Borne Alkyd Resins: Impacts of Modification on Their Coating Properties. *Chem. Afr.* **2022**, *5*, 211–225. [CrossRef]
3. Steward, P.A.; Hearn, J.; Wilkinson, M.C. An Overview of Polymer Latex Film Formation and Properties. *Adv. Colloid Interface Sci.* **2000**, *86*, 195–267. [CrossRef] [PubMed]
4. Brito, E.L.; Ballard, N. Film Formation of Hard-Core/Soft-Shell Latex Particles. *J. Polym. Sci.* **2022**, *61*, 410–421. [CrossRef]
5. Berce, P.; Skale, S.; Razboršek, T.; Slemnik, M. Influence of Coalescing Aids on the Latex Properties and Film Formation of Waterborne Coatings. *J. Appl. Polym. Sci.* **2017**, *134*, 45142. [CrossRef]
6. Berce, P.; Skale, S.; Slemnik, M. Electrochemical Impedance Spectroscopy Study of Waterborne Coatings Film Formation. *Prog. Org. Coat.* **2015**, *82*, 1–6. [CrossRef]
7. Limousin, E.; Ballard, N.; Asua, J.M. Soft Core-Hard Shell Latex Particles for Mechanically Strong VOC-Free Polymer Films. *J. Appl. Polym. Sci.* **2019**, *136*, 47608. [CrossRef]

8. Dron, S.M.; Bohorquez, S.J.; Mestach, D.; Paulis, M. Reducing the Amount of Coalescing Aid in High Performance Waterborne Polymeric Coatings. *Eur. Polym. J.* **2022**, *170*, 111175. [CrossRef]
9. Yuan, X.; Huo, D.; Qian, Q. Effect of Annealing on the Phase Structure and the Properties of the Film Formed from P(St-Co-BA)/P(MMA-Co-BA) Composite Latex. *J. Colloid Interface Sci.* **2010**, *346*, 72–78. [CrossRef]
10. Hasanzadeh, I.; Mahdavian, A.R.; Salehi-Mobarakeh, H. Particle Size and Shell Composition as Effective Parameters on MFFT for Acrylic Core–Shell Particles Prepared via Seeded Emulsion Polymerization. *Prog. Org. Coat.* **2014**, *77*, 1874–1882. [CrossRef]
11. Akindoyo, J.O.; Beg, M.D.H.; Ghazali, S.; Islam, M.R.; Jeyaratnam, N.; Yuvaraj, A.R. Polyurethane Types, Synthesis and Applications—A Review. *RSC Adv.* **2016**, *6*, 114453–114482. [CrossRef]
12. Wang, L.; Zhu, Y.; Qu, J. Preparation and Assistant-Film-Forming Performance of Aqueous Polyurethane Dispersions. *Prog. Org. Coat.* **2017**, *105*, 9–17. [CrossRef]
13. Yang, J.; Winnik, M.A.; Ylitalo, D.; DeVoe, R.J. Polyurethane−Polyacrylate Interpenetrating Networks. 1. Preparation and Morphology. *Macromolecules* **1996**, *29*, 7047–7054. [CrossRef]
14. Mehravar, S.; Ballard, N.; Tomovska, R.; Asua, J.M. Polyurethane/Acrylic Hybrid Waterborne Dispersions: Synthesis, Properties and Applications. *Ind. Eng. Chem. Res.* **2019**, *58*, 20902–20922. [CrossRef]
15. Peruzzo, P.J.; Anbinder, P.S.; Pardini, O.R.; Vega, J.; Costa, C.A.; Galembeck, F.; Amalvy, J.I. Waterborne Polyurethane/Acrylate: Comparison of Hybrid and Blend Systems. *Prog. Org. Coat.* **2011**, *72*, 429–437. [CrossRef]
16. Xu, H.; Qiu, F.; Wang, Y.; Yang, D.; Wu, W.; Chen, Z.; Zhu, J. Preparation, Mechanical Properties of Waterborne Polyurethane and Crosslinked Polyurethane-Acrylate Composite. *J. Appl. Polym. Sci.* **2011**, *124*, 958–968. [CrossRef]
17. Bizet, B.; Grau, E.; Cramail, H.; Asua, J.M. Crosslinked isocyanate-free poly(hydroxy urethane) s–Poly(butyl methacrylate) hybrid latexes. *Eur. Polym. J.* **2021**, *146*, 110254. [CrossRef]
18. Dong, A.; An, Y.; Feng, S.; Sun, D. Preparation and morphology studies of core-shell type waterborne polyacrylate–polyurethane microspheres. *J. Colloid Interface Sci.* **1999**, *214*, 118–122. [CrossRef]
19. Kim, B.K.; Lee, J.C. Modification of waterborne polyurethanes by acrylate incorporations. *J. Appl. Polym. Sci.* **1995**, *58*, 1117–1124. [CrossRef]
20. Zhu, Z.; Li, R.; Zhang, C.; Gong, S. Preparation and properties of high solid content and low viscosity waterborne polyurethane—Acrylate emulsion with a reactive emulsifier. *Polymers* **2018**, *10*, 154. [CrossRef]
21. ASTM D2354; Standard Test Method for Minimum Film Formation Temperature (MFFT) of Emulsion Vehicles. ASTM International: West Conshohocken, PA, USA, 2018.
22. ASTM D1474; Standard Test Methods for Indentation Hardness of Organic Coatings. ASTM International: West Conshohocken, PA, USA, 2008.
23. ISO 2409:2020; Paints and varnishes—Cross-cut test. International Organization for Standardization: Geneva, Switzerland, 2020.
24. Xu, J.; Jiang, Y.; Qiu, F.; Dai, Y.; Yang, D.; Yu, Z.; Yang, P. Synthesis, Mechanical Properties and Iron Surface Conservation Behavior of UV-Curable Waterborne Polyurethane-Acrylate Coating Modified with Inorganic Carbonate. *Polym. Bull.* **2018**, *75*, 4713–4734. [CrossRef]
25. Jang, T.; Kim, H.J.; Jang, J.B.; Kim, T.H.; Lee, W.; Seo, B.; Ko, W.B.; Lim, C.S. Synthesis of Waterborne Polyurethane Using Phosphorus-Modified Rigid Polyol and Its Physical Properties. *Polymers* **2021**, *13*, 432. [CrossRef] [PubMed]
26. Wang, W.; Li, L.; Jin, S.; Wang, Y.; Lan, G.; Chen, Y. Study on Cellulose Acetate Butyrate/Plasticizer Systems by Molecular Dynamics Simulation and Experimental Characterization. *Polymers* **2020**, *12*, 1272. [CrossRef] [PubMed]
27. Yen, F.S.; Hong, J.L. Hydrogen-Bond Interactions between Ester and Urethane Linkages in Small Model Compounds and Polyurethanes. *Macromolecules* **1997**, *30*, 7927–7938. [CrossRef]
28. Kamoun, E.A.; El-Betany, A.; Menzel, H.; Chen, X. Influence of Photoinitiator Concentration and Irradiation Time on the Crosslinking Performance of Visible-Light Activated Pullulan-HEMA Hydrogels. *Int. J. Biol. Macromol.* **2018**, *120*, 1884–1892. [CrossRef] [PubMed]
29. Cakić, S.M.; Špírková, M.; Ristić, I.S.; B-Simendić, J.K.; Milena, M.; Poręba, R. The Waterborne Polyurethane Dispersions Based on Polycarbonate Diol: Effect of Ionic Content. *Mater. Chem. Phys.* **2013**, *138*, 277–285. [CrossRef]
30. Zhang, Q.; Deng, Y.; Fu, Z.; Zhang, H. Effects of the molecular structure on the vibration reduction and properties of hyperbranched waterborne polyurethane–acrylate for damping coatings. *J. Appl. Polym. Sci.* **2019**, *136*, 47733. [CrossRef]
31. Dai, Q.; Wu, D.; Zhang, Z.; Ye, Q. Preparation of monodisperse poly (methyl methacrylate) particles by radiation-induced dispersion polymerization using vinyl terminus polysiloxane macromonomer as a polymerizable stabilizer. *Polymer* **2003**, *44*, 73–77. [CrossRef]
32. Liu, B.L.; Zhang, B.T.; Cao, S.S.; Deng, X.B.; Hou, X.; Chen, H. Preparation of the stable core–shell latex particles containing organic-siloxane in the shell. *Prog. Org. Coat.* **2008**, *61*, 21–27. [CrossRef]
33. Brouwer, W.M. Comparative methods for the characterization of core—Shell latex particles. *Colloid Surf.* **1989**, *40*, 235–247. [CrossRef]
34. Lei, L.; Xia, Z.; Ou, C.; Zhang, L.; Zhong, L. Effects of crosslinking on adhesion behavior of waterborne polyurethane ink binder. *Prog. Org. Coat.* **2015**, *88*, 155–163. [CrossRef]
35. Chattopadhyay, D.K.; Sreedhar, B.; Raju, K.V. Effect of chain extender on phase mixing and coating properties of polyurethane ureas. *Ind. Eng. Chem. Res.* **2005**, *44*, 1772–1779. [CrossRef]

36. Yang, W.; Cheng, X.; Wang, H.; Liu, Y.; Du, Z. Surface and Mechanical Properties of Waterborne Polyurethane Films Reinforced by Hydroxyl-Terminated Poly(Fluoroalkyl Methacrylates). *Polymer* **2017**, *133*, 68–77. [CrossRef]
37. Satguru, R.; Mcmahon, J.; Padget, J.C.; Coogan, R.G. Aqueous Polyurethanes: Polymer Colloids with Unusual Colloidal, Morphological, and Application Characteristics. *J. Coat. Technol.* **1994**, *66*, 47–55.
38. Deng, Y.; Zhou, C.; Zhang, Q.; Zhang, M.; Zhang, H. Structure and performance of waterborne polyurethane-acrylate composite emulsions for industrial coatings: Effect of preparation methods. *Colloid Polym. Sci.* **2020**, *298*, 139–149. [CrossRef]
39. Alvarez, G.A.; Fuensanta, M.; Orozco, V.H.; Giraldo, L.F.; Martín-Martínez, J.M. Hybrid waterborne polyurethane/acrylate dispersion synthesized with bisphenol A-glicidylmethacrylate (Bis-GMA) grafting agent. *Prog. Org. Coat.* **2018**, *118*, 30–39. [CrossRef]
40. Bano, S.; Iqbal, T.; Ramzan, N.; Farooq, U. Study of Surface Mechanical Characteristics of ABS/PC Blends Using Nanoindentation. *Processes* **2021**, *9*, 637. [CrossRef]
41. Chen, Y.; Pan, M.; Li, Y.; Xu, J.Z.; Zhong, G.J.; Ji, X.; Yan, Z.; Li, Z.M. Core-shell nanoparticles toughened polylactide with excellent transparency and stiffness-toughness balance. *Compos. Sci. Technol.* **2018**, *164*, 168–177. [CrossRef]
42. Luo, H.; Wei, H.; Wang, Z.; Li, H.; Chen, Y.; Xiang, J.; Fan, H. Fabrication of UV-curable waterborne polyurethane coatings with self-cleaning, anti-graffiti performance, and corrosion resistance. *Colloid Surf. A-Physicochem. Eng. Asp.* **2023**, *676*, 132177. [CrossRef]
43. Desai, S.; Thakore, I.M.; Brennan, A.; Devi, S. Thermomechanical properties and morphology of interpenetrating polymer networks of polyurethane–poly(methyl methacrylate). *J. Appl. Polym. Sci.* **2002**, *83*, 1576–1585. [CrossRef]

Disclaimer/Publisher's Note: The statements, opinions and data contained in all publications are solely those of the individual author(s) and contributor(s) and not of MDPI and/or the editor(s). MDPI and/or the editor(s) disclaim responsibility for any injury to people or property resulting from any ideas, methods, instructions or products referred to in the content.

Article

Synthesis of Room Temperature Curable Polymer Binder Mixed with Polymethyl Methacrylate and Urethane Acrylate for High-Strength and Improved Transparency

Ju-Hong Lee, Won-Bin Lim, Jin-Gyu Min, Jae-Ryong Lee, Ju-Won Kim, Ji-Hong Bae * and Pil-Ho Huh *

Department of Polymer Science and Engineering, Pusan National University, Busan 46241, Republic of Korea; dlwnghd15@pusan.ac.kr (J.-H.L.); wblim@pusan.ac.kr (W.-B.L.); jgmin@pusan.ac.kr (J.-G.M.); leejr1216@pusan.ac.kr (J.-R.L.); kjw8618@pusan.ac.kr (J.-W.K.)
* Correspondence: jhbae@pusan.ac.kr (J.-H.B.); pilho.huh@pusan.ac.kr (P.-H.H.); Tel.: +82-51-510-3637 (J.-H.B.)

Abstract: Urethane acrylate (UA) was synthesized from various di-polyols, such as poly(tetrahydrofuran) (PTMG, Mn = 1000), poly(ethylene glycol) (PEG, Mn = 1000), and poly(propylene glycol) (PPG, Mn = 1000), for use as a polymer binder for paint. Polymethyl methacrylate (PMMA) and UA were blended to form an acrylic resin with high transmittance and stress-strain curve. When PMMA was blended with UA, a network structure was formed due to physical entanglement between the two polymers, increasing the mechanical properties. UA was synthesized by forming a prepolymer using di-polyol and hexamethylene diisocyanate, which were chain structure monomers, and capping them with 2-hydroxyethyl methacrylate to provide an acryl group. Fourier transform infrared spectroscopy was used to observe the changes in functional groups, and gel permeation chromatography was used to confirm that the three series showed similar molecular weight and PDI values. The yellowing phenomenon that appears mainly in the curing reaction of the polymer binder was solved, and the mechanical properties according to the effects of the polyol used in the main chain were compared. The content of the blended UA was quantified using ultravioletvisible spectroscopy at a wavelength of 370 nm based on 5, 10, 15, and 20 wt%, and the shear strength and tensile strength were evaluated using specimens in a suitable mode. The ratio for producing the polymer binder was optimized. The mechanical properties of the polymer binder with 5–10 wt% UA were improved in all series.

Keywords: urethane acrylate; polymer binder; photopolymer; curable polymer

1. Introduction

External stimulus response technology using smart materials is emerging as an attractive research field. To reduce automobile accidents that occur at night, there is a need for research on smart materials that can be applied to road surfaces. External stimulus response technology is an important skill for securing driving stability for traffic weaknesses and drivers, and it is a technology that makes it possible to recognize real time road information [1–8]. Photo stimulation-sensitive technology can be used in various fields by mixing a photo-stimulated material and various polymer resins with the property of emitting light. A representative photo stimulation-sensitive (sunlight, street lighting, car headlights, etc.) technology is a luminescent paint binder used in road and automobile lines. The polymer binders used in luminescent paints show high efficiency and stability in dark spaces where there is no light. Continuously illuminated painted lanes create a safe traffic environment that provides information to drivers and prevents accidents. Therefore, polymer binder manufacturing technology that mixes with phosphorescent pigments to safely protect the pigments and can be used in the next lane is a technology that is attracting attention. The synthesis of binders that can protect pigments has become an important research goal. A photo-stimulated binder that exhibits excellent luminous properties requires the following mechanical properties: transmittance that does not reduce the luminous properties of the phosphorescent paint, adhesion to the road surface, weather resistance to withstand weather changes, and toughness to withstand external shocks [9–23]. Therefore, polymethyl methacrylate (PMMA) is a representative polymer used in paint manufacturing, but it has the disadvantage of inappropriate physical properties when used independently. To solve these problems, this study used PMMA and a composite series added with various acrylates. Among numerous acrylates, UA has excellent mechanical properties such as weather resistance, abrasion resistance, and alkali resistance, and structurally exhibits a network or linear structure depending on the type of polyol (diol, triol, etc.). Urethane has the advantage of varying mechanical properties (impact resistance, friction resistance, wear resistance, etc.) depending on the type of polyol and isocyanate. The use of diol and isocyanate with a linear structure can be applied to paints with excellent performance. Therefore, the UA synthesized in this study exhibits improved toughness and optical properties when blended with PMMA to produce a binder. In addition, polymer binders for road surfaces must exhibit high abrasion resistance to withstand external forces such as vehicle weight and speed, and UA is suitable because it provides the properties necessary for the purpose of use in paint. To successfully blend UA with PMMA, it is essential to synthesize them as oligomers. Their high molecular weight has the disadvantage of poor processability due to viscosity. Also, insufficient urethane bonding will cause phase separation from PMMA, so it is essential to take care. Therefore, in this study, a blend process of UA and PMMA was performed after synthesizing oligomeric UA [24–31]. PMMA is synthesized through a copolymerization process using a butyl acrylate monomer and a methyl methacrylate (MMA) monomer and has excellent optical properties, high weather resistance, and adhesive strength [32]. Urethane prepolymer is formed through a urethane bond by reacting the hydroxyl group of polyols with an NCO group of isocyanates. UA was synthesized by adding an acrylic group and capping the ends, resulting in high chemical resistance and tensile strength values. Representative polyols for synthesizing UA include polyester and polyether polyols, as well as PPG, PEG, and PTMG. Polyether polyols with the same molecular weight were used in the range of this study. Various types of isocyanates, such as methylene diphenyl diisocyanate (MDI) and toluene diisocyanate (TDI), which are used industrially, as well as HDI and 4,4'-diisocyanatodicyclohexylmethane (H_{12}MDI), were used. Urethane prepolymers were synthesized using chain structured HDI to maximize the optical properties MDI and TDI with phenyl groups were not used because of the yellowing phenomenon. As the acrylate, 2-HEMA, the most widely applied industrially, was used to react with the NCO group at the end of the urethane prepolymer [33–39].

Three series of UA were synthesized to observe changes in the physical properties of UA according to polyols. By controlling the added catalyst, reaction time, and temperature, UA having a similar molecular weight was prepared and blended with PMMA. Curing of acrylates currently used industrially proceeds with crosslinking using heat or light. In this study, however, polymeric binders made by blending PMMA and UA can crosslink only with the reaction heat at room temperature [40–43]. Benzoyl peroxide (BPO) was used as an initiator for the curing reaction at room temperature, and N,N-bis(2-hydroxyethyl)-p-toluidine (PTE) was added as a catalyst to initiate the reaction of BPO. In this study, three series of UA according to the type of polyol were synthesized, and their mechanical properties were evaluated after blending with PMMA. As a result, the highest permeability, shear strength, and tensile strength were exhibited at a UA content of 5 to 10 wt%. An optimized polymer binder was prepared by adjusting the PMMA to UA ratio while the content of the initiator and reaction promoting catalyst was fixed. In future studies, it will be mixed with luminescent paint and used as a promising photo-stimulation smart, sensitive material [44–58].

2. Material and Methods

The urethane binder was synthesized using polypropylene glycol (PPG, Mn = 1000 g/mol, Merck KGaA, Darmstadt, Germany), polyethylene glycol (PEG, Mn = 1000 g/mol, Merck

KGaA, Darmstadt, Germany), polytetrahydrofuran (PTMG, Mn = 1000 g/mol, Merck KGaA, Darmstadt, Germany) as the polyol of UA, and hexamethylene diisocyanate (HDI, Mn = 168.2 g/mol, Tokyo Chemical Industry Co., Ltd., Tokyo, Japan) as the isocyanate of the main chain. PMMA (Mn = 28,000, Jungseok Chemical Co., Ltd., Jeonju, Republic of Korea) was used to form a polymer binder by complexing with UA. MMA (Mn = 100.121 g/mol, Merck KGaA, Darmstadt, Germany) was used as a diluent to control the viscosity of the synthesized UA. Dibutyltin dilaurate (DBTDL, Mn = 631.56 g/mol, Merck KGaA, Darmstadt, Germany) was used as an organometallic catalyst to promote the urethane reaction between the –OH group of the polyol and the isocyanate –NCO group. N,N-bis(2-hydroxyethyl)-p-toluidine (PTE, Mn = 195.26 g/mol, Jungseok Chemical Co., Ltd., Jeonju, Republic of Korea) and BPO (Mn = 242.23 g/mol, Jungseok Chemical Co., Ltd., Jeonju, Republic of Korea) were used as catalysts and initiators to cure the polymer binder blended with PMMA and UA, respectively.

3. Results and Discussion

In the case of synthesized UA, the changes in mechanical properties according to the type of polyether polyol with the same molecular weight were compared. PPG/PEG/PTMG and HDI were added at a 1:2.2 molar ratio to a 250 mL four-necked flask. The urethane prepolymer manufacturing process was set to 50–60 °C under the condition of adding 0.1 wt% tin catalyst and stirred at 100 rpm using a mechanical stirrer. The synthesis process was analyzed by Fourier transform infrared (FT-IR) spectroscopy and shown in Figure 1. The OH stretching vibration around 3700 cm^{-1} and the NCO peak around 2250 cm^{-1} decreased because of the reaction between polyol and isocyanate, and a new urethane bond reaction was generated at 1725 cm^{-1}, as shown in (a), (b). UA was synthesized by reacting the NCO group at the terminal of the urethane prepolymer with the OH group of acrylates. In the case of 2-HEMA with an acryl group, a 2.2 molar ratio was added to react and remove the isocyanate. The C=O peak at 1725 cm^{-1} and C=C peak at 810 cm^{-1} generated by the addition of an acryl group were confirmed, as shown in (c), (d). High viscosity UA was difficult to blend with PMMA. Therefore, the reaction temperature was lowered to 30 °C, and 30 wt% of MMA was added as a diluent and mixed.

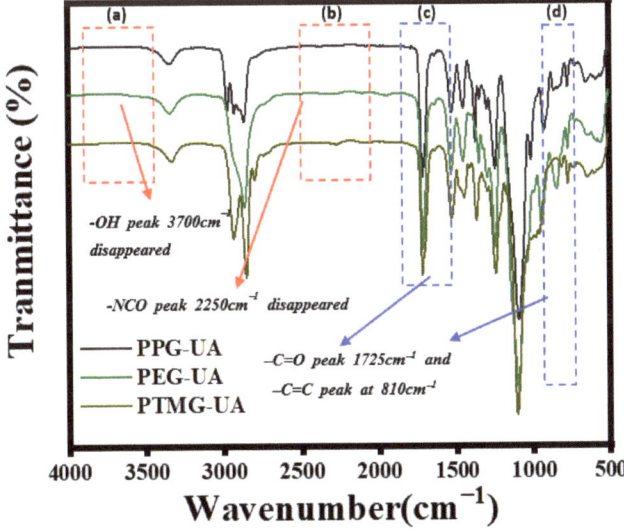

Figure 1. FT-IR spectra of the urethane acrylate according to the type of polyol. The (**a**) shows the hydroxyl group of the synthesized urethane acrylate, (**b**) representatives the isocyanate group, and (**c**,**d**) show the C=O and C=C bond peaks, respectively.

Scheme 1 shows the synthesis of UA through three polyols and the curing mechanism after blending with PMMA. Three series of polyols (PPG, PEG, and PTMG) were reacted with HDI to form urethane prepolymers, respectively. The isocyanate group of the prepolymer reacted with additionally added 2-HEMA to produce Di-UA with acrylate capped end groups. It was blended with the prepared PMMA to form a polymeric binder support. 2-HEMA added during the blending process helps improve the shear strength and tensile strength through intermolecular hydrogen bonding. PTE was added as a catalyst and blended for three hours in an ultrasonic homogenizer. BPO as an initiator was added to the successfully mixed polymer binder and sufficient friction energy was applied using a vortex mixer. The PTE promotes the initiation of the BPO, and as a result, the polymeric binder was cured by self-heating at room temperature. Typical initiation reactions occur by externally applied light and heat. It is important to note that no additional external energy is required for the synthesis of polymeric binders in this paper. The formulation of the polymer binder is shown in Table 1.

The molecular weight of the synthesized UA series was measured by gel permeation chromatography (GPC). Tetrahydrofuran (THF) was used as the GPC solvent for the measurement. In a 20 mL vial, 0.05 g of the synthesized UA and 1 mL of THF solvent were added and dissolved by ultrasound from a sonicator for 6 h. The residue was then filtered off using a 0.5 μm pore-size filter. The filtered sample was introduced into the GPC using a 1 mL needle. A target molecular weight of Mn 7–8000 was achieved, and a polydispersity index (PDI) of 1.95–2.00 was confirmed, as shown in Figure 2 and Table 2.

Table 1. Formulations of the acrylic binder series.

Functionality	Components	Content (wt%)				
		0 wt%	5 wt%	10 wt%	15 wt%	20 wt%
Base acrylate	PMMA	89	84	79	74	69
	2-HEMA	10	10	10	10	10
Di-acrylate	URETHANE	0	5	10	15	20
Catalyst	PTE	1	1	1	1	1
Initiator	BPO	2	2	2	1	2

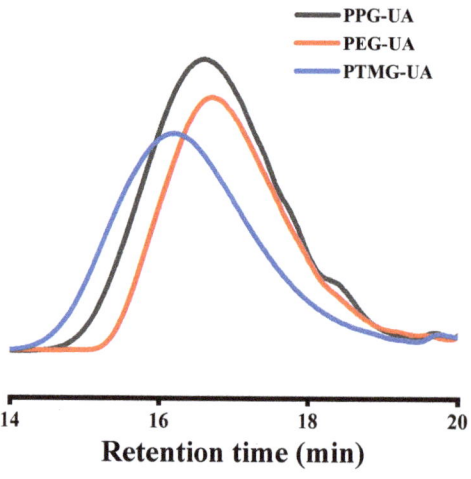

Figure 2. GPC curves of urethane acrylate binders synthesized with a similar MW formulation.

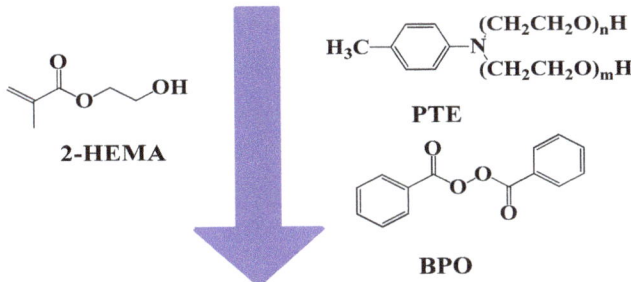

Scheme 1. Synthetic procedure to form the di-urethane acrylate/polymethyl methacrylate binder. Arrows indicate composite and blend paths.

Table 2. Molecular weights and PDI of the urethane acrylate.

	M_n	M_w	PDI
PPG-UA	7780	15,152	1.95
PEG-UA	7053	12,904	1.90
PTMG-UA	7874	15,434	2.00

Adding low molecular weight UA reduces the mechanical properties of the binder, which has the disadvantage of a weak structure due to the low molecular weight UA chains between PMMA polymer chains. By contrast, high molecular weight acrylates have problems with the curing time and rate at room temperature owing to their high viscosity, resulting in extremely low efficiency. In addition, the processability of the binder is very important for spraying on the road surface. When forming UA with a molecular weight of 10,000 or more, the viscosity due to the high molecular weight reduces workability. Therefore, UA with an appropriate molecular weight of 5–8000 for use in road paints was synthesized. Polymeric binders were prepared by varying the content ratio of UA in the synthesized oligomeric form and PMMA. To determine the difference in binder properties according to changes in acrylate content, the mechanical properties of a polymer binder blended with UA and PMMA were evaluated while fixing the type and content of the catalyst and initiator. All tests were conducted more than five times depending on the type of sample. The average value of the results excluding the maximum and minimum values is shown. The shear strength of the SUS specimens was measured in accordance with ASTM D1002 standard [59]. Apply one drop of the prepared binder to the end of the SUS specimen (width 25.4 mm, height 12.7 mm). Using another SUS specimen, overlap to the same size and confirm that curing occurs at room temperature. Curing is completed, and the measurement is performed using UTM. The shear strength according to the UA content was measured, as shown in Figure 3. UA synthesized using three different types of polyols was blended with PMMA at 5–10 wt% it showed the highest value and decreased gradually. In this experiment, it was possible to confirm the critical content of UA blended with PMMA. The polymers were physically entangled, resulting in a network structure and high shear strength. It was conducted according to the difference in the polyol type and UA content. The mixture of 5 wt% of UA using PPG as a polyol showed the highest value of 14.6 MPa, as shown in Figure 3 and Table 3. The methyl groups of the main chain were not packed in a linear structure and had a free volume. Therefore, PPG-UA was blended with PMMA because of its good compatibility. Compared to PTMG-UA, in the case of PPG-UA, relatively high results were obtained due to the short molecular unit chain length, that is, many acrylate functional groups. As a result, to obtain PMMA-based polymer binders with high mechanical properties, optimal design of the appropriate molecular chain structure length is required.

Table 3. Shear strength according to the urethane acrylate content.

(MPa)	Ref.	UA 5 wt%	UA 10 wt%	UA 15 wt%	UA 20 wt%
PMMA/PPG-UA	2.5	14.6	11.1	10.5	9.4
PMMA/PEG-UA	2.5	6.6	7.7	5.5	5.2
PMMA/PTMG-UA	2.5	3.6	5.3	3.5	3.3

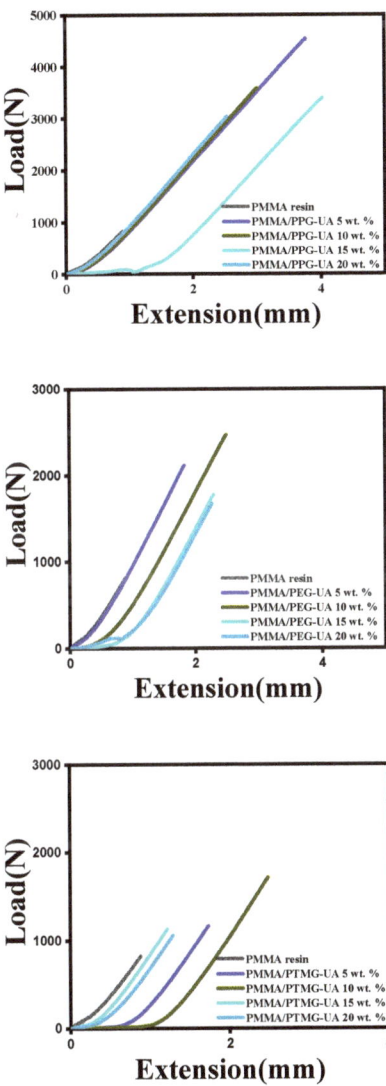

Figure 3. Shear strength test according to the type and content of urethane acrylate.

Figure 4 shows the UV-vis data of the three series of polymeric binders at a wavelength of 370 nm, which has the highest energy in the visible light region. The prepared binder for UV-vis measurement was formed to a thickness of 175 μm on a glass slide using a Baker applicator. As with other physical property tests, measurements were completed more than five times. It was confirmed as the average value of the results excluding the maximum and minimum values. The transmittance is especially important to improve the characteristics of polymer binders for road marking when mixed with phosphorescent pigments and mechanical strength. The low transmittance makes it impossible to control the luminous effect when the phosphorescent pigment and the polymer binder are mixed. PMMA is a representative polymer with high transmittance and showed similar or improved transmittance at 370 nm when 5 wt% of PPG/PEG/PTMG-UA was used. PPG-UA measured the highest light transmittance at 89.5%, as shown in Table 4. PEG and PTMG-UA showed low transmittance because of the lack of space for light to pass through due to the structural characteristics of the linear stack. On the other hand, the transmittance decreased gradually for binders in which the UA content exceeded 10 wt%. Curing the binder with the addition of diol UA resulted in lower transmittance because it turns hazy as the content increases. Compared to PTMG and PPG, as the content of PEG-based polymer binder increases, UV transmittance rapidly decreased. Structurally, the PEG chain forms a more linear structure than other comparative products. The linear structure is packed and makes it difficult for light to pass through. As a result, the polymer binder containing 5 wt% of UA showed an optimized value.

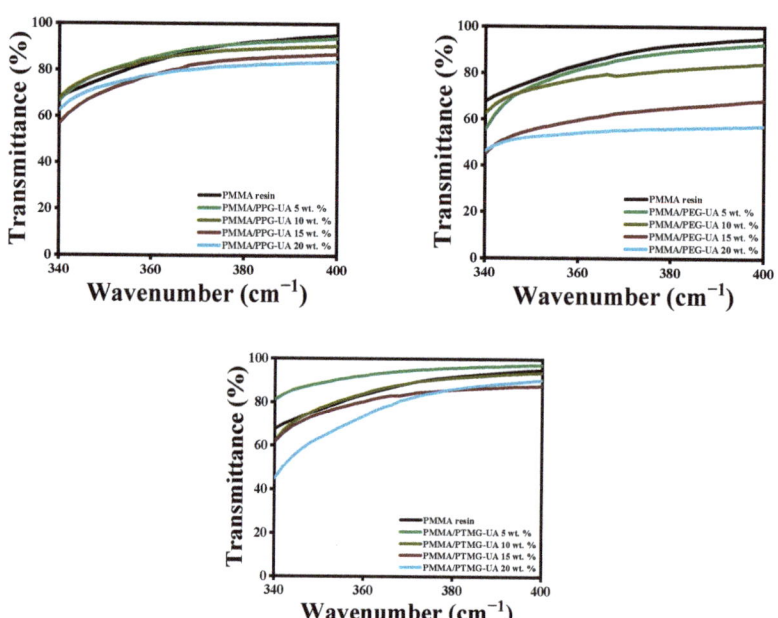

Figure 4. UV-vis transmittance according to the difference in the content of urethane acrylate.

Table 4. Transmittance at 370 nm according to urethane acrylate content.

Transmittance (%)	Ref.	UA 5 wt%	UA 10 wt%	UA 15 wt%	UA 20 wt%
PMMA/PPG-UA	88.7	89.5	87.3	82.7	80.6
PMMA/PEG-UA	88.7	86.2	79.3	63.1	55.5
PMMA/PTMG-UA	88.7	88.5	87.6	83.8	81.8

The curing process of the synthesized binder was performed in a mold manufactured according to the ASTM D638 standard and is shown in Figure 5 [60,61]. In this study, a Teflon mold was used to produce dog-bone specimens under ASTM D638 Type IV conditions because a release agent spraying process is necessary to separate the manufactured specimens when performing binder curing in an iron mold. Figure 6 presents the tensile strength of the polymer binder according to the UA content. The prepared specimen was measured at a speed of 10 mm/min using UTM. In the case of the polymer binder cured with the PMMA resin, the stress was 7.1 MPa, whereas the binder containing 5 wt% PPG-UA showed a stress of 9.7 MPa. PMMA/PEG-UA and PMMA/PTMG-UA showed 5.2 MPa and 3.5 MPa, respectively, at a 10 wt% content. Table 5 lists the stress values for each content of the UA series. PMMA/PEG-UA was expected to show high tensile strength because of the high intermolecular attraction through hydrogen bonding, but the miscibility between PEG-UA and PMMA was poor. PMMA/PTMG-UA affects the strain rather than stress because the chain length of PTMG-based UA is long. The polymer binder prepared with 5 wt% of PMMA/PPG-UA showed approximately 30% improvement in the mechanical properties compared to the polymer binder made using PMMA resin.

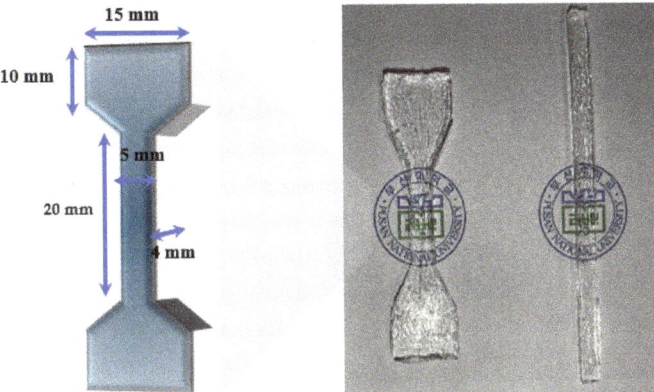

Figure 5. Specimen specification figures of PMMA/urethane acrylate binder for tensile strength.

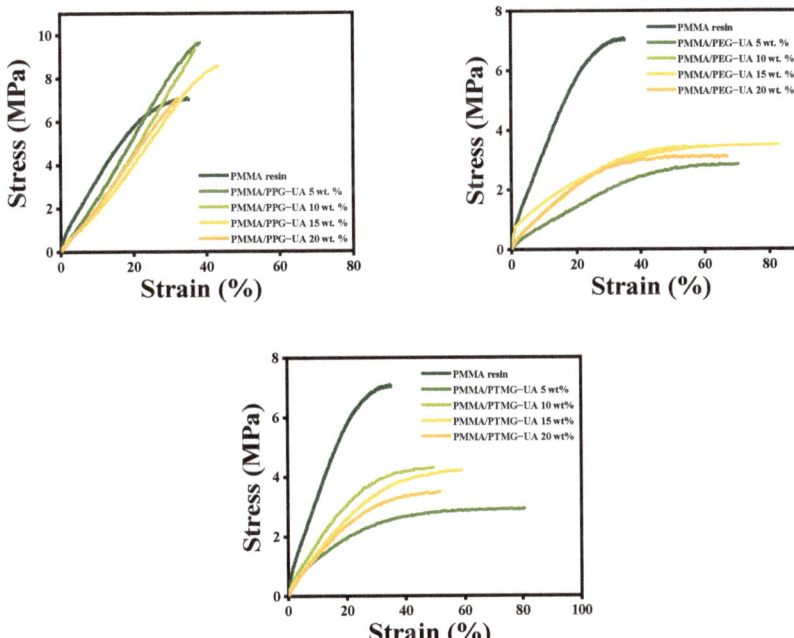

Figure 6. Tensile strength of the urethane acrylate-polymethyl methacrylate polymer binder.

Table 5. Tensile strength at 5 wt% according to urethane acrylate using different types of polyols.

(MPa)	Ref.	UA 5 wt%	UA 10 wt%	UA 15 wt%	UA 20 wt%
PMMA/PPG-UA	7.1	9.7	9.2	8.5	7.1
PMMA/PEG-UA	7.1	2.8	3.5	3.4	3.1
PMMA/PTMG-UA	7.1	2.9	4.3	4.2	3.5

4. Conclusions

In this study, a room temperature curing type binder series was designed as the most important element technology in externally stimulated polymer binder materials, and the limitations of commercialization were overcome by improving the low mechanical properties of PMMA based binders. In the design of a binder composed of a blend of PMMA and UA, UA using various types of polyols (PPG, PEG, PTMG) was synthesized to have the same or similar molecular weight to compare its performance evaluation. The proposed UA series was synthesized to have a molecular weight of 5–8000. Low molecular weights of 2–3000 showed very low mechanical properties for application to binders, and high molecular weights of 10,000 or more were difficult to apply due to processing problems. The optimal mixing ratio was suggested through various content splits so that the designed binder shows excellent mechanical properties. The prepared polymer binder showed improvement in transmittance, shear strength, and tensile strength at UA 5–10 wt%, and an improved polymer binder could be made with structural effects using PPG-UA. PMMA based binders blended with UA using other types of polyols (PEG, PTMG) also showed slightly improved mechanical properties. In this study, the designed PMMA/UA binder can be cured at room temperature without external energy (light, heat, etc.) due to the heat generated from the reaction with the initiator (BPO) and catalyst (PTE). A binder design that can be cured at room temperature is the most important element technology in externally stimulated polymer binders. In addition, in this study, to overcome the mechanical property limitations of PMMA based binders, UA was designed using various polyols and improved physical properties were confirmed. The proposed PMMA/UA binder material could be a promising candidate for future road marking polymer binders.

Author Contributions: Conceptualization, J.-H.L., J.-H.B. and P.-H.H.; methodology, J.-H.L., J.-H.B. and P.-H.H.; formal analysis, J.-H.L. and W.-B.L.; investigation, J.-H.L. and J.-G.M.; resources, J.-H.L., J.-H.B. and P.-H.H.; data curation, J.-H.L., J.-R.L. and J.-W.K.; writing—original draft preparation, J.-H.L.; writing—review and editing, J.-H.L., J.-H.B. and P.-H.H.; supervision, J.-H.B. and P.-H.H.; funding acquisition, J.-H.B. and P.-H.H. All authors have read and agreed to the published version of the manuscript.

Funding: This research was funded by the Korea Agency for Infrastructure Technology Advancement (grant No. 22POQW-B152739-04).

Institutional Review Board Statement: Not applicable.

Data Availability Statement: Data are contained within the article.

Conflicts of Interest: The authors declare no conflicts of interest.

References

1. Kozbial, A.; Guan, W.; Li, L. Manipulating the molecular conformation of a nanometer-thick environmentally friendly coating to control the surface energy. *J. Mater. Chem. A Mater.* **2017**, *5*, 9752–9759. [CrossRef]
2. Zheng, Q.; Xu, C.; Jiang, Z.; Zhu, M.; Chen, C.; Fu, F. Smart Actuators Based on External Stimulus Response. *Front. Chem.* **2021**, *9*, 650358. [CrossRef] [PubMed]
3. Mrinalini, M.; Prasanthkumar, S. Recent Advances on Stimuli-Responsive Smart Materials and their Applications. *ChemPlusChem* **2019**, *84*, 1103–1121. [CrossRef] [PubMed]

4. Ramakrishnan, T.; Kumar, S.S.; Chelladurai, S.J.S.; Gnanasekaran, S.; Sivananthan, S.; Geetha, N.K.; Arthanari, R.; Assefa, G.B. Recent Developments in Stimuli Responsive Smart Materials and Applications: An Overview. *J. Nanomater.* **2022**, *2022*, 4031059. [CrossRef]
5. Yan, D.; Wang, Z.; Zhang, Z. Stimuli-Responsive Crystalline Smart Materials: From Rational Design and Fabrication to Applications. *Acc. Chem. Res.* **2022**, *55*, 1047–1058. [CrossRef]
6. Kreye, O.; Mutlu, H.; Meier, M.A.R. Sustainable routes to polyurethane precursors. *Green Chem.* **2013**, *15*, 1431–1455. [CrossRef]
7. Yamago, S.; Yahata, Y.; Nakanishi, K.; Konishi, S.; Kayahara, E.; Nomura, A.; Goto, A.; Tsujii, Y. Synthesis of concentrated polymer brushes via surface-initiated organotellurium mediated living radical polymerization. *Macromolecules* **2013**, *46*, 6777–6785. [CrossRef]
8. Treat, N.J.; Fors, B.P.; Kramer, J.W.; Christianson, M.; Chiu, C.; Read De Alaniz, J.; Hawker, C.J. Controlled radical polymerization of acrylates regulated by visible light. *ACS Macro Lett.* **2014**, *3*, 580–584. [CrossRef]
9. Wang, K.; Lu, Z.; Zou, Y.; Zhu, Y.; Yu, J. Preparation and Performance Characterization of an Active Luminous Coating for Asphalt Pavement Marking. *Coatings* **2023**, *13*, 1108. [CrossRef]
10. Bi, Y.; Pei, J.; Chen, Z.; Zhang, L.; Li, R.; Hu, D. Preparation and characterization of luminescent road-marking paint. *Int. J. Pavement Res. Technol.* **2020**, *14*, 252–258. [CrossRef]
11. Nance, J.; Sparks, T.D. From streetlights to phosphors: A review on the visibility of roadway marking. *Prog. Org. Coat.* **2020**, *148*, 105749. [CrossRef]
12. Yoon, T.H.; Lee, Y.K.; Lim, B.S.; Kim, C.W. Degree of polymerization of resin composites by different light sources. *J. Oral Rehabil.* **2002**, *29*, 1165–1173. [CrossRef] [PubMed]
13. Arikawa, H.; Takahashi, H.; Kanie, T.; Ban, S. Effect of various visible light photoinitiators on the polymerization and color of light-activated resins. *Dent. Mater. J.* **2009**, *28*, 454–460. [CrossRef] [PubMed]
14. Son, S.W.; Yeon, J.H. Mechanical properties of acrylic polymer concrete containing methacrylic acid as an additive. *Constr. Build. Mater.* **2012**, *37*, 669–679. [CrossRef]
15. Asopa, V.; Suresh, S.; Khandelwal, M.; Sharma, V.; Asopa, S.S.; Kaira, L.S. A comparative evaluation of properties of zirconia reinforced high impact acrylic resin with that of high impact acrylic resin. *Saudi J. Dent. Res.* **2015**, *6*, 146–151. [CrossRef]
16. Hadizadeh, E.; Pazokifard, S.; Mirabedini, S.M.; Ashrafian, H. Optimizing practical properties of MMA-based cold plastic road marking paints using mixture experimental design. *Prog. Org. Coat.* **2020**, *147*, 105784. [CrossRef]
17. Fatemi, S.; Varkani, M.K.; Ranjbar, Z.; Bastani, S. Optimization of the water-based road-marking paint by experimental design, mixture method. *Prog. Org. Coat.* **2006**, *55*, 337–344. [CrossRef]
18. Mirabedini, S.M.; Zareanshahreki, F.; Mannari, V. Enhancing thermoplastic road-marking paints performance using sustainable rosin ester. *Prog. Org. Coat.* **2020**, *139*, 105454. [CrossRef]
19. Bae, J.B.; Won, J.C.; Lim, W.B.; Lee, J.H.; Min, J.G.; Kim, S.W.; Kim, J.H.; Huh, P.H. Highly Flexible and Photo-Activating Acryl-Polyurethane for 3D Steric Architectures. *Polymers* **2021**, *13*, 844–853. [CrossRef]
20. Park, H.W.; Seo, H.S.; Kwon, K.; Lee, J.H. Enhanced Heat Resistance of Acrylic Pressure-Sensitive Adhesive by Incorporating Silicone Blocks Using Silicone-Based Macro-Azo-Initiator. *Polymers* **2020**, *12*, 2410–2414. [CrossRef]
21. Gago, I.; Rio, M.D.; Leon, G.; Miguel, B. Urethane-Acrylate/Aramid Nanocomposites Based on Graphenic Materials. A Comparative Study of Their Mechanical Properties. *Polymers* **2020**, *12*, 2388–2397. [CrossRef] [PubMed]
22. Chen, H.; Lee, S.Y.; Lin, Y.M. Synthesis and Formulation of PCL-Based Urethane Acrylates for DLP 3D Printers. *Polymers* **2020**, *12*, 1500–1516. [CrossRef] [PubMed]
23. Teo, K.T.; Hassan, A.; Gan, S.N. UV-Curable Urethane Acrylate Resin from Palm Fatty Acid Distillate. *Polymers* **2018**, *10*, 1374–1389. [CrossRef] [PubMed]
24. Nasir, K.M.; Halim, N.A.; Tajuddin, H.A.; Arof, A.K.; Abidin, Z.H.Z. The effect of PMMA on physical properties of dammar for coating paint application. *Pigment Resin. Technol.* **2013**, *42*, 137–145. [CrossRef]
25. Grilli, A.; Bocci, M.; Virgili, A.; Conti, C. Mechanical Characterization and Chemical Identification of Clear Binders for Road Surface Courses. *Mater. Sci. Eng. A* **2020**, *9*, 4930646. [CrossRef]
26. Yildirim, Y. Polymer modified asphalt binders. *Constr. Build. Mater.* **2007**, *21*, 66–72. [CrossRef]
27. Airey, G.D.; Mohammed, M.H.; Fichter, C. Rheological characteristics of synthetic road binders. *Fuel* **2008**, *87*, 1763–1775. [CrossRef]
28. Kamal, I.; Bas, Y. Materials and technologies in road pavements—An overview. *Mater. Today Proc.* **2021**, *42*, 2660–2667. [CrossRef]
29. Schneider, M.; Michels, R.; Pipich, V.; Goerigk, G.; Sauer, V.; Heim, H.P.; Huber, K. Morphology of Blends with Crosslinked PMMA Microgels and Linear PMMA Chains. *Macromolecules* **2013**, *46*, 9091–9103. [CrossRef]
30. Rudtsch, S.; Hammerschmidt, U. Intercomparison of Measurements of the Thermophysical Properties of Polymethyl Methacrylate. *Int. J. Thermophys.* **2004**, *25*, 1475–1482. [CrossRef]
31. Fang, Z.; Gao, L.; Zhou, L.; Zheng, Z.; Guo, B.; Zhang, C. Synthesis and Characterization of Excellent TransparencyPoly(urethane-methacrylate). *J. Appl. Polym. Sci.* **2009**, *111*, 724–729. [CrossRef]
32. Ali, U.; Karim, K.J.A.; Buang, N.A. A Review of the Properties and Applications of Poly (Methyl Methacrylate) (PMMA). *Macromolecules* **2015**, *55*, 678–705. [CrossRef]
33. Maurya, S.D.; Kurmvanshi, S.K.; Mohanty, S.; Nayak, S.K. A Review on Acrylate-Terminated Urethane Oligomers and Polymers: Synthesis and Applications. *Polym. Plast. Technol. Eng.* **2018**, *57*, 625–656. [CrossRef]

34. Swiderski, K.W.; Khudyakov, I.V. Synthesis and Properties of Urethane Acrylate Oligomers: Direct versus Reverse Addition. *Ind. Eng. Chem. Res.* **2004**, *43*, 6281–6284. [CrossRef]
35. Hua, F.J.; Hu, C.P. Morphology and mechanical properties of urethane acrylate resin networks. *J. Appl. Polym. Sci.* **2000**, *77*, 1532–1537. [CrossRef]
36. Liu, B.; Nie, J.; He, Y. From rosin to high adhesive polyurethane acrylate: Synthesis and properties. *Int. J. Adhes. Adhes.* **2016**, *66*, 99–103. [CrossRef]
37. Oprea, S.; Vlad, S.; Stanciu, A. Poly(urethane-methacrylates). Synthesis and characterization. *Polymer* **2001**, *42*, 7257–7266. [CrossRef]
38. Sultan, M.; Atta, S.; Bhatti, H.N.; Islam, A.; Jamil, A.; Bibi, I.; Gull, N. Synthesis, Characterization, and Application Studies of Polyurethane Acrylate Thermoset Coatings: Effect of Hard Segment. *Polym. Plast. Technol. Eng.* **2017**, *56*, 1608–1618. [CrossRef]
39. Kim, B.K.; Lee, K.H.; Kim, H.D. Preparation and properties of UV-curable polyurethane acrylates. *J. Appl. Polym. Sci.* **1996**, *60*, 799–805. [CrossRef]
40. Studer, K.; Nguyen, P.T.; Decker, C.; Beck, E.; Schwalm, R. Redox and photoinitiated crosslinking polymerization. *Prog. Org. Coat.* **2005**, *53*, 126–133. [CrossRef]
41. Zhang, S.; Shi, Z.; Xu, H.; Ma, X.; Yin, J.; Tian, M. Revisiting the mechanism of redox polymerization to build the hydrogel with excellent properties using a novel initiator. *Soft Matter.* **2016**, *12*, 2575–2582. [CrossRef]
42. Fu, J.; Yu, H.; Wang, L.; Liang, R.; Zhang, C.; Jin, M. Preparation and properties of UV-curable polyurethane acrylate/SiO$_2$ composite hard coatings. *Prog. Org. Coat.* **2021**, *153*, 106121. [CrossRef]
43. Nowak, M.; Bednarczyk, P.; Mozelewska, K.; Czech, Z. Synthesis and Characterization of Urethane Acrylate Resin Based on 1,3-Propanediol for Coating Applications. *Coatings* **2022**, *12*, 1860. [CrossRef]
44. Vinçotte, A.; Beauvoit, E.; Boyard, N.; Guilminot, E. Effect of solvent on PARALOID®B72 and B44 acrylic resins used as adhesives in conservation. *Herit. Sci.* **2019**, *7*, 42. [CrossRef]
45. Basar, O.; Veliyath, V.P.; Tarak, F.; Sabet, E. A Systematic Study on Impact of Binder Formulation on Green Body Strength of Vat-Photopolymerisation 3D Printed Silica Ceramics Used in Investment Casting. *Polymers* **2023**, *15*, 3141–3158. [CrossRef]
46. Tuan, T.S.R.; Aung, M.M.; Ahmad, A.; Rayung, M.; Su'ait, M.S.; Yusof, N.A.; Lae, K.Z.W. Enhancement of Plasticizing Effect on Bio-Based Polyurethane Acrylate Solid Polymer Electrolyte and Its Properties. *Polymers* **2018**, *10*, 1142–1159. [CrossRef]
47. Xiang, H.; Wang, X.; Lin, G.; Xi, L.; Yang, Y.; Lei, D.; Dong, H.; Su, J.; Cui, Y.; Liu, X. Preparation, Characterization and Application of UV-Curable Flexible Hyperbranched Polyurethane Acrylate. *Polymers* **2017**, *9*, 552–563. [CrossRef]
48. Chang, H.H.; Tseng, Y.T.; Huang, S.W.; Kuo, Y.F.; Yeh, C.L.; Wu, C.H.; Huang, Y.C.; Jeng, R.J.; Lin, J.J.; Lin, C.P. Evaluation of Carbon Dioxide-Based Urethane Acrylate Composites for Sealers of Root Canal Obturation. *Polymers* **2020**, *12*, 482–500. [CrossRef]
49. Ghanali, S.K.; Adrus, N.; Majid, R.; Ali, F.; Jamaluddin, J. UV-LED as a New Emerging Tool for Curable Polyurethane Acrylate Hydrophobic Coating. *Polymers* **2021**, *13*, 487–497. [CrossRef]
50. Zhen, Y.; Mingguang, H.; Han, C. Synthesis and properties of hydroxy acrylic resin with high solid content. *AIP Conf. Proc.* **2017**, *1890*. [CrossRef]
51. Brown, R.A.; Coogan, R.G.; Gortier, D.G.; Reeve, M.S.; Rega, J.D. Comparing and contrasting the properties of urethane/acrylic hybrids with those of corresponding blends of urethane dispersions and acrylic emulsion. *Prog. Org. Coat.* **2005**, *52*, 73–84. [CrossRef]
52. Pieper, R.J.; Ekin, A.; Webster, D.C.; Casse, F.; Callow, J.A.; Callow, M.E. Combinatorial approach to study the effect of acrylic polyol composition on the properties of crosslinked siloxane-polyurethane fouling-release coatings. *J. Coat. Technol.* **2007**, *4*, 453–461. [CrossRef]
53. Yao, T.; Han, S.; Gong, X.; Zhang, J.; Chang, X.; Zhang, Z. Performance evaluation of a polyurethane-urea binder for asphalt pavement groove-filling. *Constr. Build. Mater.* **2022**, *315*, 125734. [CrossRef]
54. Mousaa, I.M.; Ali, N.M.; Attia, M.K. Preparation of high performance coating films based on urethane acrylate oligomer and liquid silicone rubber for corrosion protection of mild steel using electron beam radiation. *Prog. Org. Coat.* **2021**, *155*, 106222. [CrossRef]
55. Zhang, J.; Li, X.; Shi, X.; Hua, M.; Zhou, X.; Wang, X. Synthesis of core–shell acrylic–polyurethane hybrid latex as binder of aqueous pigment inks for digital inkjet printing. *Prog. Nat. Sci.* **2021**, *22*, 71–78. [CrossRef]
56. Zheng, W.; Wang, H.; Chen, Y.; Ji, J.; You, Z.; Zhang, Y. A review on compatibility between crumb rubber and asphalt binder. *Constr. Build. Mater.* **2021**, *297*, 123820. [CrossRef]
57. Verdet, M.; Salenikovich, A.; Cointe, A.; Coureau, J.L.; Galimeard, P.; Toro, W.M.; Blanchet, P.B.; Delisee, C. Mechanical Performance of Polyurethane and Epoxy Adhesives in Connections with Glued-in Rods at Elevated Temperatures. *Constr. Build. Mater.* **2016**, *11*, 8200–8214. [CrossRef]
58. Lim, W.B.; Bae, J.H.; Lee, G.H.; Lee, J.H.; Min, J.G.; Huh, P.H. Transparency- and Repellency-Enhanced Acrylic-Based Binder for Stimuli-Responsive Road Paint Safety Improvement Technology. *Materials* **2021**, *14*, 6829. [CrossRef]
59. Saleeva, L.; Kashapov, R.; Shakirzyanov, F.; Kuznetsov, E.; Kashapov, L.; Smirnova, V.; Kashapov, N.; Saleeva, G.; Sachenkov, O.; Saleev, R. The Effect of Surface Processing on the Shear Strength of Cobalt-Chromium Dental Alloy and Ceramics. *Materials* **2022**, *15*, 2987. [CrossRef]

70. Laureto, J.; Pearce, J. Anisotropic mechanical property variance between ASTM D638-14 type i and type iv fused filament fabricated specimens. *Polym. Test.* **2018**, *68*, 294–301. [CrossRef]
71. Monserrat, B.A.; Lluma, J.; Mesa, R.J.; Rodriguez, J.T. Study of the Influence of the Manufacturing Parameters on Tensile Properties of Thermoplastic Elastomers. *Polymers* **2022**, *14*, 576. [CrossRef] [PubMed]

Disclaimer/Publisher's Note: The statements, opinions and data contained in all publications are solely those of the individual author(s) and contributor(s) and not of MDPI and/or the editor(s). MDPI and/or the editor(s) disclaim responsibility for any injury to people or property resulting from any ideas, methods, instructions or products referred to in the content.

Article

Synthesis and Properties of Cationic Core-Shell Fluorinated Polyurethane Acrylate

Junhua Chen [1,2], Xiaoting Lu [1], Jinlian Chen [1], Shiting Li [1], He Zhang [1,2], Yinping Wu [1,2], Dongyu Zhu [3] and Xiangying Hao [1,2,*]

[1] School of Environmental and Chemical Engineering, Zhaoqing University, Zhaoqing 526061, China; cehjchen@yeah.net (J.C.); luxiaoting23@mails.ucas.ac.cn (X.L.)
[2] Guangdong Provincial Key Laboratory of Environmental Health and Land Resource, College of Environmental and Chemical Engineering, Zhaoqing University, Zhaoqing 526061, China
[3] School of Chemical Engineering and Light Industry, Guangdong University of Technology, Guangzhou 510006, China; zdy16@gdut.edu.cn
* Correspondence: xyinghao@zqu.edu.cn

Abstract: Vinyl-capped cationic waterborne polyurethane (CWPU) was prepared using isophorone diisocyanate (IPDI), polycarbonate diol (PCDL), N-methyldiethanolamine (MDEA), and trimethylolpropane (TMP) as raw materials and hydroxyethyl methacrylate (HEMA) as a capping agent. Then, a crosslinked FPUA composite emulsion with polyurethane (PU) as the shell and fluorinated acrylate (PA) as the core was prepared by core-shell emulsion polymerization with CWPU as the seed emulsion, together with dodecafluoroheptyl methacrylate (DFMA), diacetone acrylamide (DAAM), and methyl methacrylate (MMA). The effects of the core-shell ratio of PA/PU on the surface properties, mechanical properties, and heat resistance of FPUA emulsions and films were investigated. The results showed that when w(PA) = 30~50%, the stability of FPUA emulsion was the highest, and the particles showed a core-shell structure with bright and dark intersections under TEM. When w(PA) = 30%, the tensile strength reached 23.35 ± 0.08 MPa. When w(PA) = 50%, the fluorine content on the surface of the coating film was 14.75% and the contact angle was as high as 98.5°, which showed good hydrophobicity; the surface flatness of the film was observed under AFM. It is found that the tensile strength of the film increases and then decreases with the increase in the core-shell ratio and the heat resistance of the FPUA film is gradually increased. The FPUA film has excellent properties such as good impact resistance, high flexibility, high adhesion, and corrosion resistance.

Keywords: cationic waterborne polyurethane; core-shell structure; isocyanate reactivity; tensile strength; acrylate-based composites

1. Introduction

The substantial environmental degradation that has been occurring in recent years has focused a lot of emphasis on the study and development of environmentally friendly materials [1–3]. Waterborne polymers have been employed extensively in recent decades due to their low toxicity and environmental friendliness [4–6]. Waterborne resins have hydrophilic groups with high surface energy but low water and solvent resistance; waterborne polyurethanes are an important family of ecologically benign waterborne resins with good mechanical and physical properties [7,8]. However, cationic aqueous polyurethanes with sulfur or ammonium ions in the side chain or main chain have a strong hygroscopicity and bactericidal activity and they are frequently used to treat surfaces made of paper, leather, cloth, and glass [9]. Acrylates have good water resistance and double bonding aids in the formation of a crosslinked structure; therefore, combining them with aqueous polyurethanes can result in composite emulsions with outstanding overall performance [10]. On the other hand, the mechanical properties of single polyurethane aqueous dispersions are poor, with poor self-thickening and low solid content. Some of the promising applications of aqueous polyurethanes formed using isocyanate chemistry are electrode functionalization for electrochemical measurements [11], plant-based cellulose polymers for single-molecule studies [12], or metal oxide surface conjugation as a suitable chemical platform [13].

As a result of their strong electronegativity and low polarization, fluorine atoms can effectively lower the surface tension of polymers, allowing them to display a variety of benefits like stability, good weathering, and resistance to heat [14]. Perfluoroalkyl acrylic acid copolymers, for example, have an extremely low critical surface tension in the range of 10–11 mN/m [15]. This is because fluorocarbon side-chain-containing polymers have a high abundance of CF_3 groups on the surface. When fluorine atoms are added to polyurethane acrylate systems, they can effectively reduce their surface energy and improve their thermal stability and excellent water resistance [16–20]. Fluorine atoms have unique properties, which makes fluorinated polymers outstanding and unique [21]. Nevertheless, fluorinated monomers are not very compatible and come at a hefty price [22]. Copolymer compatibility and emulsion stability are known to diminish with higher incorporation of fluorinated monomers in copolymers [23]. Consequently, it is important to optimize the polymerization process to minimize the number of fluorinated monomers while preserving a suitable surface tension (oil/water repellency).

Due to their general benefits in film production, core-shell fluorinated copolymer emulsions have drawn a lot of interest when compared to regular emulsions without unique features [24–27]. The construction of the core-shell structure allows the acquisition of final polymer properties that are difficult to obtain by blending or random copolymerization of the two polymers [28–30]. It also improves the copolymer properties such as impact resistance, abrasion resistance, and water resistance because of the ionic bonding, grafting, or interpenetrating networks between the core-shell layers [31,32]. Typically, semi-continuous or seeded emulsion polymerization is used to create core-shell fluorinated copolymers [33]. This process involves polymerizing a mixture of monomers to form a nucleus and then post-polymerizing a different mixture of monomers on the nucleus seed to generate a shell [34]. The core-shell technique can be categorized as either copolymerized or non-copolymerized depending on how the PU and FPA chain breaks are connected.

By employing a monomer pre-emulsification synthesis technique, Zhong added hexafluorobutyl methacrylate (HFMA) to waterborne polyurethane dispersion and created crosslinked fluorinated polyurethane resin (FWPU) with a uniform particle size by using dihydrazide adipate as a chain extender [35]. By esterifying perfluorooctane chloride (PFOC) and hydroxypropyl methacrylate (HPMA), Bai created the fluorine-containing acrylate monomer PFMA [36]. Potassium persulfate (KPS) and sodium bicarbonate were then utilized as an initiator/buffer system, while sodium dodecyl sulfate (SDS)/Tween 80 was used as a mixed emulsifier. Semi-continuous core-shell polymerization was used to create the fluorinated acrylate emulsions MMA/BA/St and PFMA/MMA/BA. Ting used methyl methacrylate (MMA), butyl acrylate (BA), dodecafluoroheptyl methacrylate (DFMA), and 3-(Trimethoxysilyl)propyl methacrylate (MPS) as raw materials to create a series of self-crosslinked fluorinated polyacrylate emulsion particles with core-shell structure [37]. Investigations were performed into how the MPS dose affected the characteristics of latex film. By using the core-shell polymerization process, Hirose created a polyurethane/polyacrylate composite emulsion with an interpenetrating polymer network (IPN) [38]. By combining vinyl end-sealed polyurethane prepolymer and a combination of acrylate and perfluoroalkyl acrylate as raw ingredients, Park created a water-based fluorinated polyurethane-acrylate emulsion [14]. It was investigated as to how much PA/FPA affected the surface, thermal, and mechanical characteristics of polymer antifouling coatings. In their study, Chakrabarty prepared polystyrene nuclear emulsion particles with perfluoroalkyl acrylate as the shell layer [39]. They demonstrated that, when compared to random copolymers or emulsion blends of styrene and perfluoroalkyl acrylate, coil-shell particles are the most effective particles for lowering the surface energy of emulsion film.

Certain desired features of polyacrylate coatings can be introduced by adding fluorine-containing groups to copolymers in polyurethane acrylics [40]. Consequently, high-end self-cleaning coatings and water-/oil-repellent surface coatings for textiles, papers, and leathers have been using more and more fluorinated polyurethane-acrylate (FPUA) [41]. It has also progressively developed into a focal point for domestic and international research on fluorine-modified waterborne polyurethanes [42]. On the other hand, cationic fluorinated waterborne polyurethanes have not been the subject of as many national and international investigations. In addition, there are many accessible fluorinated monomers but their utilization rate is low, the polymerization process is hard to manage, and the reaction process is complicated.

In this study, cationic FPUA that self-emulsifies was created by adding unsaturated monomer to partially seal the PU chain break that contains an NCO end group. By combining this process with monomer pre-emulsification, hydrophobic PA/FPA monomer was expanded from the outside to the inside of PU micellar for polymerization. The polymerization process was straightforward and simple to manage. Transmission electron microscopy (TEM), a contact angle measurement device, infrared spectroscopy, and XPS were used to describe and compare the structure and surface hydrophobic characteristics of PU and modified FPUA.

2. Materials and Methods

2.1. Materials

Polycarbonate diol (PCDL, purity > 95%, M_n ~ 1000) was procured from Shanghai Huihua Industrial Co., Ltd., Shanghai, China. Isophorone diisocyanate (IPDI, purity > 99.5%, NCO% $\geq$ 37.5%) was supplied by Bayer Chemicals (Leverkusen, Germany). Dodecafluoroheptyl methacrylate was purchased from Shangfu Technology Co., Ltd., Shanghai, China. N-methyldiethanolamine (MDEA, $\geq$99.5%), 1,4-butanediol (BDO, $\geq$99.5%), methyl methacrylate (MMA, $\geq$99.5%), hydroxyethyl methacrylate (HEMA, $\geq$96%), trimethylolpropane (TMP, $\geq$99.5%), and dibutyltin dilaurate (DBTDL, $\geq$99.5%) were purchased from Macklin Inc., Shanghai, China. Dihydrazide adipate (ADH, $\geq$98%) and 2,6-tert-butyl-p-cresol (BHT, $\geq$99.5%) was supplied by Aladdin Reagent Co., Ltd., Shanghai, China. Glacial acetic acid (HAc, $\geq$90%) and ammonium persulfate (APS, $\geq$99%) were purchased from Shanghai Lingfeng Chemical Reagent Co., Ltd., Shanghai, China. Deionization is homemade in the laboratory. All the other reagents underwent further purification.

2.2. Preparation of Cationic Polyurethane Aqueous Dispersion (CWPU)

Pre-add PCDL and IPDI in a three-necked flask with a stirrer, raise the temperature to 80~85 °C, drop in catalyst DBTDL for 1~2 h, and then add TMP and BDO to keep the reaction for 2 h. When the viscosity of the system rises, inject an appropriate amount of acetone to adjust the viscosity of the system. After the system is stabilized, the temperature of the reaction system is lowered to 60 °C, and then MDEA is added. In order to reduce the viscosity of the system and to reduce the foam in the synthesis, the system is diluted by adding an appropriate amount of acetone and the acceleration of the drop is adjusted so that it is all added within one hour. The heating is then continued for two to three hours. HEMA is added to the system, the ends are blocked with double bonds, held for two hours, and then completely neutralized with glacial acetic acid. Finally, an amount of deionized water is gradually added to the prepolymer and emulsified with rapid shear. This resulted in a cationic partially double-bonded aqueous polyurethane dispersion. The crude reaction mixture was at a 75% yield. Scheme 1 is shown in the figure below, with the detailed dosages shown in Table S1.

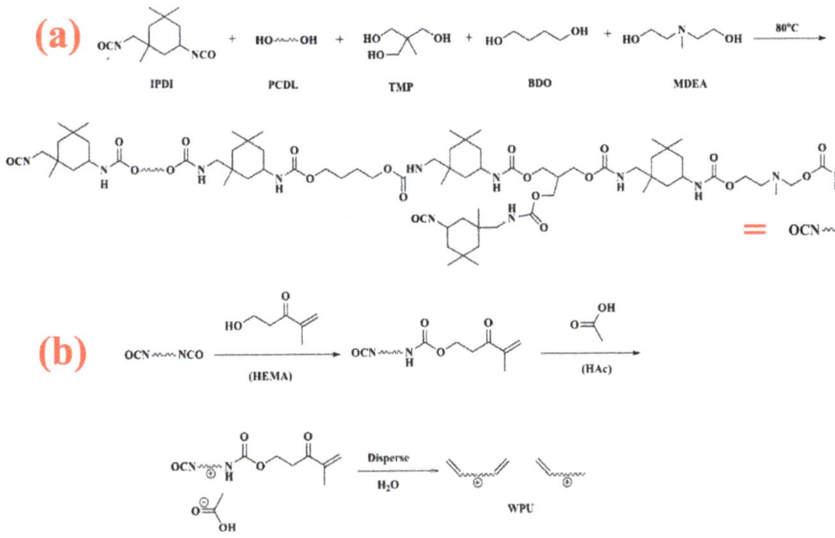

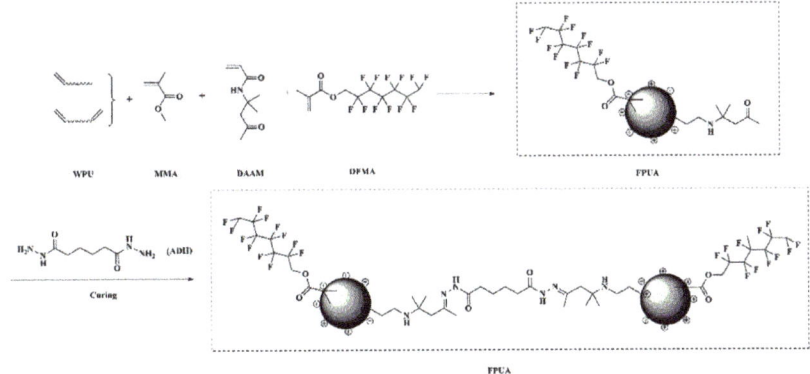

Scheme 1. Synthesis of the CWPU. (a) Isocyanate-terminated prepolymers; (b) Cationic polyurethane aqueous dispersion.

2.3. Preparation of Fluorinated Waterborne Polyurethane-Acrylate Core-Shell Emulsion (FPUA)

As one of the reaction monomers, the above-obtained polyurethane dispersion was added together with MMA, DAAM, and DFMA, each of which accounted for 30% and 50% of the total weight. DFMA made up 10% of the total weight. To create FPUA, cored shell emulsions with cored shells of 30/70 and 50/50 were used as well as 0.2 g APS initiator, which was introduced for the emulsion polymerization reaction after an even mixing step at 80 °C. The temperature was then maintained for 4 h. The crude reaction mixture was at 86% yield. In Table S1, the typical formulations are displayed. The synthetic circuit is shown in Scheme 2.

Scheme 2. Preparation of FPUA.

2.4. Preparation of FPUA Film

Place the emulsion in a Teflon mold and allow it to cure for a week at room temperature before putting it in a vacuum drying oven at 60 °C for 72 h to produce a light-yellow film with a thickness of around 1 mm.

2.5. Characterization

A Fourier transform infrared spectrometer (PERKIN ELMER 1730, Waltham, MA, USA) was used to conduct FTIR tests using the potassium bromide tablet method. The range of wave numbers is 400–4000 cm^{-1}. The membranes were prepared using a Fourier Transform Attenuated Total Reflectance Infrared Spectrometer (ATR-FTIR) manufactured by Nicolet Company, Madison, WI, USA, with a wave number range of 4000–525 cm^{-1} and 32 scans with a resolution of 4 cm^{-1}. At room temperature, the emulsion was diluted 1000 times and then the particle size and distribution of the emulsion were measured by Zetasizer Nano-ZS (Malvern, UK). The UJC-2000C1 static contact angle measuring instrument (Shanghai Zhongchen Digital Equipment Co., Ltd., Shanghai, China) measured the contact angles of the surface layer of the coating film and the air medium at room temperature with distilled water, dichloroethane, and hexadecane. The emulsion was diluted 200 times with deionized water and then thoroughly mixed using a sonicator. It was then dripped onto a conductive copper mesh and stained with 2% phosphotungstic acid and, after drying, the morphology of the stained emulsion particles was examined using a transmission electron microscope (TEM, Hitachi, Tokyo, Japan, model H-7650) with an operating voltage of 60 kV. The samples to be tested were uniformly coated on the surface of monocrystalline silicon, their surfaces were sprayed with gold after drying, and they were then characterized in a scanning electron microscope (SEM) model SU-8220 of Hitachi, Japan. The samples were uniformly coated on the monocrystalline silicon surface before testing and the samples were tested with a SPA-400 knockdown atomic force microscope (AFM) with a scanning range of 5×5 μm and a viewing scale of 10 nm. Each sample was finally presented in the form of a two-dimensional planar image as well as a three-dimensional image. A Hitachi GENESYS 180 UV–visible spectrometer was used to analyze the transmittance of the coatings. A blank slide served as the test backdrop, while a slide served as the substrate. After testing each sample group three times in parallel at three separate test sites, the average transmittance at 500 nm for the three groups was determined.

The film was heated at 80 °C for 2 h and then the surface elements of the film were analyzed qualitatively or quantitatively by using an X-ray photoelectron spectrometer (ESCALAB 250XI, Thermo Scientific, Waltham, MA, USA). The heat loss of the samples was measured using a thermogravimetric differential thermal integrated analyzer (Labsys Evo) from Setaram, Caluire-et-Cuire, France, with temperatures ranging from ambient temperature to 600 °C with an increasing rate of 10 °C/min. A dynamic thermo-mechanical analyzer model DMA 242C3 from NETZSCH, Selb, Germany, was used to test the thermo-mechanical properties of the coatings. The following parameters were used: maximum amplitude of 10 μm, maximum dynamic force of 2 N, static force of 0.5 N, temperature scanning range of −50~100 °C, temperature rise rate of 5.0 K/min, test frequencies of 1.0 Hz, 3.333 Hz, and 5.0 Hz. A multifunctional electronic tensile testing machine EKT-TS2000 (Ektron Tek Co., Ltd., Taiwan, China) was used to measure the mechanical behavior of the sample. The tensile test speed was 10 mm/min. Specimens with specifications of 80 mm × 10 mm (length × width) were used for the evaluation. The samples were cut into tensile test strips of 0.5–1.0 mm thickness. The test was repeated three times to ensure the accuracy of the measurement results. To test the corrosion resistance of the coated film, four different solutions were used: 5.0 wt% NaCl solution, 0.5 mol/L $CuSO_4$ solution, H_2SO_4 solution (pH = 0, using methyl orange staining), and NaOH solution (pH = 14, using rhodamine staining). Then, 0.5 mL droplets were taken and added to the coated and uncoated areas, respectively, and the corrosion of the respective surfaces was observed after 24 h. The corrosion of the coated and uncoated surfaces was observed after 24 h.

3. Results and Discussion

The stretching vibration absorption peak of –NH– and –OH is at 3343 cm^{-1} in Figure 1. The characteristic absorption peak of –CH_3 and –CH_2– is at 2946 cm^{-1}, the stretching vibration absorption peak of carbonyl group is at 1739 cm^{-1}, and the absorption peak of C=C is at 1651 cm^{-1}. It was established that to maximize the degree of crosslinking, the chemical bonding between HEMA and prepolymer was crucial to the synthesis of CWPU. Additionally, there is no absorption peak between 2000 and 2500 cm^{-1}, which further suggests that the –NCO reaction is complete and results in the formation of the carbamate structure. The distinctive absorption peak of N–H in –CONH has a wave number at 1532 cm^{-1}.

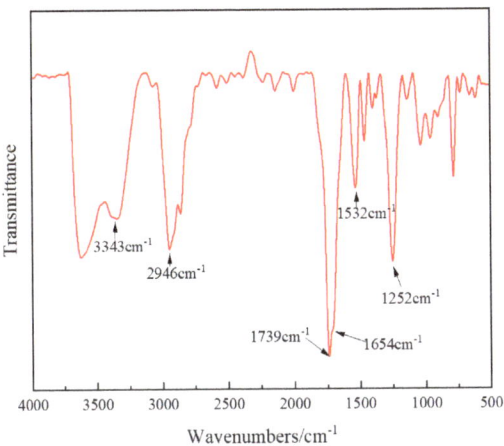

Figure 1. FT-IR spectrum of CWPU.

The curves (a) and (b) of Figure 2 depict the stretching vibration peaks of –NH– and –OH at 3340 cm^{-1}, the methyl-methyl absorption peak at about 2910 cm^{-1} and 2864 cm^{-1}, and the absence of any absorption peaks between 2000 and 2500 cm^{-1}. It was demonstrated that throughout the synthesis process, –NCO underwent complete reactivity and changed into a carbamate structure. The carbonyl group has a stretching vibration peak at 1738 cm^{-1}, a double bond has a characteristic absorption peak at 1645 cm^{-1}, and N–H in –CONH has a characteristic absorption peak at 1532 cm^{-1}, 1462 cm^{-1}, and 1245 cm^{-1}. The curve also reveals the deformation absorption peak of CF_2 at 696 cm^{-1}, the stretching vibration absorption peak of C–F at 1143 cm^{-1}, the stretching vibration absorption peak of CF_3 at 846 cm^{-1}, and the absorption peak of CF_3 at 794 cm^{-1}, all of which signify the inclusion of fluorine in fluorinated polyurethane. Additionally, it can be seen from a comparison of curves (a) and (b) in Figure 2 that there is a variation in the strength of the same band peak when the PA/PU ratio is 50/50 and 30/70. The height of the 30/70 peak can be clearly seen to be higher than the 50/50 peak among them at 1738 cm^{-1} and 1532 cm^{-1}, demonstrating that when PA/PU is 30/70, the absorption peak of the carbonyl group and carbamate bond is rather strong. Moreover, the absorption peak of 50/50 is stronger than that of 30/70 at 1143 cm^{-1} and 794 cm^{-1}, further supporting the idea that when PA/PU is 50/50, the proportion of fluorine-containing acrylate increases, increasing the strength of the C–F and CF_3 stretching vibration peak. This also further confirms the core-shell structure of the self-encapsulation behavior.

Figure 3 shows that the particle size distributions of the CWPU and 30 wt% and 50 wt% FPUA emulsions are unimodal, indicating that the physicochemical distributions of the emulsions are fairly uniform. Among all the emulsions, PU has the smallest particle size and the narrowest particle size distribution, which indicates that the latex particles have good storage stability. Secondly, the modified 30 wt% FPUA emulsion had smaller particles and a narrower distribution compared to the 50 wt% FPUA emulsion. The amphiphilic CWPUs were self-assembled to form aggregates, which were surrounded by ionizable groups produced by the MDEA units. Relatively large aggregates were observed in FWPU-Seed solutions containing CWPU and one-third of the monomer. Considering the lack of hydrophilic portion of MMA and DFMA, it is assumed that these monomers migrate into the CWPU aggregates to reach a steady state and thus are responsible for the formation of larger aggregates. Water-soluble crosslinked DAAMs are dispersed on the surface of the polymer particles after polymerization to form chemical bonds and are susceptible to crosslinking during curing. In FWPU solutions containing CWPU and all monomers, the particle diameter further increases to nearly 100 nm.

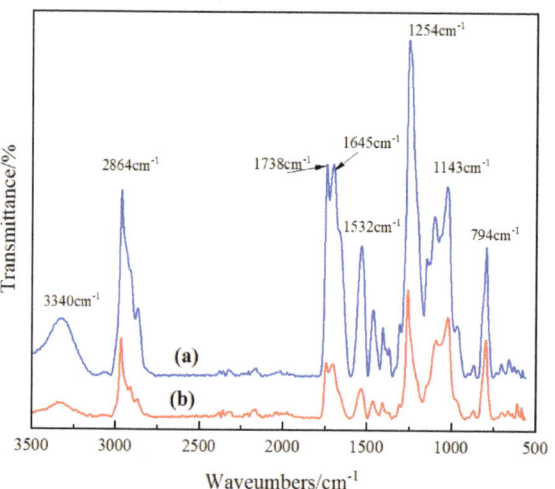

Figure 2. ATR spectrum of FPUA. (a) PA/PU = 50/50; (b) PA/PU = 30/70.

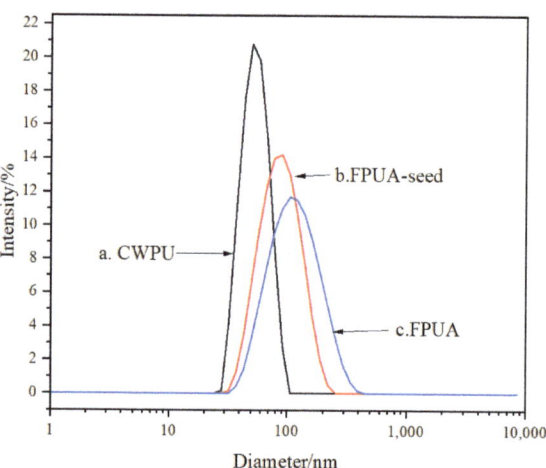

Figure 3. Particle size distribution. (a) CWPU; (b) 30 wt% FPUA; (c) 50 wt% FPUA.

Figure 4a,c depicts TEM images of FPUA emulsion particles demonstrating a core-to-shell ratio of 30/70 at particle sizes of 2 µm and 500 nm, respectively. For 500 nm sizes, Figure 4b portrays TEM images of FPUA emulsion particles with a core-to-shell ratio of 50/50. Initially, the TEM image exhibited a black-and-white pattern linked to light and dark upon negative staining of the emulsion with phosphotungstic acid. The polymeric particles were distinctly visible as spherical white spots on the dark background. An apparent core-shell architecture was discernible amongst them, with the hydrophilic PU chain segment functioning as the spherical shell of the particles and the hydrophobic fluoroacrylate monomer enveloped by PU forming the nucleus of particle. After comparison of TEM photographs of varied core-shell ratios, it was observed that in the 50/50 ratio of FPUA emulsion, numerous PA particles did not swell into PU for polymerization but copolymerized externally due to the incompatibility between the hydrophilic shell PU and hydrophobic PA. Consequently, the size of the core-shell's internal and external halves was once diminished, simplifying the formation of a wrapped core-shell structure. However, a more uniform and distinguishable transtypic core-shell structure can be generated by FPUA emulsions exhibiting a core-to-shell ratio of 30/70.

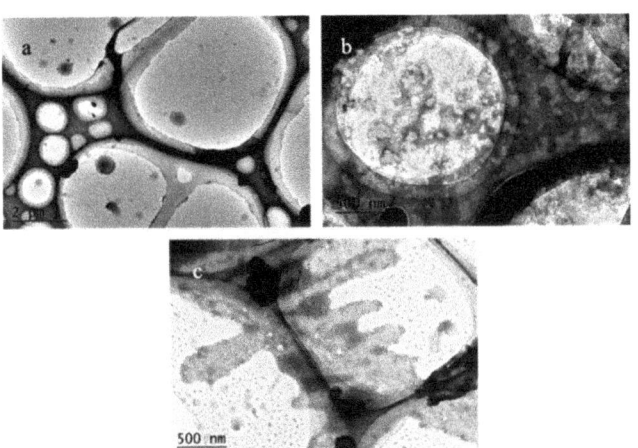

Figure 4. The TEM photograph of FPUA composite. (**a**) size of 2 μm; (**b**) size of 500 nm; (**c**) size of 500 nm.

Figure 5 provides a graphical representation of how the core-shell ratio impacts the mechanical properties of latex film. For a more detailed insight into the data, refer to Table S2. The introduction of fluorine into the film increases its hardness, as is evident in the table. This is primarily due to the strong polar bond C–F in the fluorine-containing side chain of the hard segment of the polyurethane chain, which can generate NH···F hydrogen bonds. As the core-shell ratio increases, the tensile strength of the film initially increases before eventually decreasing. From 6.3 ± 0.05 to 23.35 ± 0.08 MPa, the tensile strength of FPUA rises. However, Young's modulus and elongation at the break of the film gradually decline, indicating that as the core-shell ratio increases, the flexibility of the film decreases, making it more brittle. This observation highlights two key aspects. Firstly, fluorinated waterborne polyurethane has significantly higher hardness compared to unfluorinated polyurethane. However, it is less tough and flexible. Secondly, excessive fluorine content also impacts the mechanical properties of the film. When the core-shell ratio is increased to 50/50, the tensile strength of the film decreases from 23.35 ± 0.08 MPa to 18.61 ± 0.04 MPa. Alongside this, Young's modulus reduces by 17.2%, while the elongation at break decreases by 86.5%. These findings indicate that as the proportion of fluorinated polyurethane adhesive (FPA) increases, the polarity of the composite emulsion gradually increases, leading to low compatibility between the hydrophilic group and the hydrophobic FPA segment. Consequently, the movement of the segment is inhibited, resulting in a significant reduction in flexibility and elongation at the break of FPUA and a softer more brittle film.

The fluorine-containing acrylate modification of FPUA film yields notable enhancements in adhesion, impact resistance, hardness, and flexibility, as seen in Figure 6 and Table S3. This implies that the acrylate monomer addition raises the FPUA film's crosslinking density, assisting in the formation of a dense network structure and improving the mechanical characteristics of the film.

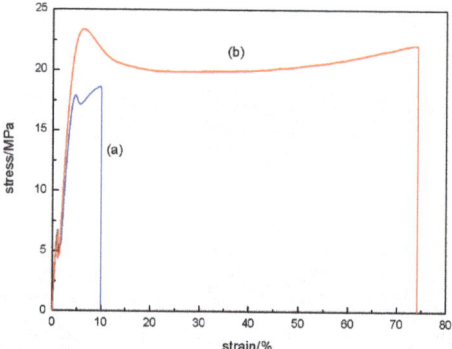

Figure 5. The tensile stress–strain curve of FPUA. (a) PA/PU = 50/50; (b) PA/PU = 30/70.

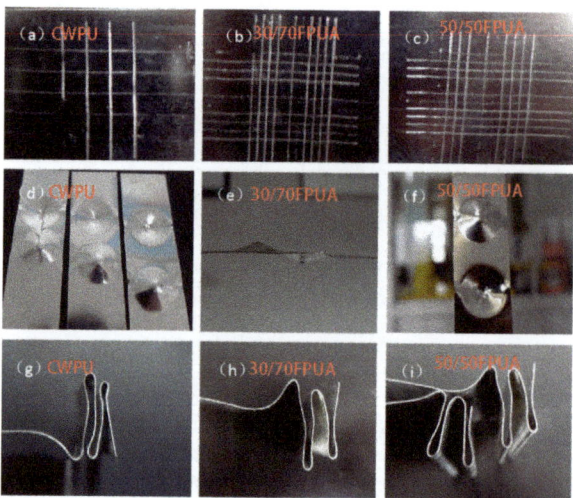

Figure 6. Adhesion (**a–c**), impact (**d–f**), and flexibility (**g–i**) tests of FPUA film.

Figure 7 demonstrates the relationship between mechanical loss (tan δ), energy storage modulus, and temperature for a cured FPUA film. Tan δ is a measure of the energy dissipation in a material under cyclic loading and its maximum value corresponds to the glass transition temperature (Tg) of the polymer. Tg represents the temperature at which the polymer transitions from a glassy state to a rubbery or viscous state. The modulus at Tg + 60 °C provides information about the cross-linking density of the film. Cross-linking density refers to the number of chemical bonds formed between polymer chains, which affects the stiffness and rigidity of the material. In this case, the cross-linking density and stiffness are directly related to both Tg and the modulus of the cross-linked polymer. Table S4 presents crucial data obtained from these calculations, specifically comparing the cross-linking densities of two different FPUA films: 50/50FPUA and 30/70FPUA. The results indicate that the cross-linking density of the 30/70FPUA film is significantly higher than that of the 50/50FPUA film, suggesting that the former film is more rigid. However, it is interesting to note that there are no significant differences in Tg and modulus between the 50/50FPUA and 30/70FPUA films in the glassy state. This implies that while the cross-linking density influences the rigidity of the films, it does not have a significant impact on their glass transition behavior. Therefore, the core-shell structure in this case manifests through variations in cross-linking density, which in turn affects the mechanical properties such as stiffness. The glass transition temperature and modulus serve as important parameters for understanding the behavior of these cross-linked polymer films.

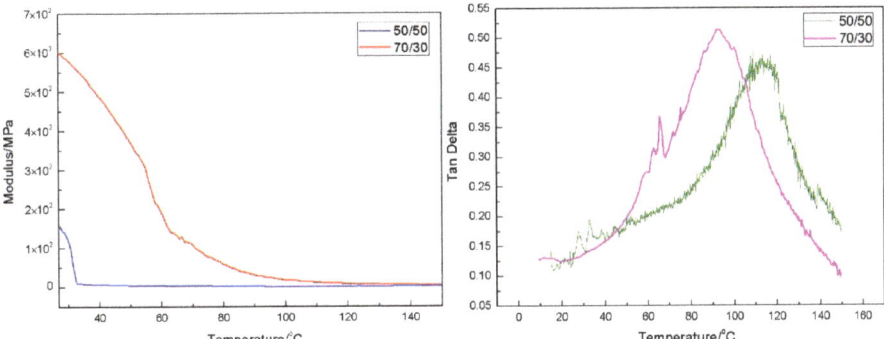

Figure 7. Storage modulus and tan δ for FPUA films.

Before and after wearing, the surface elements of the polished FPUA films with two core-shell ratios were investigated. As illustrated in Figure 8, the primary constituents on the surface of films are C, N, O, and F. Fluorine elements are more likely to migrate and enrich at the film/air surface following high-temperature heat treatment because their signal intensity is higher before grinding than after. Since the surface of the film is destroyed during grinding, the fluorine content decreases. Furthermore, comparing Figure 8, it can be shown that the enrichment of fluorine on the surface of the FPUA film is lower when the PA/PU ratio is 30/70 than when the PA/PU ratio is 50/50. Table S5 illustrates this quantitative growth more clearly.

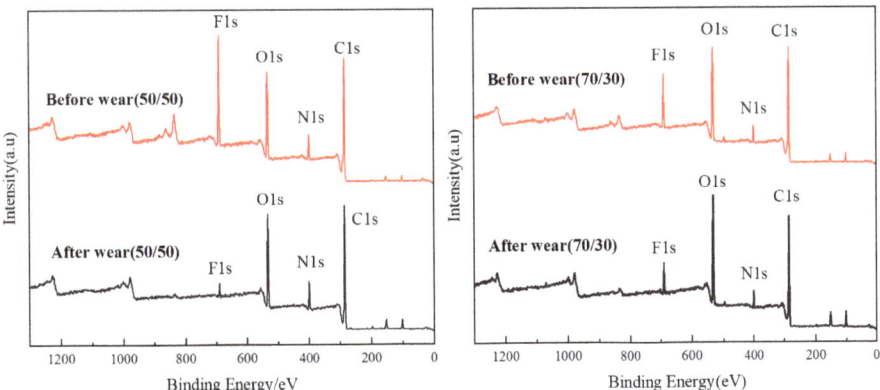

Figure 8. XPS wide spectra of the FPUA film.

The surface tension of the film is directly correlated with the magnitude of the contact angle. Typically, the lesser the contact angle and the lower the surface tension, the coarser the surface of film. Figure 9 and Table S6 demonstrate that, when specimens are moistened with identical solvent, the surface contact angle escalates in relation to fluoride content of samples. Owing to the strong electronegativity and diminished polarity of fluorine atoms, the structure of the backbone is cloaked in fluoro-bearing groups, which shield the interior hydrophilic groups of polymer molecules, thereby endowing the surface of the FPUA film with superior hydrophobic and oleophobic properties. The penetration capacity of various solvents into the surface of identical specimens also varies. Different solvents have different capacities for penetrating the surfaces of similar specimens. Hexadecane efficiently impregnates the sample's surface more heavily than diiodomethane and water because of hexadecane's lower polarity than water and its ability to deposit on the specimen surface at a moderate contact angle.

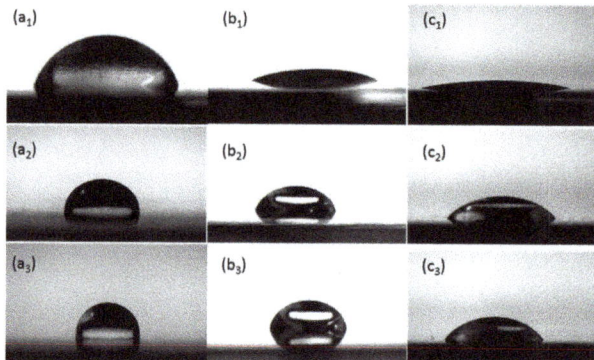

Figure 9. Images of droplets and the contact angles of water (a_1–a_3), diiodomethane (b_1–b_3), and hexadecane (c_1–c_3) on the coating.

Figure S1 shows the SEM of the coated film with a homogeneous surface without significant phase separation. Figure 10 shows the top view of the AFM plane of the coated film surface and its three-dimensional (3D) morphology image. From the figure, it can be observed that the phase region of the CWPU adhesive film in Figure 10A,a is small, relatively uniform, fluctuates within the range of 3 nm, and the surface flatness is high. On the other hand, the surface of fluorine-modified polyurethane is relatively high in roughness because of the incompatibility of the soft and hard segments, which will cause the phenomenon of micro-phase separation. Figure 10B,b,C,c shows the height map and 3D stereogram of FPUA gel film when the addition of FPA monomer is 30 wt% and 50 wt%, respectively. At 30 wt% FPA, the surface of the film is relatively flat and the roughness fluctuates around 7 nm, while at 50 wt% FPA, the surface of the film is the roughest and the fluctuation range is around 20 nm. This indicates that the fluorine content increases with the increase in the core-shell ratio and that the fluorine chain segments have the tendency to migrate to the air-adhesive film interface spontaneously during high-temperature curing, at this time, due to the low surface energy of fluorine; to reduce the surface area of the fluorine, it will form the bumps, which leads to the increase in surface roughness.

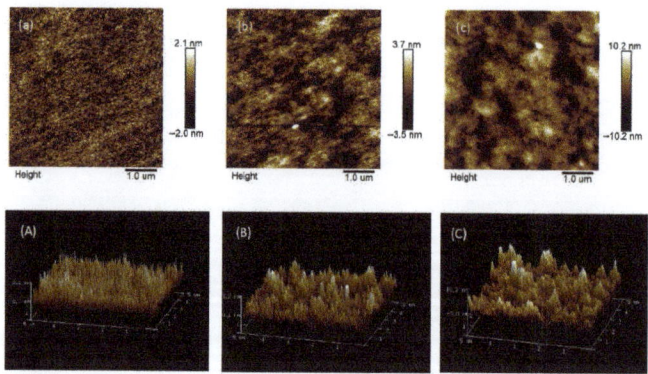

Figure 10. The AFM of CWPU/FPUA films. (**a**–**c**) CWPU/30 wt%/50 wt% FPUA; (**A**–**C**) CWPU/ 30 wt%/50 wt% FPUA.

The TGA curves of FPUA-coated films in nitrogen and their thermogravimetric analyses (DTG) are shown in Figure 11, respectively. The residual carbon content (Y_c) and the temperature of maximum loss rate (T_{max}) of the samples after 600 °C are shown in Table S7. The thermal decomposition temperatures of the polymers do not differ significantly at the initial stage and at 20% weight loss, the decomposition temperatures of the two are comparable, even the decomposition temperature of 50/50FPUA is slightly lower than that of 30/70FPUA, which may be attributed to the poor thermal stability of the MMA chain segments. The subsequent decomposition process can be divided into two main stages, corresponding to the thermal decomposition of the soft and hard phases of the polymer. The decomposition temperature of the hard phase of 50/50FPUA is 254 °C (the C–N bonding energy of the hard phase is low) and the decomposition end temperature is 393 °C, with the maximum decomposition rate occurring at 320.1 °C, while the initial decomposition temperature of the soft phase is 410 °C and the decomposition end temperature is 475 °C. The decomposition stages of the soft and hard phases of 30/70 are not much different from those of 50/50FPUA but the maximum decomposition rate is greater than that of 30/70FPUA and the maximum decomposition rate is higher than that of 50/70FPUA, which is more obvious. The decomposition stage of the 30/70 soft and hard phases is similar to that of the 30/70 phase but the maximum decomposition rate is greater than that of the 50/50 FPUA, which shows that the heat resistance of FPUA increases with the increase in the proportion of FPA. This is because the addition of fluorinated monomers further improves the heat resistance of the coating film due to the larger bond energy and higher stability of the C–F bond. The higher the DFMA content in the shell layer in the core-shell structure, the higher the C–F bond binding energy, the more stable the structure, and the better the heat resistance.

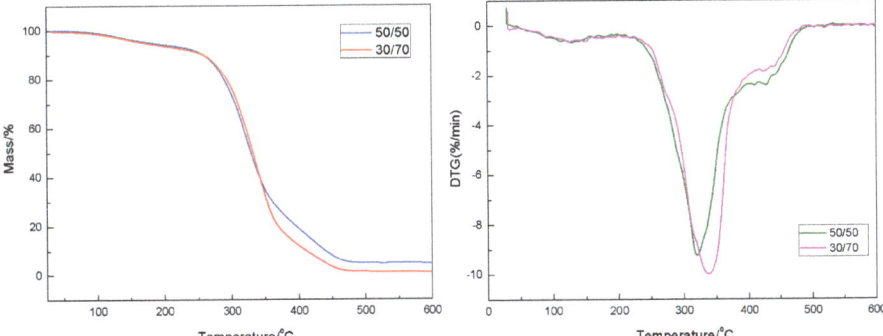

Figure 11. TGA curves and their corresponding DTG curves of FPUA films.

The bulk of the uncoated tinplates have corroded to discoloration because of a chemical reaction between the solution on their surface and the iron sheet substrate, as seen in Figure 12. The tinplate surface coated with FPUA emulsion, however, still maintains high surface integrity and very little corrosion after 15 h of acid–alkali salt corrosion. This demonstrates that the modified FPUA film has good corrosion resistance and can withstand the interference of a harsh external environment.

The transmittance curves for different light wavelengths are shown in Figure 13 after the glass plate has been coated. As the wavelength of the incoming light increases, the transmittance of the film shows a trend of slight rise at first, followed by a reduction, as seen in Figure 13a. Under the two ratios, the transmittance of film in the visible spectrum (380–780 nm) is greater than 93%. The quantity of light passing through the slide film displaying the clear and transparent text before and after coating was found to be negligible to nonexistent in natural light, as shown in Figure 13b. This reveals that the system is extremely stable and that the particle size of the emulsion is consistent. It also shows that the produced FPUA emulsion has a high degree of transparency and a smooth granular-free film. It is crucial for electrical and film goods that have to adhere to stringent transparency standards.

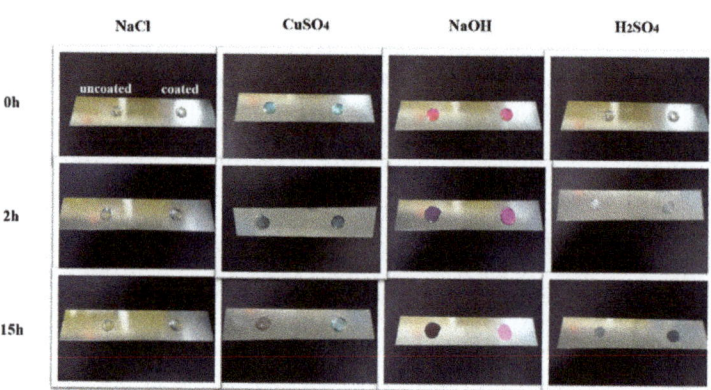

Figure 12. Corrosion resistance of FPUA films.

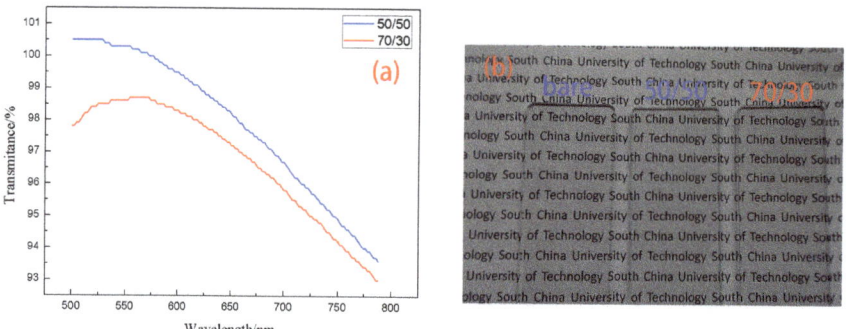

Figure 13. Transmittance test of FPUA film. (a) Transmittance Curve; (b) Corresponding coatings are tested for transparency.

4. Conclusions

In summary, a core-shell fluorinated polyurethane-acrylate composite emulsion was prepared by using synthetic cationic waterborne polyurethane as the seed emulsion and pre-emulsions were made by solubilizing with acrylic monomers, such as DFMA, MMA, and DMMA, to prepare a core-shell fluorinated polyurethane-acrylate composite emulsion; then, ADH was introduced as the post-crosslinking agent and keto hydrazine cross-linking was used to make the polymer form a chemically crosslinked film between the core-shell and the core-shell. The experiments mainly investigated the effects of fluorinated acrylate modification before and after the modification as well as the effects of two different core-shell ratios, 30/70 and 50/50, on the properties of FPUA emulsions and adhesive films. The results showed that the emulsion storage stability was better when the addition of fluorinated acrylic acid (PA) was in the range of 30–50 wt%. It was observed by TEM that the polymer particles showed a clearly visible core-shell structure, with the hydrophilic PU chain segments as the shell and the hydrophobic PA chain segments as the core. After high temperature heat treatment, the fluorine content on the surface of FPUA film increases with the increase in the core-shell ratio, up to 14.75%, which is also verified by the AFM test of the film. The contact angle test on the surface of the film shows that when w(PA) is 50%, the surface contact angle reaches 98.5°, which is hydrophobic to a certain extent. The tensile strength of the films was found to increase and then decrease with the increase in the core-shell ratio and the maximum was 23.35 ± 0.08 MPa at a core-shell ratio of 30/70; however, with the increase in the core-shell ratio to 50/50, the number of fluorine-containing chain segments increased and the density of the polar groups became larger and larger, which impeded the free movement of the chain segments and thus made the FPUA films hard and brittle. With the increase in the core-shell ratio, the heat resistance of FPUA films gradually increased. The fluorinated acrylate-modified FPUA film has excellent adhesion, impact resistance, hardness, flexibility, and chemical resistance. The light transmittance of the prepared films is above 93%, which has high application value. In the future, we will conduct in-depth research on the process of fluorinated acrylate-modified cationic waterborne polyurethanes, including the effective promotion of the compatibility of the two phases of PU and PA, the enhancement of the conversion rate of fluorinated chain segments, and the reduction in process costs.

Supplementary Materials: The following supporting information can be downloaded at: https://www.mdpi.com/article/10.3390/polym16010086/s1, Figure S1: SEM images of the films. (a) Monocrystalline silicon coated sheet; (b) CWPU; (c) 30/70FPUA; (d) 50/50 FPUA; Table S1: Typical dosage for cationic polyurethane aqueous dispersion (CWPU); Table S2: Tensile strength, Elongation at break and Young's modulus of FPUA; Table S3: The adhesion, impact resistance and flexibility of FPUA; Table S4: The T_g, modulus and crosslink density of FPUA; Table S5: Atomic percentage of FPUA film surfaces before and after wearing test; Table S6: The contact angle of water, diiodomethane and hexadecane droplets on the films; Table S7: Thermal stability of FPUA films.

Author Contributions: Conceptualization, J.C. (Junhua Chen); methodology, J.C. (Junhua Chen) and Y.W.; software, H.Z.; validation, H.Z.; formal analysis, X.L. and J.C. (Jinlian Chen); investigation, X.L. and J.C. (Jinlian Chen); resources, D.Z.; data curation, S.L.; writing—original draft preparation, J.C. (Junhua Chen); writing—review and editing, X.H.; visualization, J.C. (Junhua Chen) and Y.W.; supervision, X.H.; project administration, X.H.; funding acquisition, D.Z. All authors have read and agreed to the published version of the manuscript.

Funding: This study was financially supported by the Youth Enhancement Project of the Natural Science Foundation of Guangdong Province (Dongyu Zhu, 202341515030135).

Institutional Review Board Statement: Not applicable.

Data Availability Statement: The data are contained within the article.

Conflicts of Interest: The authors declare no conflicts of interest.

References

1. Sharmin, E.; Zafar, F.; Akram, D.; Alam, M.; Ahmad, S. Recent advances in vegetable oils-based environment friendly coatings: A review. *Ind. Crops Prod.* **2015**, *76*, 215–229. [CrossRef]
2. Hormaiztegui, M.E.V.; Mucci, V.L.; Santamaria-Echart, A.; Corcuera, M.Á.; Eceiza, A.; Aranguren, M.I. Waterborne polyurethane nanocomposites based on vegetable oil and microfibrillated cellulose. *J. Appl. Polym. Sci.* **2016**, *133*, 44207. [CrossRef]
3. Luo, Q.; Wen, X.; Xu, R.; Liu, Z.; Xiang, H.; Li, Z.; Liu, X. Preparation and Properties of Novel Modified Waterborne Polyurethane Acrylate. *Coatings* **2022**, *12*, 1135. [CrossRef]
4. Gong, S.; Xiang, S.; Wang, T.; Cai, D. Towards Solar-Driven Formation of Robust and Self-Healable Waterborne Polyurethane Containing Disulfide Bonds via in-situ Incorporation of 2D Titanium Carbide MXene. *J. Renew. Mater.* **2023**, *11*, 1063–1076. [CrossRef]
5. Lei, W.; Zhou, X.; Fang, C.; Song, Y.; Li, Y. Eco-friendly waterborne polyurethane reinforced with cellulose nanocrystal from office waste paper by two different methods. *Carbohydr. Polym.* **2019**, *209*, 299–309. [CrossRef] [PubMed]
6. De Smet, D.; Wéry, M.; Uyttendaele, W.; Vanneste, M. Bio-Based Waterborne PU for Durable Textile Coatings. *Polymers* **2021**, *13*, 4229. [CrossRef] [PubMed]
7. Li, Y.; Chen, S.; Shen, J.; Zhang, S.; Liu, M.; Lv, R.; Xu, W. Preparation and Properties of Biobased, Cationic, Waterborne Polyurethanes Dispersions from Castor Oil and Poly (Caprolactone) Diol. *Appl. Sci.* **2021**, *11*, 4784. [CrossRef]
8. Deng, Y.; Zhang, C.; Zhang, T.; Wu, B.; Zhang, Y.; Wu, J. Study of a Novel Fluorine-Containing Polyether Waterborne Polyurethane with POSS as a Cross-Linking Agent. *Polymers* **2023**, *15*, 1936. [CrossRef]
9. Phunphoem, S.; Saravari, O.; Supaphol, P. Synthesis of cationic waterborne polyurethanes from waste frying oil as antibacterial film coatings. *Int. J. Polym. Sci.* **2019**, *2019*, 2903158. [CrossRef]

10. Wang, R.; Zhang, Z.; Bai, X.; Xu, Z.; Zheng, J.; Pan, F.; Yuan, C. Preparation and Properties of UV and Aziridine Dual–Cured Polyurethane Acrylate Emulsion. *Coatings* **2022**, *12*, 1293. [CrossRef]
11. Fang, C.H.; Liu, P.I.; Chung, L.C.; Shao, H.; Ho, C.H.; Chen, R.S.; Fan, H.T.; Liang, T.M.; Chang, M.C.; Horng, R.Y. A flexible and hydrophobic polyurethane elastomer used as binder for the activated carbon electrode in capacitive deionization. *Desalination* **2016**, *399*, 34–39. [CrossRef]
12. Marcuello, C.; Foulon, L.; Chabbert, B.; Molinari, M.; Aguié-Béghin, V. Langmuir–Blodgett Procedure to Precisely Con-trol the Coverage of Functionalized AFM Cantilevers for SMFS Measurements: Application with Cellulose Nanocrystals. *Langmuir* **2018**, *34*, 9376–9386. [CrossRef]
13. Tardio, S.; Abel, M.L.; Carr, R.H.; Watts, J.F. The interfacial interaction between isocyanate and stainless steel. *Int. J. Adhes. Adhes.* **2019**, *88*, 1–10. [CrossRef]
14. Zhou, J.; Zhu, C.; Liang, H.; Wang, Z.; Wang, H. Preparation of UV-curable low surface energy polyurethane acrylate/fluorinated siloxane resin hybrid coating with enhanced surface and abrasion resistance properties. *Materials* **2020**, *13*, 1388. [CrossRef] [PubMed]
15. Boschet, F.; Ameduri, B. (Co) polymers of chlorotrifluoroethylene: Synthesis, properties, and applications. *Chem. Rev.* **2014**, *114*, 927–980. [CrossRef]
16. Vitale, A.; Bongiovanni, R.; Ameduri, B. Fluorinated oligomers and polymers in photopolymerization. *Chem. Rev.* **2015**, *115*, 8835–8866. [CrossRef] [PubMed]
17. Shin, M.S.; Lee, Y.H.; Rahman, M.M.; Kim, H. Synthesis and properties of waterborne fluorinated polyurethane-acrylate using a solvent-/emulsifier-free method. *Polymer* **2013**, *54*, 4873–4882. [CrossRef]
18. Xia, W.; Zhu, N.; Hou, R.; Zhong, W.; Chen, M. Preparation and characterization of fluorinated hydrophobic UV-crosslinkable thiol-ene polyurethane coatings. *Coatings* **2017**, *7*, 117. [CrossRef]
19. Wang, S.; Liu, W.; Tan, J. Synthesis and properties of fluorine-containing polyurethane based on long chain fluorinated polyacrylate. *J. Macromol. Sci. Part A Pure Appl.Chem.* **2016**, *53*, 41–48. [CrossRef]
20. Huang, H.; Fang, S.; Luo, S.; Hu, J.; Yin, S.; Wei, J.; Yu, Q. Multiscale modification on acrylic resin coating for concrete with silicon/fluorine and graphene oxide (GO) nanosheets. *Constr. Build. Mater.* **2021**, *305*, 124297. [CrossRef]
21. Ghazali, S.K.; Adrus, N.; Majid, R.A.; Ali, F.; Jamaluddin, J. UV-LED as a new emerging tool for curable polyurethane acrylate hydrophobic coating. *Polymers* **2021**, *13*, 487. [CrossRef] [PubMed]
22. Zhu, C.; Yang, H.; Liang, H.; Wang, Z.; Dong, J.; Xiong, L.; Zhou, J.; Ke, J.; Xu, X.; Xi, W. A novel synthetic UV-curable fluorinated siloxane resin for low surface energy coating. *Polymers* **2018**, *10*, 979. [CrossRef] [PubMed]
23. Chen, L.J.; Wu, F.Q. Structure and properties of novel fluorinated polyacrylate latex prepared with reactive surfactant. *Polym. Sci. Ser. B.* **2011**, *53*, 606–611. [CrossRef]
24. Xu, W.; Wang, S.; Hao, L.; Wang, X. Preparation and characterization of trilayer core–shell polysilsesquioxane–fluoroacrylate copolymer composite emulsion particles. *J. Appl. Polym. Sci.* **2017**, *134*, 44845. [CrossRef]
25. Sahu, A.; Kumar, D. Core-shell quantum dots: A review on classification, materials, application, and theoretical modeling. *J. Alloys Compd.* **2022**, *942*, 166508. [CrossRef]
26. Duan, S.; Wu, R.; Xiong, Y.H.; Ren, H.M.; Lei, C.; Zhao, Y.Q.; Zhang, X.Y.; Xu, F.J. Multifunctional antimicrobial materials: From rational design to biomedical applications. *Prog. Mater. Sci.* **2022**, *125*, 100887. [CrossRef]
27. Cheaburu-Yilmaz, C.N.; Ozkan, C.K.; Yilmaz, O. Synthesis and Application of Reactive Acrylic Latexes: Effect of Particle Morphology. *Polymers* **2022**, *14*, 2187. [CrossRef]
28. Guo, X.; Ge, S.; Wang, J.; Zhang, X.; Zhang, T.; Lin, J.; Zhao, C.X.; Wang, B.; Zhu, G.; Guo, Z. Waterborne acrylic resin modified with glycidyl methacrylate (GMA): Formula optimization and property analysis. *Polymer* **2018**, *143*, 155–163. [CrossRef]
29. Zeng, Y.; Yang, W.; Xu, P.; Cai, X.; Dong, W.; Chen, M.; Du, M.; Liu, T.; Lemstra, P.J.; Ma, P. The bonding strength, water resistance and flame retardancy of soy protein-based adhesive by incorporating tailor-made core–shell nanohybrid compounds. *Chem. Eng. J.* **2022**, *428*, 132390. [CrossRef]
30. Yang, J.; Mu, Y.; Li, X. Morphology of multilayer core-shell emulsion: Influence of crosslinking agent. *Mater. Lett.* **2022**, *322*, 132493. [CrossRef]
31. Lee, S.W.; Lee, Y.H.; Park, H.; Kim, H.D. Effect of total acrylic/fluorinated acrylic monomer contents on the properties of waterborne polyurethane/acrylic hybrid emulsions. *Macromol. Res.* **2013**, *21*, 709–718. [CrossRef]
32. Cheaburu-Yilmaz, C.N.; Yilmaz, O.; Darie-Nita, R.N. The effect of different soft core/hard shell ratios on the coating performance of acrylic copolymer latexes. *Polymers* **2021**, *13*, 3521. [CrossRef] [PubMed]
33. Zhong, X.; Lin, J.; Wang, Z.; Xiao, C.; Yang, H.; Wang, J.; Wu, X. Preparation of a crosslinked coating containing fluorinated water-dispersible polyurethane particles. *Prog. Org. Coat.* **2016**, *99*, 216–222. [CrossRef]
34. Bai, S.; Zheng, W.; Yang, G.; Fu, F.; Liu, Y.; Xu, P.; Lv, Y.; Liu, M. Synthesis of core-shell fluorinated acrylate copolymers and its application as finishing agent for textile. *Fibers Polym.* **2017**, *18*, 1848–1857. [CrossRef]
35. Lü, T.; Qi, D.; Zhang, D.; Liu, Q.; Zhao, H. Fabrication of self-cross-linking fluorinated polyacrylate latex particles with core-shell structure and film properties. *React. Funct. Polym.* **2016**, *104*, 9–14. [CrossRef]
36. Hirose, M.; Kadowaki, F.; Zhou, J. The structure and properties of core shell type acrylic-polyurethane hybrid aqueous emulsions. *Prog. Org. Coat.* **1997**, *31*, 157–169. [CrossRef]

7. Park, Y.G.; Lee, Y.H.; Rahman, M.M.; Park, C.C.; Kim, H.D. Preparation and properties of waterborne polyurethane/self-cross-linkable fluorinated acrylic copolymer hybrid emulsions using a solvent/emulsifier-free method. *Colloid Polym. Sci.* **2015**, *293*, 1369–1382. [CrossRef]
8. Chakrabarty, A.; Singha, N.K. Tailor-made polyfluoroacrylate and its block copolymer by RAFT polymerization in miniemulsion; improved hydrophobicity in the core–shell block copolymer. *J. Colloid Interface Sci.* **2013**, *408*, 66–74. [CrossRef]
9. Wang, Y.; Qiu, F.; Xu, B.; Xu, J.; Jiang, Y.; Yang, D.; Li, P. Preparation, mechanical properties and surface morphologies of waterborne fluorinated polyurethane-acrylate. *Prog. Org. Coat.* **2013**, *76*, 876–883. [CrossRef]
10. Jiang, G.; Tuo, X.; Wang, D.; Li, Q. Preparation, characterization, and properties of fluorinated polyurethanes. *J. Polym. Sci. Part A Polym. Chem.* **2009**, *47*, 3248–3256. [CrossRef]
11. Yu, Y.; Wang, J.; Zong, J.; Zhang, S.; Deng, Q.; Liu, S. Synthesis of a fluoro-diol and preparation of fluorinated waterborne polyurethanes with high elongation at break. *J. Macromol. Sci Part A Pure Appl.Chem.* **2018**, *55*, 183–191. [CrossRef]
12. Liu, Y.; Zhao, Y.J.; Teng, J.L.; Wang, J.H.; Wu, L.S.; Zheng, Y.L. Research progress of nano self-cleaning anti-fouling coatings. *IOP Conf. Ser. Mater. Sci. Eng.* **2018**, *284*, 012016. [CrossRef]

Disclaimer/Publisher's Note: The statements, opinions and data contained in all publications are solely those of the individual author(s) and contributor(s) and not of MDPI and/or the editor(s). MDPI and/or the editor(s) disclaim responsibility for any injury to people or property resulting from any ideas, methods, instructions or products referred to in the content.

Article

Investigation of Novel Solid Dielectric Material for Transformer Windings

Aysel Ersoy [1,*], Fatih Atalar [1] and Alper Aydoğan [2]

[1] Department of Electrical and Electronic Engineering, Istanbul University-Cerrahpasa, Bağlariçi St. No. 7, Avcılar, 34320 Istanbul, Turkey; fatih.atalar@iuc.edu.tr
[2] Computing & Information Services Office, Muğla Sıtkı Koçman University, Kötekli Campus, 48000 Muğla, Turkey; alperaydogan@mu.edu.tr
* Correspondence: aersoy@iuc.edu.tr

Citation: Ersoy, A.; Atalar, F.; Aydoğan, A. Investigation of Novel Solid Dielectric Material for Transformer Windings. *Polymers* 2023, 15, 4671. https://doi.org/10.3390/polym15244671

Academic Editors: Zhuohong Yang and Chang-An Xu

Received: 22 November 2023
Revised: 8 December 2023
Accepted: 10 December 2023
Published: 11 December 2023

Copyright: © 2023 by the authors. Licensee MDPI, Basel, Switzerland. This article is an open access article distributed under the terms and conditions of the Creative Commons Attribution (CC BY) license (https://creativecommons.org/licenses/by/4.0/).

Abstract: Improvement techniques aimed at enhancing the dielectric strength and minimizing the dielectric loss of insulation materials have piqued the interest of many researchers. It is worth noting that the electrical breakdown traits of insulation material are determined by their electrochemical and mechanical performance. Possible good mechanical, electrical, and chemical properties of new materials are considered during the generation process. Thermoplastic polyurethane (TPU) is often used as a high-voltage insulator due to its favorable mechanical properties, high insulation resistance, lightweight qualities, recovery, large actuation strain, and cost-effectiveness. The elastomer structure of thermoplastic polyurethane (TPU) enables its application in a broad range of high-voltage (HV) insulation systems. This study aims to evaluate the feasibility of using TPU on transformer windings as a solid insulator instead of pressboards. The investigation conducted through experiments sheds light on the potential of TPU in expanding the range of insulating materials for HV transformers. Transformers play a crucial role in HV systems, hence the selection of suitable materials like cellulose and polyurethane is of utmost importance. This study involved the preparation of an experimental setup in the laboratory. Breakdown tests were conducted by generating a non-uniform electric field using a needle–plane electrode configuration in a test chamber filled with mineral oil. Various voltages ranging from 14.4 kV to 25.2 kV were applied to induce electric field stress with a step rise of 3.6 kV. The partial discharges and peak numbers were measured based on the predetermined threshold values. The study investigated and compared the behaviors of two solid insulating materials under differing non-electric field stress conditions. Harmonic component analysis was utilized to observe the differences between the two materials. Notably, at 21.6 kV and 25.2 kV, polyurethane demonstrated superior performance compared to pressboard with regards to the threshold value of leakage current.

Keywords: pressboard; polyurethane; harmonics; transformers mineral oil; breakdown

1. Introduction

The maintenance of dielectric strengths for solid insulators used in transformers is crucial in ensuring the continuity of energy supply. Solid insulators operate in the same environment as mineral oil, a petroleum-based liquid dielectric, but may experience a decrease in insulation performance over time [1,2]. Despite periodic maintenance being of great importance, the deterioration of solid insulators is considered an irreversible condition [3]. Hence, ensuring healthy operation of the transformer mainly relies on appropriate solid insulator selection and testing. Pressboard materials based on cellulose are typically used as solid insulators [4,5] in oil-type transformers [6]. Nevertheless, transformer stability can be achieved by utilizing insulators with high dielectric strength like polyurethane. Therefore, it is advisable to consider conducting appropriate tests on polyurethane materials to enhance insulation resistance.

In their studies, Aruna et al. investigated several options for paper insulators. They compared the electrical properties of cellulose film with those of bacterial cellulose and also examined their advantages, disadvantages, and nuclear magnetic resonances. Moreover, they found that electrical loading, operating temperatures, and water level in oil have a significant impact on the deterioration of cellulose insulation [7]. The study analyzed the heat behavior of different papers in oil-impregnated paper insulators under heat-induced aging. Morooka et al. determined that kraft paper's flexibility decreases over time and its modulus of elasticity reduces, which jeopardizes the insulation system [8].

The insulation characteristic of a transformer can be evaluated using impulse, direct current (DC) or alternative current (AC) voltages. Nedjar carried out a reliability analysis of paper insulators using both AC and DC voltages by applying step voltage. The results showed that discharges occur when the electric field applied to the pressboards exceeds the critical value, and these discharges occur between oil and pressboard. A conductive path is formed in the process, which has been observed to degrade the insulation system. It has been reported that the resistance of oil-impregnated pressboards to DC voltage is superior to that of AC voltage, according to a reliability analysis based on step voltage practices [9]. The modelling of lightning strikes in transformers was undertaken by Florkowski and et al. They discovered that the strength of the windings is influenced by the waveform of the applied voltage and the standard voltage pulse used. It has been established that overvoltage mainly impacts the insulation system windings and ground electrode. According to reports, overvoltage has reached its maximum value. The insulator has been found to be adversely affected by this situation at standard impulse voltage. However, there were no observed effects on the insulator interface at discrete impulse voltages [10].

The investigation into the exterior design of transformers is ongoing. Chaw et al. conducted a comparison between the rectangular and circular winding distribution transformers at the 1000 kVA level. Consequently, models were created for 1000 kVA 50 Hz 11/0.4 kV three-phase, two-winding, triangular star-connected core distribution transformers. It is predicted that efficiency will improve as transformer losses decrease. The magnetic voltage is anticipated to differ between the proposed transformers and the target design, as the losses in the transformers are reduced [11].

It is crucial to consider the variability of experimental methods when assessing the condition of a transformer, along with the significance of the recommended oils and insulation materials used in transformer insulation. The condition of the oil plays a critical role, as demonstrated in research conducted by Malik Yadav, Mishra, and Mehto, investigating transformer breakdown and maintenance scenarios. The study compared traditional methods against the neuro-fuzzy technique [12].

Pradhan and Yew conducted a frequency field analysis of data obtained from their experimental study and evaluated it using harmonic values [13]. Sun et al. examined the efficacy of fiber coating materials in power transformers and demonstrated the superiority of ethylene tetrafluoroethylene (ETFE), polytetrafluoroethylene (PTFE), and polyamide 12 (PA 12) over greaseproof paper [14]. Substitutions in electromagnetic field distribution, particularly in windings, are currently intensively studied in high voltage transformers and distribution transformers. Sun et al.'s modelling study indicated that the increase in pressboard thickness from 5 mm to 10 mm could result in a change of up to 30% in electric field decrease [15].

This study compares the effectiveness of pressboard, a cellulose derivative, and polyurethane, a common material in the electrical industry, as a transformer insulator. We measured the peaks of leakage currents flowing from the earth electrode after applying voltage to the prepared insulating plates, and we calculated the harmonics as well. In our study, we have experimentally researched and provided data on the breakdown behavior when using polyurethane and polymers as alternatives to paper insulators. To accurately analyze surface catastrophic events, we used scanning electron microscopy (SEM) images of each case and identified the superior cases of TPU.

2. Measurement Methodology

The study measured leakage currents on the earth electrode using the Fluke ScopeMeter 199C series oscilloscope, which belongs to the Fluke brand and is rated as CAT IV Class. These oscilloscopes have a sampling rate of 2.5 GS/s (giga samples/second) and a bandwidth of 200 MHz. Measurements were recorded graphically and evaluated for a 6 s duration, with a total of 30,000 data points. The FlukeView® SW160 software, with a multi-function counter interface, was purchased and used to process the data in MATLAB 2022b software.

Various assessment techniques and norms are available to evaluate the partition between solid and liquid media within transformers as well as to comprehend the partial discharge that may arise. Specific tests have to be executed, predominantly in the case of cellulose-based pressboards. A test mechanism was established in this investigation, making use of pertinent standards as a basis [16,17]. The pressboard is of 100 × 100 mm dimension and it has fiber density of 1.2 g/cm^3, a relative dielectric constant of $\varepsilon_r = 4.1$ and the loss factor is 0.0057. The relative dielectric constant of mineral oil is $\varepsilon_r = 2.2$. Also, the PU has same dimension of pressboard. Its relative dielectric constant of $\varepsilon_r = 12.25$ and the loss factor is 0.0021. The investigation involved the initial recording of leakage current signals. Disturbance points in the leakage current signal waveform are attributed to partial discharge. The study analyzed and interpreted the peak values at these points. The testing equipment and measurement system used in this study are identical to those employed in our earlier research [18]. The main objective of the present study is to compare two dissimilar types of solid insulating materials. The electrode system setup, used for testing the proposed PU and pressboard, is displayed in Figure 1. The thicknesses of the dielectric materials examined in the investigation are 1 mm, 2 mm, and 3 mm. In addition, the mineral oil level remained unvaried throughout the research. The voltage levels were varied, though. Four voltage levels ranging from 14.4 kV to 25.2 kV were applied to the insulators. Accordingly, the leakage current threshold values during partial discharge were determined. The peaks of the measured values that exceeded this threshold value for the applied voltage level indicate the commencement of partial discharge and result in signal deterioration [19–22].

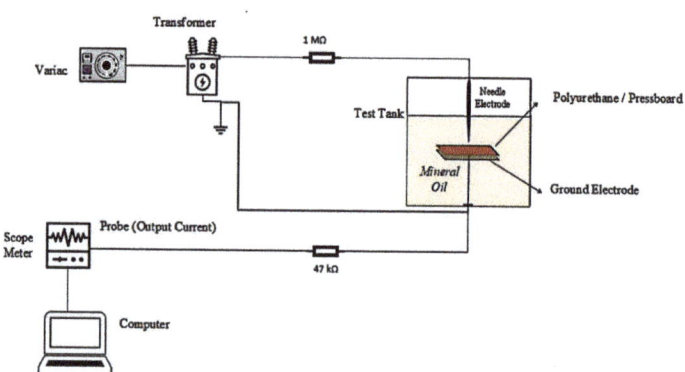

Figure 1. Electrode system for surface electric discharges, with solid dielectric made of pressboard and polyurethane.

3. Results and Discussion

3.1. Leakage Current and Harmonic Analysis

The number of current peaks and values were calculated at all voltage levels for the pressboard of 3 different thicknesses. Values for these data are given in Tables 1–3, respectively, for 1 mm, 2 mm, and 3 mm thicknesses.

Table 1. Pressboard and Polyurethane Threshold and Peak Values (1 mm).

Material	Voltage (kV)	Peak Value Positive (A)	Peak Count Positive	Threshold Value (A)
Pressboard	14.4	0.44	55	0.42
Polyurethane	14.4	0.46	36	0.44
Pressboard	18	0.54	8	0.52
Polyurethane	18	0.56	8	0.54
Pressboard	21.6	0.68	0	0.68
Polyurethane	21.6	1.04	29	0.68
Pressboard	25.2	2.04	16	0.96
Polyurethane	25.2	1.44	70	1.2

Table 2. Pressboard and Polyurethane Threshold and Peak Values (2 mm).

Material	Voltage (kV)	Peak Value Positive (A)	Peak Count Positive	Threshold Value (A)
Pressboard	14.4	0.58	14	0.56
Polyurethane	14.4	0.64	15	0.62
Pressboard	18	0.54	0	0.54
Polyurethane	18	0.54	8	0.52
Pressboard	21.6	0.86	31	0.84
Polyurethane	21.6	1.08	14	0.96
Pressboard	25.2	1.76	40	0.96
Polyurethane	25.2	1.52	92	1.2

Table 3. Pressboard and Polyurethane Threshold and Peak Values (3 mm).

Material	Voltage (kV)	Peak Value Positive (A)	Peak Count Positive	Threshold Value (A)
Pressboard	14.4	0.44	7	0.42
Polyurethane	14.4	0.44	3	0.42
Pressboard	18	0.58	89	0.54
Polyurethane	18	0.54	76	0.52
Pressboard	21.6	0.72	25	0.68
Polyurethane	21.6	0.68	0	0.68
Pressboard	25.2	0.96	2	0.88
Polyurethane	25.2	0.88	0	0.88

Threshold values of flowing currents are selected with similar values for each solid insulator. As can be seen in Table 1, pressboard has more peak points even though its threshold is slightly less than polyurethane at the applied lowest voltage level. When it comes to the situation at the highest voltage level, polyurethane shows worse insulation performance. It can be concluded from this point that pressboard can be selected for higher voltage levels at the smallest electrode gap. However, polyurethane has more strength when the voltage level is 14.4 kV. So, new investigations are needed to clarify interpretation with new electrode gaps. Table 2 consists of the data of two solid insulators at the thickness of 2 mm.

As shown in Table 2, there is no significant difference between two solid insulators at the voltage level of 14.4 kV and 18 kV. PU has higher peak current value at 21.6 kV. At this voltage level, PU is more advantageous to select as an insulator. However, at the highest voltage level, even though pressboard has higher peak current value, its peak points are less than PU. So, we have to select a new electrode gap to make more reasonable comparison for selecting between PU and pressboard. Table 3 shows the same manner of data of PU and pressboard at the thickness of 3 mm.

As can be seen from Table 3, there is no significant difference between the two insulators at the 14.4 kV voltage level. At this level, only at the point where the current signal is disturbed, the peak number of signs with partial discharge is higher in the pressboard. This makes the polyurethane material a step above the pressboard. When the 18 kV and 21.6 kV voltage levels are examined, it is seen that the pressboard leakage current is 40 mA more than that of the polyurethane. There are also serious differences between the discharge peak points for these levels. At 25.2 kV voltage level, there is 80 mA difference between the leakage currents measured from PU and pressboard.

The peak values and number of these points of leakage current are not enough to make a more comprehensive evaluation in terms of the dielectric strength behavior of PU and pressboard. So, in this study, harmonic current values between 50–1000 Hz frequency band are calculated for every special situation. Another purpose of harmonic current performance is to determine the dominant harmonic component of current signal. With the help of this method, there is an opportunity to make a clearer evaluation in dielectric behavior differences. The harmonic current values of PU and pressboard at the 1 mm electrode gap are summarized in Table 4.

Table 4. Harmonic values of leakage current at 1 mm material thickness.

Frequency (Hz)	Harmonic Values (A)							
	Polyurethane				Pressboard			
	14.4 kV	18 kV	21.6 kV	25.2 kV	14.4 kV	18 kV	21.6 kV	25.2 kV
50	0.0193	0.0049	0.0484	0.1654	0.0074	0.0040	0.0468	0.0429
100	0.1920	0.2385	0.2874	0.4649	0.1877	0.2324	0.2843	0.4316
150	0.0009	0.0015	0.0013	0.0216	0.0007	0.0010	0.0011	0.0226
200	0.0028	0.0031	0.0046	0.0081	0.0024	0.0026	0.0042	0.0059
250	0.0008	0.0012	0.0021	0.0019	0.0007	0.0011	0.0022	0.0049
300	0.0136	0.0181	0.0221	0.0308	0.0133	0.0175	0.0213	0.0316
350	0.0010	0.0016	0.0014	0.0029	0.0010	0.0012	0.0019	0.0039
400	0.0099	0.0130	0.0125	0.0229	0.0101	0.0116	0.0142	0.0186
450	0.0007	0.0008	0.0019	0.0028	0.0010	0.0008	0.0015	0.0020
500	0.0073	0.0107	0.0131	0.0216	0.0091	0.0107	0.0151	0.0198
550	0.0015	0.0017	0.0014	0.0038	0.0016	0.0011	0.0016	0.0022
600	0.0058	0.0059	0.0085	0.0124	0.0051	0.0058	0.0080	0.0108
650	0.0004	0.0008	0.0016	0.0035	0.0003	0.0011	0.0012	0.0023
700	0.0056	0.0060	0.0062	0.0178	0.0058	0.0051	0.0088	0.0102
750	0.0012	0.0011	0.0009	0.0035	0.0010	0.0008	0.0015	0.0028
800	0.0056	0.0071	0.0089	0.0128	0.0048	0.0060	0.0060	0.0121
850	0.0005	0.0009	0.0007	0.0037	0.0009	0.0007	0.0012	0.0038
900	0.0033	0.0048	0.0071	0.0096	0.0033	0.0044	0.0045	0.0108
950	0.0007	0.0011	0.0020	0.0024	0.0009	0.0010	0.0014	0.0026
1000	0.0014	0.0006	0.0028	0.0033	0.0003	0.0011	0.0019	0.0083

There is an interesting common situation for both solid insulators in terms of dominant harmonic values. It can be clearly seen in Table 4 that 2nd and 6th harmonic values have higher values than the others at the all-voltage levels for PU and pressboard. These values were calculated very close to each other for both materials. Therefore, it can be said that the leakage currents for 1 mm thickness are similar for both insulators. In this case, it is necessary to examine the harmonic values for 2 mm thick materials, as given in Table 5.

From Table 5, it is seen that there is similar behavior in terms of dominant harmonic values with 1 mm material thickness. So, it can be said that the electric field creates similar stresses on the materials of both 1 mm and 2 mm thickness. However, at the voltage levels of 21.6 kV and 25.2 kV, the fundamental components (50 Hz) values of PU are higher than the counterpart of pressboard. It can be said that harmonics, except from the fundamental component, are seen more intensely on pressboard. So, when the 2 mm thickness material is used at higher voltage levels, the PU selection is more reasonable than pressboard. This situation is supported by higher peak current value of pressboard at the highest voltage level, as seen in the Table 2 data.

Table 5. Harmonic values of leakage current at 2 mm material thickness.

	Harmonic Values (A)							
	Polyurethane				Pressboard			
Frequency (Hz)	14.4 kV	18 kV	21.6 kV	25.2 kV	14.4 kV	18 kV	21.6 kV	25.2 kV
50	0.0392	0.0046	0.1137	0.1759	0.0151	0.0040	0.0329	0.0510
100	0.2707	0.2324	0.4049	0.4777	0.2548	0.2338	0.3791	0.4364
150	0.0038	0.0015	0.0125	0.0205	0.0043	0.0012	0.0182	0.0285
200	0.0034	0.0030	0.0069	0.0072	0.0015	0.0030	0.0057	0.0071
250	0.0018	0.0005	0.0034	0.0018	0.0017	0.0014	0.0048	0.0060
300	0.0188	0.0172	0.0262	0.0309	0.0198	0.0170	0.0286	0.0344
350	0.0026	0.0014	0.0042	0.0040	0.0035	0.0012	0.0046	0.0064
400	0.0123	0.0122	0.0193	0.0214	0.0112	0.0141	0.0160	0.0164
450	0.0016	0.0008	0.0018	0.0024	0.0005	0.0012	0.0035	0.0026
500	0.0125	0.0113	0.0168	0.0211	0.0106	0.0114	0.0149	0.0164
550	0.0032	0.0011	0.0048	0.0030	0.0021	0.0016	0.0028	0.0033
600	0.0062	0.0045	0.0099	0.0122	0.0059	0.0092	0.0057	0.0081
650	0.0019	0.0011	0.0018	0.0029	0.0003	0.0014	0.0011	0.0019
700	0.0085	0.0061	0.0095	0.0147	0.0066	0.0086	0.0084	0.0090
750	0.0024	0.0010	0.0024	0.0029	0.0008	0.0018	0.0014	0.0023
800	0.0071	0.0063	0.0108	0.0140	0.0068	0.0056	0.0089	0.0092
850	0.0015	0.0011	0.0032	0.0035	0.0019	0.0009	0.0024	0.0021
900	0.0064	0.0053	0.0107	0.0137	0.0068	0.0029	0.0113	0.0135
950	0.0012	0.0013	0.0017	0.0030	0.0008	0.0007	0.0021	0.0037
1000	0.0020	0.0003	0.0035	0.0034	0.0042	0.0007	0.0063	0.0051

Harmonic values at the 3 mm thickness are summarized in Table 6. At the voltage level of 14.4 kV, the 3rd and 7th harmonics were measured as 23 mA and 10.93 mA for pressboard. No double harmonics have been observed in the leakage current components. The number of peaks exceeding the peak value threshold is higher in the pressboard than in polyurethane. While examining harmonics for 18 kV voltage level, 3rd harmonics are measured as 3.1 mA for both pressboard and PU. The calculated 5th harmonic component is 17.9 mA for pressboard and 15.7 mA for polyurethane; pressboard's 5th harmonics are higher than polyurethane. The 13th harmonic value is similarly higher on the pressboard; this level is 9 mA for pressboard and 3.6 mA for polyurethane. The leakage current values for the 15th and 17th harmonics were measured as 5 mA and 2.8 mA, respectively, for the polyurethane, and 8.9 mA and 5.6 mA, respectively, for the pressboard. At these levels, the 15th and 17th harmonic components are higher than PU.

No comparable 3rd harmonic component was observed for the voltage level of 21.6 kV. The 5th harmonic value was measured as 21.8 mA on the pressboard and 21.4 mA on the polyurethane. This apparent difference offers the possibility of evaluation and comparison in terms of partial discharge.

The surface condition of insulators is shown in Figure 2 at the thickness of 3 mm. It can be clearly seen that the diameter of the surface carbonization pattern of pressboard is wider than the polyurethane. This situation can be a determinant of the dielectric strength of these solid insulators in the oil. PU has more durability against the formation of surface tracking formation. There is no comparable surface condition for 1 mm and 2 mm material thickness. It will be given a detailed surface morphology examination in the following SEM section.

Table 6. Harmonic values of leakage current at 3 mm material thickness.

Frequency (Hz)	Harmonic Values (A)							
	Polyurethane				Pressboard			
	14.4 kV	18 kV	21.6 kV	25.2 kV	14.4 kV	18 kV	21.6 kV	25.2 kV
50	0.1876	0.2322	0.281	0.3301	0.1871	0.233	0.2844	0.3304
100	0.001	0.001	0.0017	0.0025	0.0008	0.0009	0.0009	0.0022
150	0.0026	0.0031	0.005	0.0064	0.0023	0.003	0.0045	0.0072
200	0.0011	0.0008	0.002	0.0011	0.0011	0.0012	0.0021	0.0014
250	0.0139	0.0157	0.0214	0.0243	0.0132	0.0179	0.0218	0.023
300	0.001	0.0016	0.0014	0.0016	0.0009	0.0012	0.0014	0.002
350	0.0096	0.0114	0.0134	0.0153	0.0109	0.0136	0.016	0.0204
400	0.0008	0.0008	0.0016	0.0016	0.0008	0.0011	0.0015	0.0019
450	0.008	0.0113	0.0142	0.0154	0.0096	0.0118	0.0156	0.0161
500	0.0016	0.0014	0.0016	0.0011	0.0014	0.0013	0.0016	0.0025
550	0.0066	0.0087	0.008	0.0089	0.0067	0.0082	0.0093	0.0121
600	0.0008	0.001	0.0011	0.0013	0.0006	0.0014	0.0011	0.0008
650	0.0064	0.0036	0.0096	0.0106	0.0077	0.009	0.0114	0.0119
700	0.0012	0.0009	0.0025	0.0009	0.0013	0.0018	0.002	0.0012
750	0.005	0.0089	0.0063	0.0078	0.0037	0.005	0.0071	0.0099
800	0.0005	0.0014	0.0007	0.001	0.0005	0.0013	0.0017	0.0017
850	0.0033	0.0056	0.0046	0.0047	0.0021	0.0028	0.0036	0.0045
900	0.0005	0.001	0.001	0.0009	0.0005	0.0006	0.0012	0.0009
950	0.0013	0.0013	0.0024	0.0026	0.0009	0.0012	0.0007	0.0022
1000	0.0008	0.0004	0.0009	0.0018	0.0008	0.0008	0.0011	0.0008

Figure 2. Surface breakdown points at 25.2 kV ((a) pressboard and (b) polyurethane).

3.2. Scanning Electron Microscopy (SEM) Analysis

Initially, we applied breakdown stresses on materials with similar thickness in the experimental setup and compared the outcomes. Subsequently, we analyzed and interpreted size variations of the same materials along with the reactions and behavior patterns of different materials with similar sizes through SEM photos of the resultant material.

When examining the 1 mm pressboard material with a 1 mm electrode gap, as Figure 3a illustrates, we see a drilled hole running through the center of the pressboard, which constitutes a compressed paper structure. The point of breakdown indicates darkening, though no layered structure distortion was noted. Furthermore, it was observed that a 400-micron hole had been drilled in the pressboard's interior. The burn does not progress through infiltration, as evident from these findings. The carbonized path was anticipated to move inward, but it has yet to materialize, as evidenced by its current location. The expected structure has not been formed as a result. The experiment was promptly terminated after the breakdown occurred.

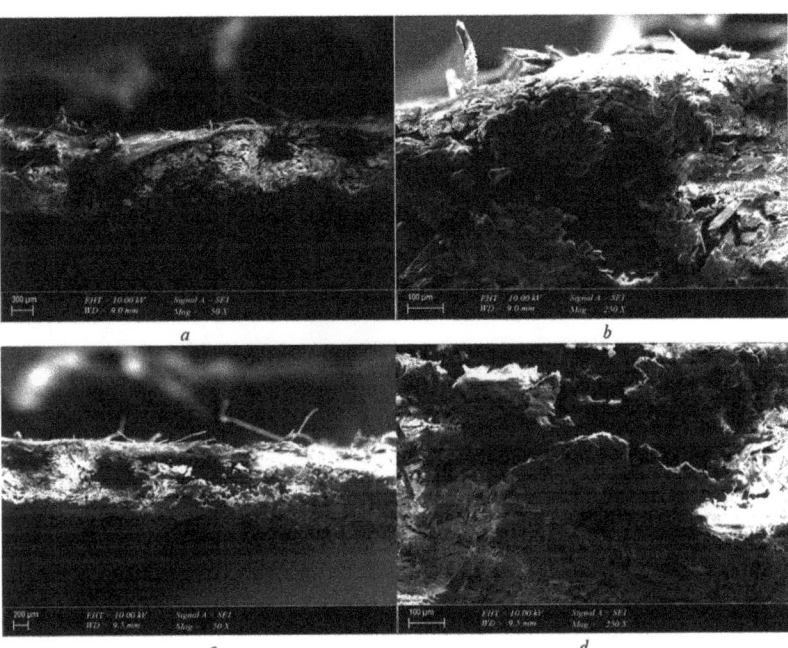

Figure 3. 1 mm pressboard breakdown in 25.2 kV SEM photos (scales are (**a**) = 300 μm-50×, (**b**) = 100 μm-250×, (**c**) = 200 μm-50×, and (**d**) = 100 μm-250×).

As demonstrated in Figure 3b, a 400-micron hole was created in the 1 mm pressboard insulator following a breakdown at the specified voltage. The carbonized path did not propagate throughout the pressboard, leading to immediate termination of the experiment upon breakdown. In addition, a weak spot in the pressboard caused an explosion after the breakdown test. The dielectric behavior observed did not disrupt other layers since the pressboard (Figure 3c), which possesses a cellulosic structure, lacks deceleration between layers due to its fibrous structure. Consequently, no significant structural changes occurred.

The cross-section obtained through the pressboard reveals explosion-shaped zones indicating micro-level physical deformation formed on the paper after electrical breakdown. However, the deformation did not exceed the layers found in the fibrous structure of the pressboard, and, as such, no significant structural changes occurred.

It is clearly seen in Figure 4a that polyurethane is a structurally hollow material. The hollow state of the structure is remarkable in the tests carried out with polyurethane samples, which are also used as insulators in cable headers commercially. It is obvious that it will show an effective behavior in the face of volumetric stresses in polyurethane material just like in the same thickness pressboard.

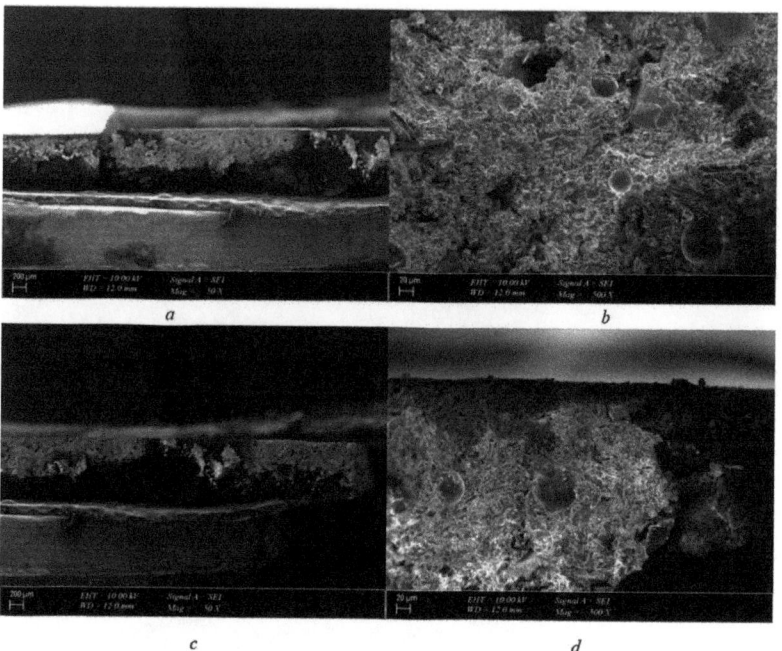

Figure 4. 1 mm polyurethane breakdown in 25.2 kV SEM photo (scales are (**a**) = 200 μm-50×, (**b**) = 20 μm-500×, (**c**) = 200 μm-50×, and (**d**) = 20 μm-500×).

Pressure is formed on the pressboard, which is located inside the oil. It is not easy to see that the transformer has bombed due to this pressure. This is due to the fact that the paper absorbs pressure. In the same case, it is also seen that, due to the convex structure found in polyurethane, it works as actively as the pressboard in absorbing this pressure and its availability instead of the pressboard is observed.

When the polyurethane material was approached and examined with a 20 μm scale, the carbonized region of the structure in the polyurethane material was not disturbed. Just like in the pressboard material, it seems that the breakdown ends with a regional burn.

There are no gaps or traces of carbon along the channel. The breakdown is limited to this area, as in the case of a 1 mm pressboard, so, in the case of a 1 mm polyurethane insulator, the material is structurally very thin. Therefore, if we interpret the breakdown, it is seen both more quickly and as a point distortion. If the structure of the material was thicker, the possibility that the breakdown would progress and form a carbonized path could have been considered, but this did not happen for a 1 mm structure.

The 200 μm scaled cross-section shows that the breakdown occurred at a single point, and the convex structure of the polyurethane material absorbed the resulting pressure, preventing the sudden expansion and progression of the breakdown. As can be seen in Figure 4d, when 20 μm was approached to the polyurethane from a different cross-section, it was seen that the progress after the breakdown did not lead to a carbonization.

As can be seen from Figure 5a, the inner part of the 2 mm pressboard structure is chipped. After the breakdown, a burn was formed. As a result of combustion, a duct was formed. There is progress along this channel; why did combustion occur along a channel for a 2 mm material when there is only a point discharge for a 1 mm pressboard material? The answer to this question is as follows. This material will not break down immediately and will not tear. As can be seen from Figure 5b, the structure is carbonized from the traces formed on the 2 mm pressboard compared to the 1 mm pressboard.

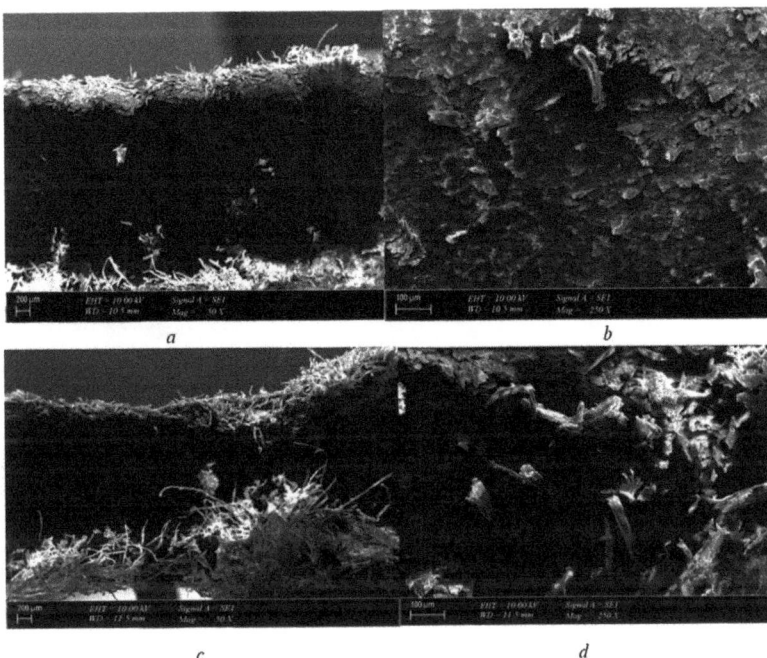

Figure 5. 2 mm pressboard breakdown in 25.2 kV SEM photo (scales are (**a**) = 200 μm-50×, (**b**) = 100 μm-250×, (**c**) = 200 μm-50×, and (**d**) = 100 μm-250×).

In Figure 5c, when a different cross-section of the pressboard formed after the breakdown voltage is examined, the fibrous structure of the paper formation and cellulosic material is clearly seen. The combustion caused by the breakdown inside the pressboard and the carbonized path formed at the end of it have been clearly revealed. As can be seen in (Figure 5d), the burns that occur in the cellulosic structure are soon clearly visible. The deterioration of the structure as a result of carbonization gives us an observation idea about the resistance of the material to breakdown.

2 mm polyurethane 200 μm cross-section is presented in Figure 6a. A similar progression that occurs on a 2 mm pressboard is also observed on polyurethane. As a result of the breakdown, the burn formed a carbonized path, but its difference from the pressboard is clearly visible. Structurally, there is a wider distortion of the pressboard. In polyurethane, the channel is much thinner. When 2 mm pressboard and polyurethane material are compared, it is seen that the structural deterioration of cellulosic paper is worse than polyurethane.

After the applied voltage is terminated, it is seen that polyurethane is structurally more robust when we enlarge it according to the 20 μm scale to examine the structure. Unlike the pressboard, it seems that polyurethane with the same mm scale does not have a hole channel as large as the one on the paper. In addition, there are no structural breakdowns. Structural bubbles contained in polyurethane help to preserve their structure due to the fact that they are trapped under pressure and become old again in a free state. When the scale is reduced (Figure 6c), the burns and channels on the polyurethane are more clearly visible. But it seems that it is not as large as the channels located on the pressboard of the same thickness. Despite this, the carbonized structure formed by burns is more clearly present here.

When 2 mm polyurethane taken from a different cross-section of the material is examined (Figure 6d), it is seen that the structural integrity is maintained and it is better than the pressboard. But, at the same time, the carbonized breakdown formed as a result of combustion can be seen in this section.

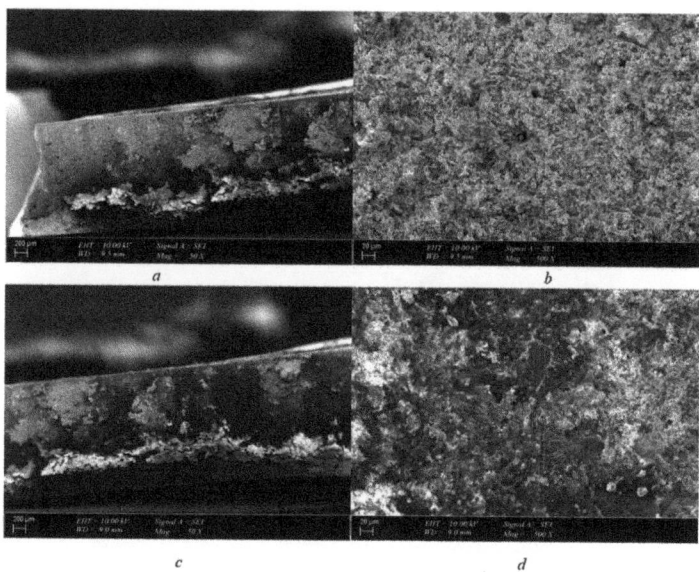

Figure 6. 2 mm polyurethane breakdown in 25.2 kV SEM photo (scales are (**a**) = 200 µm-50×, (**b**) = 20 µm-500×, (**c**) = 200 µm-50×, and (**d**) = 20 µm-500×).

As can be seen from Figure 7a, there is no breakdown on the outside of the pressboard, but there is burning on the inside. Disturbances have occurred in the internal structure. Due to the increase in thickness of the pressboard, partial discharges are observed on the inside.

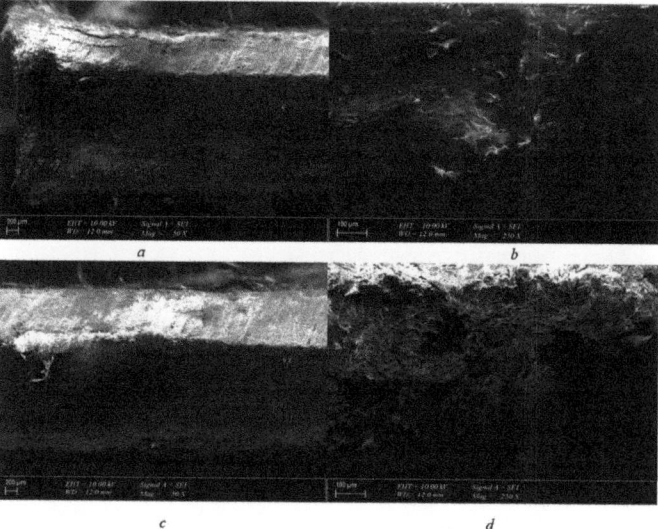

Figure 7. 3 mm pressboard breakdown in 25.2 kV SEM photo (scales are (**a**) = 200 µm-50×, (**b**) = 100 µm-250×, (**c**) = 200 µm-50×, and (**d**) = 100 µm-250×).

In (Figure 7b), it is seen that combustion occurs when the cross-sectional scale for the pressboard is enlarged, but the cellulosic structure is not as deteriorated as in the 1 mm and 2 mm pressboards. The fibers of the paper structure are most degraded on the 1 mm pressboard; for a 2 mm pressboard, less distortion is visible. The conclusion that can be drawn from this is that as the thickness of the pressboard is increased, the cellulosic fiber structure is less impaired in breakdown resistance.

It can be said that although the breakdown is not observed from the outside in the section examined for the 3 mm pressboard (Figure 7c), holes are formed as a result of burning on the inside, and, as a result, a breakdown may be present from the inside out. It has been found that while there is no breakdown outside the cellulosic fibrous structure on the pressboard, the high voltage that causes the breakdown inside can penetrate through the hole in the smallest cellulosic structure and cause explosions inside.

When the cross-section of the area (Figure 7d) is examined, it is seen that combustion disrupts the cellulosic structure inside. It has been determined that the reason for this formation is that the breakdown tension does not proceed from a single point by damaging the fibrous structure, but creates combustion at multiple points independent of each other.

As can be seen when the polyurethane material is approached by scaling 20 µm in Figure 8a, large holes formed as a result of combustion are not visible as in 2 mm polyurethane, as the material thickness increases. But it can be said that fragmentary small discharges are formed and are independent of each other, and the breakdown voltage causes more than one distortion in the weak places of the material.

When the size of the scale is reduced to 200 µm (Figure 8b), there is an opportunity to study the structure of the material from a broader point of view. At the end of this, we see that the structure remains more stable overall. It can be said that the bubbles of polyurethane are permanently damaged by breakdown, but large holes are not formed.

Figure 8. 3 mm polyurethane breakdown in 25.2 kV SEM photo (scales are (**a**) = 20 µm-500×, (**b**) = 200 µm-50×, and (**c**) = 20 µm-500×).

The SEM photo of 3 mm polyurethane is shown in (Figure 8c). This is better to see that the structure's domes are broken here. It is expected that the dome structure will better repel the pressure created by the breakdown voltage created as the material thickness increases and protects the dome structure. But, as can be seen from the figure, the convex structure is best in 1 mm polyurethane, while 2 mm is slightly more distorted in polyurethane; on the other hand, the most distorted state has appeared in 3 mm polyurethane.

4. Conclusions

The effective dielectric performance of the solid insulator in mineral oil is critical for ensuring transformers operate safely and without interruption. A high level of insulation resistance is necessary to prevent any potential short circuits and protect the transformer windings. This study investigates the electrical properties of PU and pressboard dielectric materials at thicknesses of 1 mm, 2 mm, and 3 mm. The experiments were conducted using a needle–plane electrode configuration in oil, under a non-uniform electric field. The harmonic values, peak points, and current values were measured at material thicknesses of 1 mm and 2 mm, but no clear selection criteria were identified for choosing between PU and pressboard. Even though these thicknesses lack precise interpretation, their consideration within the philosophy of science marks a new contribution to the literature. At a thickness of 3 mm, it can be stated that polyurethane (PU) offers superior insulation resistance in comparison to pressboard. As the thickness of the material increases, the dielectric performance of pressboard decreased. Although cellulose structure provides an advantage in the absorption of transformer oil, it can be said that the discharges occurring within the structure rapidly propagate and separate into channels. In this case, the leakage current flows faster in these channels and reduces the strength of the material. For thicker samples, the area in which electrons can move and form channels increases. The experimental study shows that this situation leads to more destructive effects on the pressboard. Experimental studies have shown that pressboard exhibits higher harmonic components at the same voltage levels for a 3 mm thickness. It is currently not possible to conclude that the use of polyurethane is definitively better than pressboard. Nevertheless, it is important to consider the interpretation of harmonics in the calculation of material lifespan. Although pressboard effectively removes dirt from oil due to its cellulosic structure, laboratory experiments have demonstrated that its electrical performance is inferior to that of polyurethane.

Author Contributions: A.E.: Conceptualization, methodology, software, data curation, writing—original draft preparation, investigation, reviewing and editing. F.A.: Methodology, supervision, data curation, validation, writing, reviewing and editing. A.A.: Supervision, validation, writing, reviewing and editing. All authors have read and agreed to the published version of the manuscript.

Funding: This work was supported by the Istanbul University-Cerrahpaşa Scientific Research Projects Unit with the project code FBA-2023-37284.

Institutional Review Board Statement: Not applicable.

Data Availability Statement: All data generated or analyzed during this study can be shared by the corresponding author when there is a demand from researchers.

Acknowledgments: The authors would like to thank the Istanbul University-Cerrahpaşa Scientific Research Projects Unit for financial support. The authors also thank Ketenci Elektrik company for material support.

Conflicts of Interest: The authors declare no conflict of interest.

References

1. Alencar, R.J.N.; Bezerra, U.H. Power transformer differential protection through gradient of the differential current. *J. Control. Autom. Electr. Syst.* **2013**, *24*, 162–173. [CrossRef]
2. Patel, D.; Chothani, N.G.; Mistry, K.D.; Raichura, M. Design and development of fault classification algorithm based on relevance vector machine for power transformer. *IET Electric Power Appl.* **2018**, *12*, 557–565. [CrossRef]
3. Bourreau, D.; Péden, A. Solid and non-solid dielectric material characterization for millimeter and sub-millimeter wave applications. In Proceedings of the 50th European Microwave Conference (EuMC), Utrecht, The Netherlands, 12–14 January 2021; pp. 909–912. [CrossRef]
4. Liu, J.; Cao, Z.; Fan, X.; Zhang, H.; Geng, C.; Zhang, Y. Influence of Oil–Pressboard Mass Ratio on the Equilibrium Characteristics of Furfural under Oil Replacement Conditions. *Polymers* **2020**, *12*, 2760. [CrossRef] [PubMed]

5. Gao, B.; Yu, R.; Hu, G.; Liu, C.; Zhuang, X.; Zhou, P. Development Processes of Surface Trucking and Partial Discharge of Pressboards Immersed in Mineral Oil: Effect of Tip Curvatures. *Energies* **2019**, *12*, 554. [CrossRef]
6. Koziol, M.; Nagi, L. Comparative analysis of measurement methods for partial discharges generated on the surface of solid dielectrics in insulating oil. In Proceedings of the 20th International Scientific Conference on Electric Power Engineering (EPE), Kouty nad Desnou, Czech Republic, 15–17 May 2019; pp. 1–4. [CrossRef]
7. Aruna, M.; Pattanshetti, V.V.; Ravi, K.N.; Vasudev, N. Power transformer winding insulation: A review of material characteristics. *Int. J. Mod. Eng. Res. (IJMER)* **2011**, *1*, 312–318.
8. Morooka, H.; Yamagishi, A.; Kawamura, K.; Kojima, H.; Hayakawa, N. Thermal-degradation mechanism of mineral-oil-immersed pressboards and influence of aging on pressboard properties. In Proceedings of the 2018 Condition Monitoring and Diagnosis (CMD), Perth, WA, Australia, 23–26 September 2018; pp. 1–4.
9. Nedjar, M. Weibull statistics in dielectric strength of oil-impregnated pressboard under ramped AC and DC voltages. *J. Energy Power Eng.* **2013**, *7*, 2388.
10. Florkowski, M.; Furgał, J.; Kuniewski, M.; Pająk, P. Investigations of transformer winding responses to standard full and chopped lightning impulses. In Proceedings of the 2018 IEEE International Conference on High Voltage Engineering and Application (ICHVE), Athens, Greece, 10–13 September 2018; pp. 1–4.
11. Chaw, E.E.; Tun, M.T.; Aung, H.W. Design comparison for rectangular and round winding distribution transformer (1000 kVA). *Int. J. Sci. Eng. Appl.* **2018**, *7*, 375–380.
12. Malik, H.; Yadav, A.K.; Mishra, S.; Mehto, T. Application of neuro-fuzzy scheme to investigate the winding insulation paper deterioration in oil-immersed power transformer. *Int. J. Electr. Power Energy Syst.* **2013**, *53*, 256–271. [CrossRef]
13. Pradhan, M.K.; Yew, K.J.H. Experimental investigation of insulation parameters affecting power transformer condition assessment using frequency domain spectroscopy. *IEEE Trans. Dielectr. Electr. Insul.* **2012**, *19*, 1851–1859. [CrossRef]
14. Sun, L.; Fan, X.; Jiang, S.; Wang, B.; Liu, Y.; Gao, S.; Meng, L. Study on electrical aging characteristics of fiber sheath materials in power transformer oil. *J. Electr. Eng. Technol.* **2019**, *14*, 323–330. [CrossRef]
15. Sun, W.; Yang, L.; Zare, F.; Xia, Y.; Cheng, L.; Zhou, K. 3D modeling of an HVDC converter transformer and its application on the electrical field of windings subject to voltage harmonics. *Int. J. Electr. Power Energy Syst.* **2020**, *117*, 105581. [CrossRef]
16. Calcara, L.; Sangiovanni, S.; Pompili, M. Standardized methods for the determination of breakdown voltages of liquid dielectrics. *IEEE Trans. Dielectr. Electr. Insul.* **2019**, *26*, 101–106. [CrossRef]
17. Calcara, L.; Sangiovanni, S.; Pompili, M. Partial Discharge Inception Voltage in Insulating Liquids Dependence from the Definition Used. In Proceedings of the 2020 IEEE 3rd International Conference on Dielectrics (ICD), Valencia, Spain, 5–31 July 2020; pp. 736–739. [CrossRef]
18. Atalar, F.; Uğur, E.; Bilgin, S.; Ersoy Yilmaz, A.; Uğur, M. Investigation of the oil/pressboard ratio effect on dielectric behaviour. In Proceedings of the International Scientific Conference (UNITECH), Gabrovo, Bulgaria, 15–16 November 2019; pp. 12–18.
19. Schurch, R.; Munoz, O.; Ardila-Rey, J.; Donoso, P.; Peesapati, V. Identification of Electrical Tree Aging State in Epoxy Resin Using Partial Discharge Waveforms Compared to Traditional Analysis. *Polymers* **2023**, *15*, 2461. [CrossRef] [PubMed]
20. Aydogan, A.; Atalar, F.; Ersoy Yilmaz, A.; Rozga, P. Using the method of harmonic distortion analysis in partial discharge assessment in mineral oil in a non-uniform electric field. *Energies* **2020**, *13*, 4830. [CrossRef]
21. Atalar, F.; Ersoy, A.; Rozga, P. Investigation of effects of different high voltage types on dielectric strength of insulating liquids. *Energies* **2022**, *15*, 8116. [CrossRef]
22. Ullah, I.; Ullah, R.; Amin, M.; Rahman, R.A.; Khan, A.; Ullah, N.; Alotaibi, S. Lifetime Estimation and Orientation Effect Based on Long-Term Degradation Analysis of Thermoset and Thermoplastic Insulators. *Polymers* **2022**, *14*, 3927. [CrossRef] [PubMed]

Disclaimer/Publisher's Note: The statements, opinions and data contained in all publications are solely those of the individual author(s) and contributor(s) and not of MDPI and/or the editor(s). MDPI and/or the editor(s) disclaim responsibility for any injury to people or property resulting from any ideas, methods, instructions or products referred to in the content.

Article

Aqueous Cationic Fluorinated Polyurethane for Application in Novel UV-Curable Cathodic Electrodeposition Coatings

Junhua Chen [1,2,3], Zhihao Zeng [1], Can Liu [1], Xuan Wang [1], Shiting Li [1], Feihua Ye [1,2,3], Chunsheng Li [1,2,3] and Xiaoxiao Guan [4,*]

[1] School of Environmental and Chemical Engineering, Zhaoqing University, Zhaoqing 526061, China; cehjchen@yeah.net (J.C.); zhihaozeng0203@163.com (Z.Z.); gmczunper@163.com (C.L.); pearships@163.com (X.W.); lishitingqc@163.com (S.L.); yefeihua@zqu.edu.cn (F.Y.); lichunsheng@zqu.edu.cn (C.L.)
[2] Guangdong Provincial Key Laboratory of Environmental Health and Land Resource, College of Environmental and Chemical Engineering, Zhaoqing University, Zhaoqing 526061, China
[3] Green Fine Chemical Joint Laboratory, Qingyuan 511542, China
[4] China Electronic Product Reliability and Environmental Testing Research Institute, Guangzhou 511370, China
* Correspondence: guanxiaoxiao.ok@163.com

Citation: Chen, J.; Zeng, Z.; Liu, C.; Wang, X.; Li, S.; Ye, F.; Li, C.; Guan, X. Aqueous Cationic Fluorinated Polyurethane for Application in Novel UV-Curable Cathodic Electrodeposition Coatings. *Polymers* **2023**, *15*, 3725. https://doi.org/10.3390/polym15183725

Academic Editors: Zhuohong Yang and Chang-An Xu

Received: 28 June 2023
Revised: 15 August 2023
Accepted: 21 August 2023
Published: 11 September 2023

Copyright: © 2023 by the authors. Licensee MDPI, Basel, Switzerland. This article is an open access article distributed under the terms and conditions of the Creative Commons Attribution (CC BY) license (https://creativecommons.org/licenses/by/4.0/).

Abstract: Aqueous polyurethane is an environmentally friendly, low-cost, high-performance resin with good abrasion resistance and strong adhesion. Cationic aqueous polyurethane is limited in cathodic electrophoretic coatings due to its complicated preparation process and its poor stability and single performance after emulsification and dispersion. The introduction of perfluoropolyether alcohol (PFPE-OH) and light curing technology can effectively improve the stability of aqueous polyurethane emulsions, and thus enhance the functionality of coating films. In this paper, a new UV-curable fluorinated polyurethane-based cathodic electrophoretic coating was prepared using cationic polyurethane as a precursor, introducing PFPE-OH capping, and grafting hydroxyethyl methacrylate (HEMA). The results showed that the presence of perfluoropolyether alcohol in the structure affected the variation of the moisture content of the paint film after flash evaporation. Based on the emulsion particle size and morphology tests, it can be assumed that the fluorinated cationic polyurethane emulsion is a core–shell structure with hydrophobic ends encapsulated in the polymer and hydrophilic ends on the outer surface. After abrasion testing and baking, the fluorine atoms of the coating were found to increase from 8.89% to 27.34%. The static contact angle of the coating to water was 104.6 ± 3°, and the water droplets rolled off without traces, indicating that the coating is hydrophobic. The coating has excellent thermal stability and tensile properties. The coating also passed the tests of impact resistance, flexibility, adhesion, and resistance to chemical corrosion in extreme environments. This study provides a new idea for the construction of a new and efficient cathodic electrophoretic coating system, and also provides more areas for the promotion of cationic polyurethane to practical applications.

Keywords: cationic polyurethane; UV-curable; perfluoropolyether alcohol; cathodic electrodeposition coatings

1. Introduction

Due to the unique microstructure produced between the hard and soft segments, polyurethane (PU) offers exceptional mechanical qualities. Nevertheless, the typical applications of solvent-based PU are constrained by environmental laws regarding the use of volatile organic compounds (VOCs) [1]. Traditional solvent-based PU is rapidly being replaced by environmentally friendly water-based PU as demand for environmentally friendly materials rises. A high-performance resin, known as waterborne polyurethane, utilizes water as a solvent in its formulation. In comparison to traditional solvent-based polyurethane, it offers superior properties such high hardness, wear resistance, strong adhesion, non-flammability, and low cost. It may also significantly lower the content of volatile organic compounds (VOCs) [2–7]. The textile, leather, furniture, paint, and construction industries all utilize it extensively [7–11]. Based on the nature of the ionization of hydrophilic groups in water, waterborne polyurethanes are classed as anionic, cationic, and nonionic. Ionic aqueous polyurethanes have been extensively investigated, and the anionic variety is now being produced commercially [12–17]. Cationic aqueous polyurethanes are typically synthesized by utilizing diols containing tertiary amine groups as chain extenders and quaternizing with alkylating agents or appropriate acids to produce water-soluble ionic groups. The mechanical properties of cationic aqueous polyurethanes that are not apparent in anionic materials are due to the capacity of the polar groups to form intermolecular hydrogen bonds with high intermolecular contact forces [18,19]. And cationic aqueous polyurethanes can be used for quaternary amination of the opposing ions contribute to hydrogen bonding and other ionic effects, but due to the complicated preparation process and poor stability after emulsification and dispersion, quaternary ammonium salt-based cationic aqueous polyurethanes are currently only sparingly studied [18,20–24].

Colloids known as cathodic electrophoretic coatings (EPD) are formed when charged particles and molecules that are suspended in a solvent move under the influence of an electric field and ultimately deposit on electrodes with opposing charges to create a dense surface film [25]. Cathodic electrophoretic coatings provide several advantages over traditional solvent-based coatings, including a straightforward procedure, high controllability, low cost, high permeability, electrochemical activity, and effective anti-corrosion properties [26–28]. Epoxy and acrylic cathodic electrophoretic coatings are actively being investigated. Through electrophoretic deposition of SH-SiO$_2$ nanoparticles and resin binders, Zhang et al. created durable superhydrophobic SiO$_2$ epoxy coatings with strong corrosion resistance but low weathering resistance [29].

Gong et al. created a cathodic electrophoretic solution utilizing cationic acrylate and EPD technology in order to create a conductive coating with high density and features that are hydrophobic and oleophobic [30]. The coating made using polyurethane electrophoresis solution in the traditional sense has relatively poor overall performance, such as corrosion and heat resistance, due to the introduction of hydrophilic groups, and further modification is needed for practical application in electrophoretic paints [31].

Fluorinated compounds are an excellent material for modifying. Liu et al. created a series of fluorinated aqueous polyurethane coatings by altering the content of fluorinated polymers to give them low surface energy and significant hydrophobic and oleophobic properties. The incorporation of C-F bonds can also augment the thermal stability and tensile properties of the coatings [32–36]. Unsaturated double bonds are added by the grafting of hydroxyethyl methacrylate (HEMA), which increases their thermal stability and offers active sites for light curing [37]. Finally, Liu et al. used a simple mixture of cationic aqueous polyurethanes and other UV-curing materials to prepare a new coating that improves its curing rate, reduces energy consumption, VOC emissions and reduces reactive diluents, and improves the overall physical properties of the coating, which can be applied to a variety of substrates, particularly heat-sensitive substrates [38–43].

In this experiment, cationic aqueous polyurethane was used as the electrophoresis fluid to introduce fluorinated polyether polyol (PFPE-OH) to provide low surface energy while grafting HEMA to provide a method to control the degree of cross-linking of the coating and improve the thermal stability of the coating. Finally, a novel UV-curable fluorinated cationic polyurethane-based cathodic electrophoretic coating was innovatively prepared using electrophoresis and UV curing technology. This study aims to investigate and evaluate the particle size, contact angle, water absorption rate, corrosion resistance, thermomechanical properties, and fundamental physical characteristics of the coating.

2. Experimental

2.1. Materials

Perfluoropolyether alcohol (PFPE-OH) was supplied by Hunan Nonferrous Chenzhou Fluorine Chemical Co. Ltd., Chenzhou, China. Isophorone diisocyanate (IPDI, >99.5%, NCO% $\geq$ 37.5%) was purchased from Bayer Chemicals (Leverkusen, Germany). Polycarbonate diol (PCDL, M_n = 1000), N-methyldiethanolamine (MDEA), trimethylolpropane (TMP), 1,4-butanediol (BDO), hydroxyethyl methacrylate (HEMA), 2-hydroxy-2-methyl-1-phenyl-1-propanone (Darocur 1173), glacial acetic acid (HAc), dibutyltin dilaurate (DBTDL) and 2,6-di-tert-butyl-p-cresol (DBHT) were purchased from Aladdin Reagent Co. Ltd., Shanghai, China. All organic solvents will be de-watered using 4 A° molecular sieve before use. All alcohols used were heated 120 °C in advance and de-watered under reduced pressure for about 2 h.

2.2. Cationic Aqueous Fluorinated Polyurethane Emulsion Preparation (WCFPU)

Fluorinated polyurethane polymer was synthesized via step-growth addition polymerization displayed in Figure 1. In a three-neck flask equipped with stirring paddle, thermometer, and nitrogen protection, PCDL and TMP were added after dehydration treatment, and then the temperature was increased to 80~85 °C. The temperature was then decreased to 60 °C after adding IPDI and DBTDL. BDO and MDEA butanone solution (40 wt%) were progressively added to extend the chain length and maintain the reaction for two hours. The molar functional group ratio of NCO/OH of the system at this stage was controlled to be 1.4. The system was adjusted with the proper quantity of butanone dilution when the viscosity became too high. The reaction was held for two hours to complete after the temperature was increased to 80 °C and PFPE-OH was introduced. The viscosity of the system was monitored at all times during the reaction, and butanone was added at any time. The temperature was then reduced to 60 °C. Butanone solution dissolved with HEMA and DBHT was slowly added dropwise, the remaining system NCO was capped with double bonding, and the prepolymer of CFPU was made by continuing the insulation for 2 h. The temperature was then reduced to 40 °C, and a small amount of HAc was added to neutralize the system for 30 min. Lastly, a small amount of deionized water was gently added to the prepolymer while the system was reverse-phase-emulsified at high speed for 30 min. Decompression at a negative pressure of −0.1 KPa and a temperature of 40 °C was used to remove the butanone solvent from the system. The final creamy white WCFPU emulsion with a solid content of 30 wt% was created.

2.3. UV-Cured Cathodic Electrodeposition Coating Preparation

A total of 30 wt% WCFPU emulsion was mixed with 5 wt% photoinitiator before being diluted in deionized water into a UV electrophoresis solution with a solid content of 15% and injected into the electrophoresis tank of the electrophoresis equipment. The tinplate was covered with the negative electrode, the pole plate was treated with phosphate, the tinplate was submerged in the electrophoresis tank, and there was a 10 cm space between the pole plates. The applied DC voltage was 60 V, and the electrophoresis period was 45 s. The cathode plate surface was then cleaned with a substantial volume of deionized water before being placed in an oven with a flash time of 3 min at 80 °C to eliminate the water. To achieve a uniform and smooth electrophoretic coating, the electrode plate was subjected to 100 mW/cm^2 UV curing for a duration of 60 s.

2.4. Characterization

Fourier transform infrared spectra (FT-IR) were acquired using the KBr press method on a Bruker model VERTEX 70 spectrometer (Waltham, MA, USA), covering the wavelength range of 400–4000 cm^{-1}. The hydrogen spectra (^{1}H-NMR) of the compounds were obtained on a Bruker model AVANCE III HD 400 spectrometer with CDCl$_3$ as solvents.

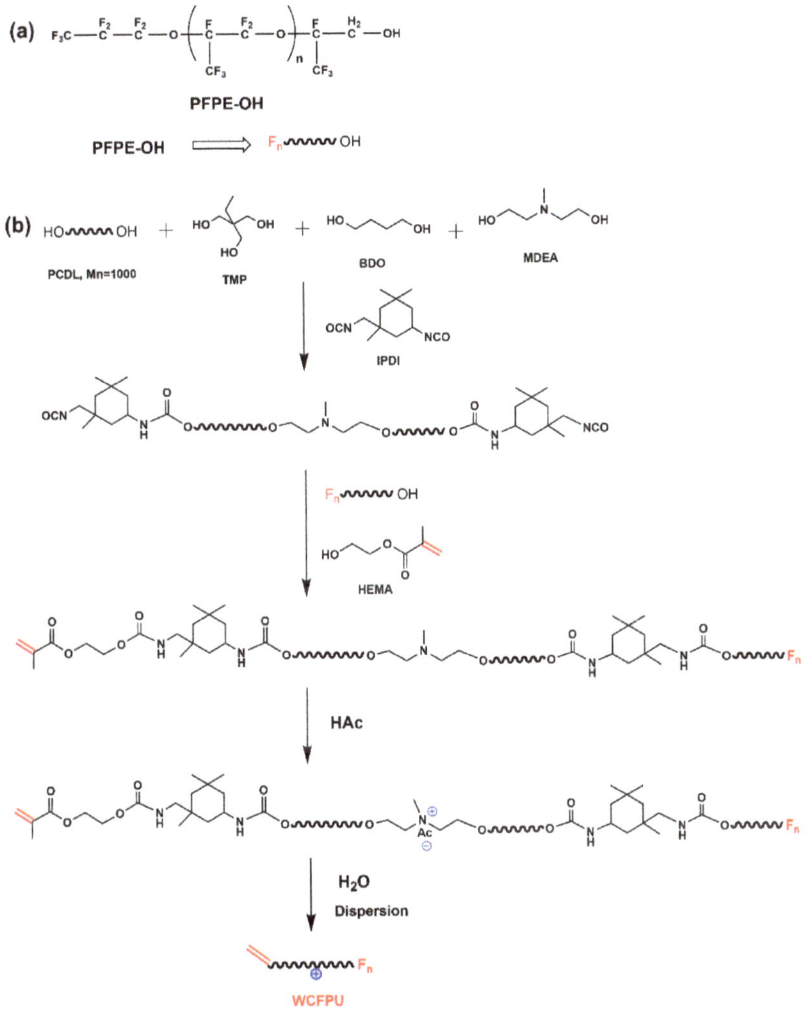

Figure 1. Waterborne cationic fluorinated polyurethane synthetic approaches (WCFPU). (**a**) The structural formula of perfluoropolyether alcohol (PFPE-OH); (**b**) the procedure for synthesis of modified polyurethane.

A volume of 0.1 mL of the sample was diluted with deionized water to a final volume of 50 mL and subjected to sonication for 30 min. The particle size distribution of the electrophoretic dispersion was determined by dynamic light scattering (DLS, Zetasizer Nano-ZS90, Malvern Instruments Ltd., Malvern, UK). A suitable amount of WCFPU emulsion was diluted 1000 times and sonicated for 30 min before a few drops of the dispersion were added onto a copper mesh using a pipette. The copper mesh was then dried in an infrared oven, followed by staining with a 5 wt% phosphotungstic acid solution for 5 min. Finally, the particle size morphology of the emulsion was examined using a transmission electron microscope (model JEM-100CX II from Japan Electronics, Beijing, China). The sample was coated with sputtered gold powder prior to examination under a scanning electron microscope (model SU-8220) capable of magnification ranging from 35 to 10,000 times.

Thermo Instruments ESCALAB 250 with XR-4 double anodes (Al/Mg) was used to perform X-ray photoelectron spectroscopy (XPS) studies under vacuum. The treated coated films were mounted on stainless steel with Cu tape on both sides and put in a sealed chamber at 10–8 mbar pressure overnight before being moved to the analytical chamber (pressure of 10–9 mbar) [2]. The XPS spectra of all materials were evaluated at a sampling depth of 6.6 nm and an electron emission angle of 45°, with the C1s line calibrated at 285.0 eV. A JC 2000C contact angle meter was used to measure the static contact angle of the coating, where the test droplets were water, diiodomethane, and hexadecane, respectively, and the volume of the droplets was 5 μL, and each sample was tested five times to take the average value. A homemade right-angled triangular test platform was designed so that the angle of the acute angle could be adjusted by lifting. The coated sample was placed on the hypotenuse of the right triangle, then the test droplet was placed on the sample, the angle was adjusted, and the slope angle when the droplet just started to move was observed and recorded as the rolling contact angle. The test droplets are water, diiodomethane and hexadecane, and the volume of the droplets is 30 μL, and each sample is tested three times to take the average value.

Differential scanning calorimetry (DSC) was used to evaluate the samples using a DSC 200F3 from Netzsch, Selb, Germany. In an aluminum sample tray, 10 mg of dried sample was sealed. To erase thermal history, the thermal cycle was conducted from 25–150 °C at a rate of 10 °C min^{-1} and maintained for 2 min. The glass transition temperature (Tg) and melting point (Tm) of polymers were then measured using a new thermal cycle that went from −70 to 250 °C at a rate of 10 °C min^{-1}. A NETZSCH model TG209 was used to determine the thermal stability of UV-cured electrodeposited films. The test atmosphere was nitrogen and the thermal scan temperature range was 25 to 600 °C with a heating rate of 10 °C/ min. The thermomechanical characteristics of the coatings were tested using a dynamic thermomechanical analyzer model DMA 242C3 from NETZSCH, Germany. The greatest dynamic force was 2 N, the maximum static force was 0.5 N, and the maximum amplitude was 10 m. The test frequencies were 1.0 Hz, 3.333 Hz, and 5.0 Hz, with the temperature range being −50~100 °C. The temperature ramping rate was 5.0 K/min.

The tensile properties were assessed using a multifunctional electronic strength tester TS2000 (Beijing Chuangcheng Zijia Technology Co., Ltd., Beijing, China) at 10 mm/min, with test bars measuring 80 × 10 mm (length × width) and thicknesses ranging from 0.5 to 1.0 mm. Each measurement was repeated at least three times.

Three coated films with dimensions of 120 mm in length, 10 mm in width, and 0.5 mm in height were submerged in deionized water and kept at 25 °C room temperature for 72 h. Before weighing the samples, absorbent cotton paper was employed to remove the water from their surfaces. The equation below may be used to calculate how much water the samples absorbed [44]:

$$\text{Absorption percentage} = \frac{W_2 - W_1}{W_1} \times 100\%$$

where W_2 and W_1 are the weight of the sample at 72 h water solubility and the weight of the original dry film, respectively.

The German QNIX 4200 (QNix, Koln, Germany) was used to gauge the coating's thickness. The gloss of the coating at 60° was measured using an ETB-0686 gloss meter in accordance with the national standard GB 9754-88 [45]. The test was performed three times. Using a Faber Castell 9000 pencil, the hardness of the coating surface was evaluated in accordance with national standard GB/T6739-1996 [46]. QFH-HG600 was used to assess the adherence of the coating to the substrate in accordance with GB/T1720-1989 [47]. The impact resistance of the paint film surface was assessed using the impact tester QCJ-50 in line with national standard GB/T 1732-93 [48], the impact tester QCJ-50 was used to evaluate the impact resistance of the paint film surface. A Shanghai QTX tester was used to measure the painted board's coating flexibility in line with GB/T 1731-93 [49]. The corrosion resistance of the electrophoretic coating was evaluated using four different solutions: 5.0 wt% NaCl, 0.5 mol/L $CuSO_4$, H_2SO_4 (pH = 0, methyl orange staining), and NaOH (pH = 14, rhodamine staining). After 24 h, corrosion of the corresponding surfaces was seen after the addition of 0.5 mL drops to the coated and untreated areas, respectively.

3. Results and Discussion

The IR spectrum of PFPE-OH is seen in Figure S1. The distinctive absorption of the alcohol hydroxyl group was shown by the wide peak at 3412.31 cm^{-1}, while the C-O-C and C-F characteristic absorption peaks were located between 1000 and 1400 cm^{-1}. Among these, the distinctive absorption peak of $-CF_3$ was 1232.62 cm^{-1}, whereas that of $-CF_2$ was 1123.58 cm^{-1}. The nuclear magnetogram of CFPU, which had a prior pretreatment to get rid of the organic solvent, is shown in Figure 2. The -C=C- bond was represented by the faint peak at 6.11 ppm, while the $-CH_3$ on the double bond side group was represented by the sharp peak at 2.01 ppm. To verify the success of double bond grafting and the presence of fluorinated monomers in the system, the IR spectra of CFPU are shown in Figure 3, and it is evident that it has a faint peak at 1635.4 cm^{-1} that is indicative of unsaturated double bond stretching vibration absorption. Also visible are the distinctive absorption peaks of C-O-C and C-F at 1100–1350 cm^{-1}. The medium broad peak at 3321 cm^{-1} is the N-H stretching vibration. The splitting peaks at 2939 cm^{-1} and 2864 cm^{-1} correspond to the stretching vibrations of $-CH_3$ and $-CH_2$-, respectively, which are mainly the methylene absorption peaks on the cationic chain extender (MDEA) and polyester structure. A strong and sharp characteristic peak at 1729 cm^{-1} is the stretching vibration of C=O. The characteristic carbamate N-H bending vibration absorption peak at was 1541 cm^{-1}. The FPU prepolymer was effectively produced, as seen by the IR and NMR pictures, and the unsaturated double bond serves as the active site for light curing. As a result, the ATR measurement of the coated film following UV electrophoresis demonstrated in Figure S2 that the unsaturated double bond at 1635.4 cm^{-1} vanishes, indicating that WCFPU has fully cured.

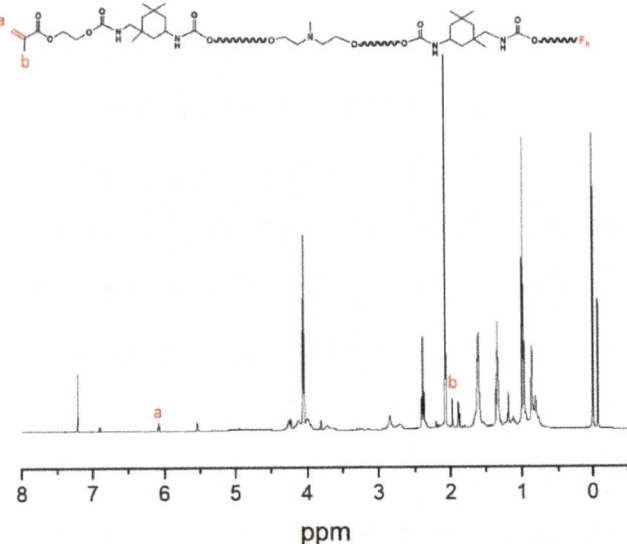

Figure 2. ^{1}H-NMR spectra of pure CFPU. The position of a is located at ~6.11 ppm weak peak is -C=C- bond, while the sharp peak of b at 2.01 ppm is due to $-CH_3$ on the side group of the double bond.

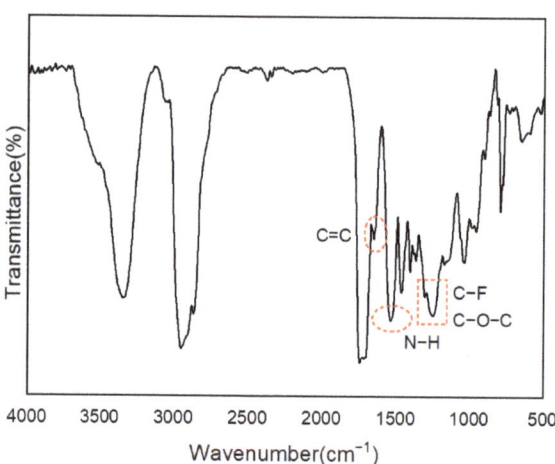

Figure 3. FT-IR spectrum of pure CFPU.

Figure 4a shows the average particle size distribution of WCFPU emulsions, which is around 91.3 nm smaller than that of the usual WCPU non-fluorinated emulsion. This indicates that the particle size of polyurethane emulsions modified with fluorinated polyether alcohols increased, and their hydrophobic ends were encapsulated within the polymer, while the hydrophilic chain segments of cationic tertiary amine groups were dispersed on the surface of the macromolecule, forming a core–shell-like structure, so their particle size became larger.

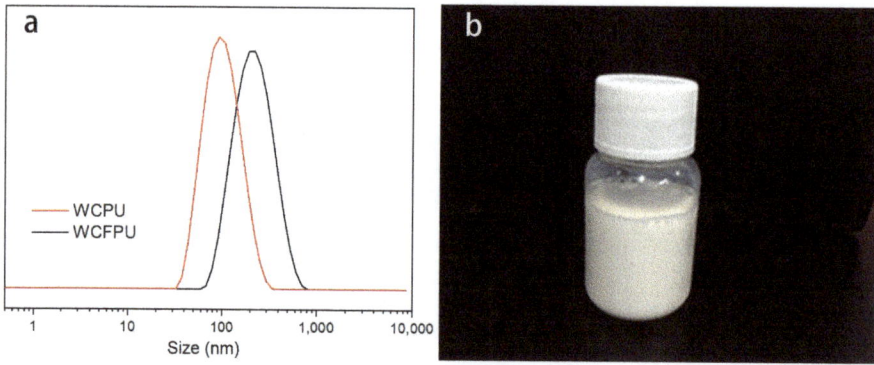

Figure 4. The DLS trace of WCFPU emulsion (**a**) and the corresponding picture of the latex (**b**).

The analysis of the WCFPU latex particle morphology is shown in Figure 5 and is enlarged to a viewing distance of 2 μm. The photos show that the grain size determined in Figure 6 is equivalent to the WCFPU latex grains, which are rather homogenous round spheres with a grain diameter of around 200 nm. The outside dark half was most likely a chain segment of an aqueous cationic chain extender, whereas the inner brilliant piece was most likely the fluorinated polyurethane chain segment with a low electron cloud density [50]. The thickness of the shell layer is about 19 nm. The photographs also further illustrate the nucleoshell structure with hydrophilic as the shell and hydrophobic as the nucleus.

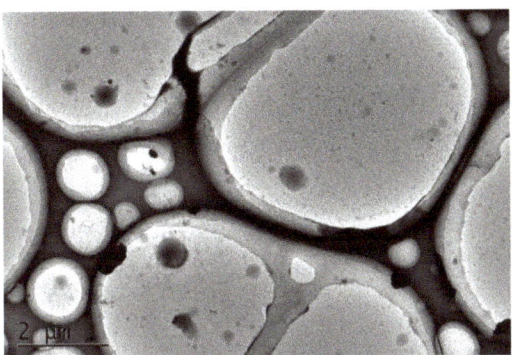

Figure 5. TEM image of WCFPU dispersions.

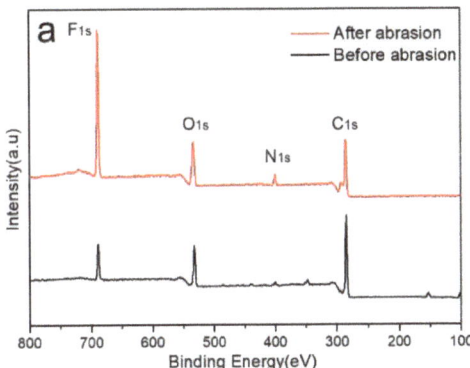

Figure 6. XPS spectra of UV-curable EPD films before and after the abrasion test (**a**) and corresponding self-made friction device (**b**).

Figure S3 show the results of an investigation into the surface morphology of the electrophoretic coating. Each area had a uniform shape and a smooth surface. It demonstrated that the system was evenly distributed, that there was no macroscopic phase separation, and that during UV curing, a highly cross-linked network formed.

To understand the fluorine chain segment movement, the XPS test was utilized to examine the surface chemistry of the UV-curable electrophoresis film. The films were polished with 500 g weights for 50 cycles, and then baked in an oven at 140 °C for 30 min. The XPS spectra before and after the coating rubbing are shown in Figure 6. It showed that the electrophoretic coating was mostly composed of C, N, O, and F as the main four elements. The fluorine atom signal intensity of the sanded and baked adhesive coating was greater than that of the unpolished one, indicating that fluorine-containing chain segments with low surface energy were more likely to migrate to and enrich the surface of the film/air. This resulted in a flat coating with a specified amount of hydrophobicity and decreased the interfacial energy on the surface of electrophoretic coating. Table S1 displays the mass percentages of each atom as determined by XPS measurements.

The Table S1 demonstrated that the fluorine atom content increased from 9.89% before sanding to 27.34% after sanding and baking, whereas the predicted average fluorine atom concentration is 4.82%. It implied that fluorine-containing chain segments with low surface energies moved to the surface of coating preferentially and were enriched during UV curing film production. The moderate amount of heat facilitated polymer chain segment migration and hydrophobic properties.

We examined the static and rolling contact angles of water, diiodomethane, and hexadecane droplets on the surface of the coating to assess the wettability of the fluorinated cationic polyurethane electrodeposition coating. The corresponding surface tensions of droplets were found to be 72 mN·m^{-1}, 50 mN·m^{-1}, and 27 mN·m^{-1} [51]. As shown in Figure 7, the static contact angles of droplets in 5 mL on the coating and the instantaneous contact angles of these droplets on the coated surface are 104.6 ± 3°, 83.6 ± 2°, and 61.3 ± 1°, respectively. This resulted from UV-curing the fluorine-containing long-chain segments, which improved the resistance of coating to water and solvents as well as the formation of regular, low-surface-energy liquid crystal forms on the surface of the latex film.

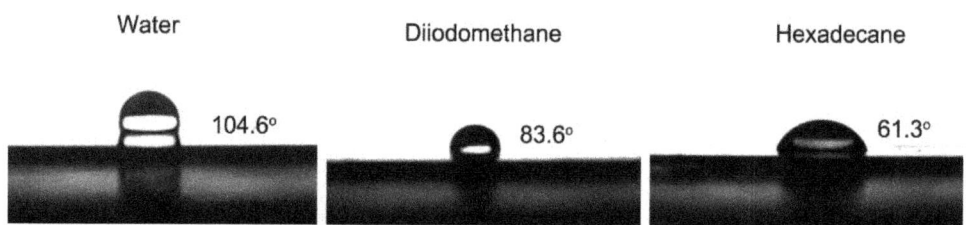

Figure 7. Images of droplets and the contact angles of water, diiodomethane, and hexadecane on the coating.

Droplets of water, diiodomethane, and cetane rolling on the coating in a volume of 50 mL are shown in Table 1, as well as their contact angles. Water droplets rolled off the surface neatly; although they had a smaller rolling angle, dii-odomethane and cetane tended to leave traces and did poorly as anti-fouling agents against oil-based pens. This was mainly due to the poor compatibility of the fluorine-containing chain segments in the coating resin structure, resulting in reduced film denseness, and the relatively insufficient content of CF_3 groups on the coating surface. The electrophoretic coating film appeared to have excellent hydrophobicity, which was consistent with the XPS test results. The average water absorption of our test coating was 2.82%, and the surface layer was enriched with many fluorine-containing chain segments, which lowered its surface energy.

Table 1. Sliding angle of water, diiodomethane, and hexadecane droplets for different volumes on the coating.

Test Droplets	Sliding Angle (50 µL)
Water	37.8 ± 1.5°
Diiodomethane	11.2 ± 0.5°
Hexadecane	9.7 ± 0.4°

Because of the higher organofluorine polymer content and the addition of C-F polar groups, as well as the increasing relative molecular mass of the polymer (Figure 8 show the DSC curves of UV cured electrodeposition coatings), the glass transition temperature of the soft segment of polyurethane in WCFPU was 53.9 °C. The figure indicated that the higher glass transition temperature, which was between 89 and 120 °C [52], which caused by the carbamate group of the hard chain section. It can be clearly observed from the figure that, as with the Tg of the soft segment, the Tg and hydrogen-bonding interactions of the hard segment decreased with the increasing length of the (-CH$_2$-CH$_2$-O-)$_n$ chain segment, and there was no significant phase separation.

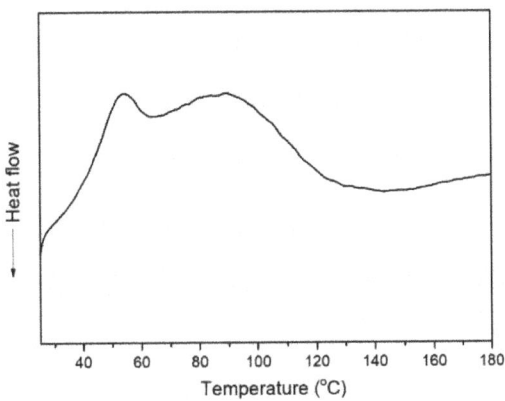

Figure 8. DSC curves of UV-curable EPD coatings.

The thermogravimetric curves of UV-cured electrophoretic coatings are shown in Figure 9. At a temperature of about 143 °C, the coating started to break down, as seen in the photograph, which was probably the result of minute molecules dissolving within the system. The carbamate structure in the structure gradually disintegrated during the second breakdown stage, which took place between 250 and 380 °C [53,54]. Between 428 and 452 °C, the breakdown entered its third stage. The improvement in the thermal stability of the UV-cured electrophoretic coating was due to the inclusion of C-F with higher bond energies, the creation of shielding protection for the interior of the polymer, and the rise in cross-link density following double bond curing.

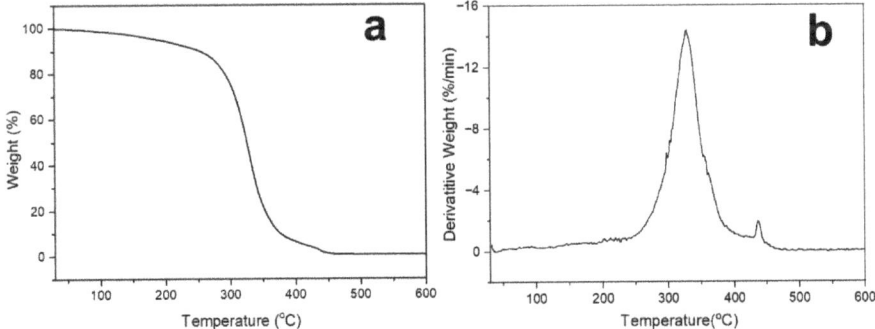

Figure 9. TGA curves (**a**) and their corresponding DTG curves (**b**) of UV-curable EPD coatings.

The stress–strain curve for the UV electrodeposition coating is shown in Figure 10. According to this curve, the fluorinated polyurethane-coated film had a tensile strength of 6.59 MPa and an elongation at break of 61.11%. The tensile strength of the original fluorine-free polyurethane coating sample was 5.68 MPa and the elongation at break was 20.14%. PCDL was used as polyester in the WCFPU structure because it provided good tensile properties. If a smaller amount of PFPE-OH was added, it led to an increase in the number of fluorinated groups introduced into the hard chain segments of the polymer, which further affected the increase in tensile strength and decrease in elongation at the break of the coated film. Furthermore, the strain-hardening phenomenon became more pronounced when fluorinated groups were introduced, which enhanced the tensile strength and reduced the flexibility of the electrophoretic film. The perfluorinated long-chain structure of PFPE-OH imparted a high degree of hydrophobicity to the membrane, increasing the content of fluorinated groups and the ratio of hydrophobic to hydrophilic chain segments, thus increasing the particle size of the emulsion and decreasing the density variation of the surface. The increase in emulsion particle size would lead to a decrease in the denseness of the emulsion after film formation, and therefore a decrease in mechanical properties [55].

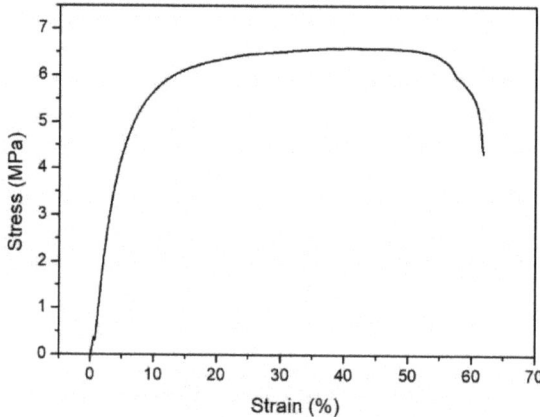

Figure 10. Stress–strain curves of UV-curable EPD coatings.

The thermomechanical behavior of the UV-curable EPD coatings is depicted in Figure 11, and the figure illustrates the great compatibility of the polyurethane system. If there was no splitting, one tan δ appeared on the graph, and conversely, two tan δ appeared if the structure was split. It can be observed from the graph that the energy storage modulus of the coating decreased with increasing temperature. The glass transition temperature (Tg) of the soft section corresponded to 64.7 °C, while the glass transition temperature of the hard section was less pronounced. Generally, the higher the glass transition temperature, the more easily molecular chain movement was hindered and the more rigid the polymer. The internal trifunctional TMP cross-linking site that restricted the mobility of the molecular chain, along with hydrogen bonding and the formation of a cross-linked network in the structure, was most likely to blame for the 2.84 damping factor.

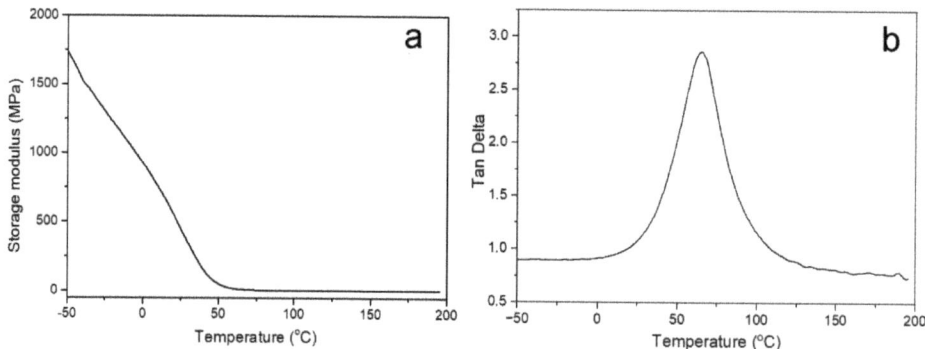

Figure 11. Temperature dependence of storage modulus (**a**) and tan δ (**b**) for UV-curable EPD films.

The UV electrodeposition coating offered high physical qualities as well as good hydrophobicity and thermal stability. As seen in Figure 12a, when the coated tinplate was struck with a 1000 g impact hammer from a height of 50 cm in both the forward and backward directions, the surface did not break or peel off. In Figure 12b, the test plate was bent 180° with a rod of 1 mm in diameter, and the coated surface did not crack at any point. Figure 12c shows the results of the adhesion test. The adhesion test uses the scratch technique described in GB/T 9286-1998 [56]. There are six degrees of adhesion, with level 5 being the worst. From level 0 to level 5, coating adherence diminishes. According to Figure 12c, the modified coating has a 0–1 level of adherence. The scribed section did not come off when the 3M tape was removed, suggesting strong adherence. This was due to the use of polycarbonate diol (PCDL) as an excellent polyester raw material for polyurethane light-curing coatings, providing excellent flexibility, adhesion, abrasion resistance, and chemical resistance to the coating.

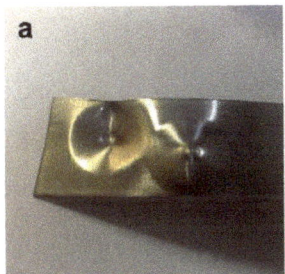 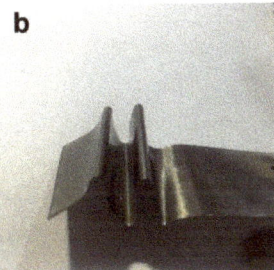 

Figure 12. Impact resistance (**a**), flexibility (**b**), and adhesion (**c**) test on the UV-curable EPD coatings.

Figure 13 shows the chemical corrosion of the electrophoretic coating surface using different 5.0 wt% NaCl solution, 0.5 mol/L $CuSO_4$ solution, H_2SO_4 solution (pH = 0, stained with methyl orange), and NaOH solution (pH = 14, stained with rhodamine) for different times. The droplets shrunk on the coated surface, while on the uncoated surface, they were spread flat. We found that when the polished uncoated tinplate came into touch with the copper sulfate solution, it immediately became covered in copper due to primary cell corrosion. After 4 h, due to partial evaporation of water, the volume of liquid was reduced and the surface of the coating appeared without corrosion or damage, but these liquids had caused varying degrees of corrosion to the uncoated tinplate. After 24 h, the water evaporated completely and the surface of the coating remained intact, while the uncoated surface showed severe chemical corrosion. It was indicated that the coating had good corrosion resistance and great potential for application in various corrosion-resistant industries.

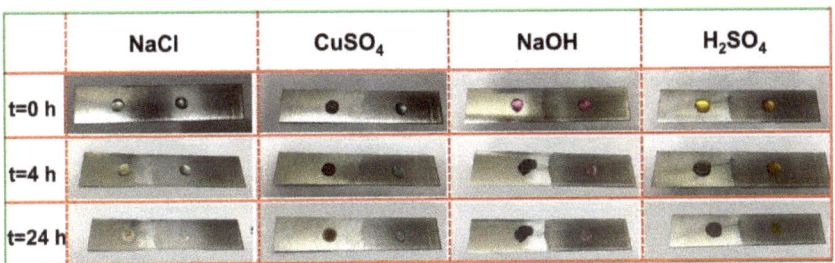

Figure 13. Effect of salt solution, strong acid solution, strong alkali solution, and corrosion cell on surface of the coatings.

4. Conclusions

In this study, a terminal fluoroalkyl cationic unsaturated polyurethane emulsion (WCFPU) was made by introducing fluorinated polyether polyol (PFPE-OH) as a side chain and grafting hydroxyethyl methacrylate (HEMA) to provide unsaturated double bonds to IPDI and polyol, and then dispersed and prepared into UV electrophoretic coatings with a solid content of 15%. The average particle size of the WCFPU emulsion was found to be about 190 nm, while the particle size of the normal non-fluorinated WCPU was 91.3 nm. Observing the TEM of the emulsion particles, we could speculate that the fluorinated cationic polyurethane emulsion was a core–shell structure with hydrophobic ends encapsulated within the polymer and hydrophilic ends on the outer surface. After wear testing and baking, it was found that the fluorine atom content of the coating increased abruptly from 8.89 to 27.34%. The static contact angle of the coating to water was $104.6 \pm 3°$, and the water droplets rolled off without traces, indicating that the coating had very good hydrophobicity. This UV electrophoretic coating had excellent thermal stability and tensile properties, and its thermomechanical behavior indicated a high polymer crosslink density. The coating also underwent the tests of impact resistance, flexibility, adhesion, and resistance to chemical corrosion in extreme environments.

Supplementary Materials: The following supporting information can be downloaded at: https://www.mdpi.com/article/10.3390/polym15183725/s1.

Author Contributions: J.C. designed and conducted the experiments and wrote the paper. S.L., X.W. and C.L. (Chunsheng Li) conducted the experiments. Z.Z., C.L. (Can Liu) and F.Y. contributed significantly to analysis and manuscript preparation. X.G. provided the whole supervision and guidance. All authors have read and agreed to the published version of the manuscript.

Funding: This research was funded by Guangdong University Innovation Project (No. 2021KQNCX106), Zhaoqing University Science Fund (No. 202006), Guangdong Basic and Applied Basic Research Foundation (No. 2020A1515111156) and Guangdong Province Enterprise Science and Technology Special Representative (No. GDKTP2021023000).

Data Availability Statement: Not available.

Acknowledgments: This research was financially supported by Guangdong University Innovation Project (No. 2021KQNCX106), Zhaoqing University Science Fund (No. 202006), Guangdong Basic and Applied Basic Research Foundation (No. 2020A1515111156) and Guangdong Province Enterprise Science and Technology Special Representative (No. GDKTP2021023000).

Conflicts of Interest: The authors declare no conflict of interest.

References

1. Lu, Y.; Larock, R.C. Soybean oil-based, aqueous cationic polyurethane dispersions: Synthesis and properties. *Prog. Org. Coat.* **2010**, *69*, 31–37. [CrossRef]
2. Tzoumani, I.; Soto Beobide, A.; Iatridi, Z.; Tzoumani, I.; Soto Beobide, A.; Iatridi, Z.; Voyiatzis, G.A.; Bokias, G.; Kallitsis, J.K. Glycidyl Methacrylate-Based Copolymers as Healing Agents of Waterborne Polyurethanes. *Int. J. Mol. Sci.* **2022**, *23*, 8118. [CrossRef]
3. Krol, P. Synthesis methods, chemical structures and phase structures of linear polyurethanes. Properties and applications of linear polyurethanes in polyurethane elastomers, copolymers and ionomers. *Prog. Mater. Sci.* **2007**, *52*, 915–1015. [CrossRef]
4. Chattopadhyay, D.K.; Raju, K.V.S.N. Structural engineering of polyurethane coatings for high performance applications. *Prog. Polym. Sci.* **2007**, *32*, 352–418. [CrossRef]
5. Tatiya, P.D.; Hedaoo, R.K.; Mahulikar, P.P.; Gite, V.V. Novel Polyurea Microcapsules Using Dendritic Functional Monomer: Synthesis, Characterization, and Its Use in Self-healing and Anticorrosive Polyurethane Coatings. *Ind. Eng. Chem. Res.* **2013**, *52*, 1562–1570. [CrossRef]
6. Xu, J.; Gao, F.; Wang, H.; Dai, R.; Dong, S.; Wang, H. Organic/inorganic hybrid waterborne polyurethane coatings with self-healing properties for anticorrosion application. *Prog. Org. Coat.* **2023**, *174*, 107244. [CrossRef]
7. Ma, X.; Chen, J.; Zhu, J.; Yan, N. Lignin-based polyurethane: Recent advances and future perspectives. *Macromol. Rapid Commun.* **2021**, *42*, 2000492. [CrossRef]
8. Lin, Z.; Sun, Z.; Xu, C.; Zhang, A.; Xiang, J.; Fan, H. A self-matting waterborne polyurethane coating with admirable abrasion-resistance. *RSC Adv.* **2021**, *11*, 27620–27626. [CrossRef]
9. Wang, C.; Mu, C.; Lin, W.; Xiao, H. Functional-modified polyurethanes for rendering surfaces antimicrobial: An overview. *Adv. Colloid Interface Sci.* **2020**, *283*, 102235. [CrossRef]
10. Xu, D.; Liang, G.; Qi, Y.; Gong, R.; Zhang, X.; Zhang, Y.; Liu, B.; Kong, L.; Dong, X.; Li, Y. Enhancing the Mechanical Properties of Waterborne Polyurethane Paint by Graphene Oxide for Wood Products. *Polymers* **2022**, *14*, 5456. [CrossRef]
11. Liu, X.; Hong, W.; Chen, X. Continuous production of water-borne polyurethanes: A review. *Polymers* **2020**, *12*, 2875. [CrossRef]
12. Gite, V.V.; Tatiya, P.D.; Marathe, R.J.; Mahulikar, P.P.; Hundiwale, D.G. Microencapsulation of quinoline as a corrosion inhibitor in polyurea microcapsules for application in anticorrosive PU coatings. *Prog. Org. Coat.* **2015**, *83*, 11–18. [CrossRef]

3. Jiang, L.; Ren, Z.Y.; Liu, W.T.; Liu, H.; Zhu, C. Synthesis and molecular interaction of tung oil-based anionic waterborne polyurethane dispersion. *J. Appl. Polym. Sci.* **2020**, *137*, 49383. [CrossRef]
4. Zhang, Y.; Zhang, W.; Wang, X.; Dong, Q.; Zeng, X.; Quirino, R.L.; Lu, Q.; Wang, Q.; Zhang, C. Waterborne polyurethanes from castor oil-based polyols for next generation of environmentally-friendly hair-styling agents. *Prog. Org. Coat.* **2020**, *142*, 105588. [CrossRef]
5. Stefanović, I.S.; Džunuzović, J.V.; Džunuzović, E.S.; Brzić, S.J.; Jasiukaitytė-Grojzdek, E.; Basagni, A.; Marega, C. Tailoring the properties of waterborne polyurethanes by incorporating different content of poly(dimethylsiloxane). *Prog. Org. Coat.* **2021**, *161*, 106474. [CrossRef]
6. Gite, V.V.; Chaudhari, A.B. Incorporation of Modified Nano Montmorillonite (MMT) in Polyurethane Coatings Based on Acrylic Copolymer and Trimer of Isophorone Diisocyanate. *Mater. Sci. Forum* **2013**, *757*, 99–109. [CrossRef]
7. Saalah, S.; Abdullah, L.C.; Aung, M.M.; Salleh, M.Z.; Awang Biak, D.R.; Basri, M.; Jusoh, E.R.; Mamat, S.; Osman Al Edrus, S.S. Chemical and Thermo-Mechanical Properties of Waterborne Polyurethane Dispersion Derived from Jatropha Oil. *Polymers* **2021**, *13*, 795. [CrossRef]
8. Borah, K.; Palanisamy, A.; Narayan, R.; Satapathy, S.; Dileepkumar, V.; Misra, S. Structure-property relationship of thermoplastic polyurethane cationomers carrying quaternary ammonium groups. *React. Funct. Polym.* **2022**, *181*, 105447. [CrossRef]
9. Fakhar, A.; Maghami, S.; Sameti, E.; Shekari, M.; Sadeghi, M. Gas separation through polyurethane—ZnO mixed matrix membranes and mathematical modeling of the interfacial morphology. *SPE Polym.* **2020**, *1*, 113–124. [CrossRef]
10. Gong, R.; Cao, H.; Zhang, H.; Qiao, L.; Wang, F.; Wang, X. Terminal Hydrophilicity-Induced Dispersion of Cationic Waterborne Polyurethane from CO_2-Based Polyol. *Macromolecules* **2020**, *53*, 6322–6330. [CrossRef]
11. Sukhawipat, N.; Raksanak, W.; Kalkornsurapranee, E.; Saetung, A.; Saetung, N. A new hybrid waterborne polyurethane coating synthesized from natural rubber and rubber seed oil with grafted acrylate. *Prog. Org. Coat. Int. Rev. J.* **2020**, *141*, 105554. [CrossRef]
12. Król, P.; Pielichowska, K.; Król, B.; Nowicka, K.; Walczak, M.; Kowal, M. Polyurethane cationomers containing fluorinated soft segments with hydrophobic properties. *Colloid Polym. Sci.* **2021**, *299*, 1011–1029. [CrossRef]
13. Zou, C.Y.; Zhang, H.M.; Qiao, L.J.; Wang, X.H.; Wang, F. Near neutral waterborne cationic polyurethane from CO_2-polyol, a compatible binder to aqueous conducting polyaniline for eco-friendly anti-corrosion purposes. *Green Chem.* **2020**, *22*, 7823–7831. [CrossRef]
14. Gong, R.; Cao, H.; Zhang, H.; Qiao, L.; Wang, X. UV-curable cationic waterborne polyurethane from CO_2-polyol with excellent water resistance. *Polym. Int. J. Sci. Technol. Polym.* **2021**, *218*, 123536. [CrossRef]
15. Saji, V.S. Electrophoretic-deposited superhydrophobic coatings. *Chem. Asian J.* **2021**, *16*, 474–491. [CrossRef]
16. Clifford, A.; Pang, X.; Zhitomirsky, I. Biomimetically modified chitosan for electrophoretic deposition of composites. *Colloids Surf. A Physicochem. Eng. Asp.* **2018**, *544*, 28–34. [CrossRef]
17. Guo, X.; Li, X.; Lai, C.; Li, W.; Zhang, D.; Xiong, Z. Cathodic electrophoretic deposition of bismuth oxide (Bi_2O_3) coatings and their photocatalytic activities. *Appl. Surf. Sci.* **2015**, *331*, 455–462. [CrossRef]
18. Zhitomirsky, D.; Roether, J.; Boccaccini, A.; Zhitomirsky, I. Electrophoretic deposition of bioactive glass/polymer composite coatings with and without HA nanoparticle inclusions for biomedical applications. *J. Mater. Process. Technol.* **2009**, *209*, 1853–1860. [CrossRef]
19. Zhang, X.F.; Jiang, F.; Chen, R.J.; Chen, Y.Q.; Hu, J.M. Robust superhydrophobic coatings prepared by cathodic electrophoresis of hydrophobic silica nanoparticles with the cationic resin as the adhesive for corrosion protection. *Corros. Sci.* **2020**, *173*, 108797. [CrossRef]
20. Gong, Z.; Zhao, W.; Chen, L. Synthesis and characterization of cationic acrylic resin used in cathodic electrodeposition coatings. *J. Polym. Res.* **2022**, *29*, 303. [CrossRef]
21. Brewer, G.E.F. Electrophoretic painting. *J. Appl. Electrochem.* **1983**, *13*, 269–275. [CrossRef]
22. Liu, K.; Su, Z.; Miao, S.; Ma, G.; Zhang, S. UV-curable enzymatic antibacterial waterborne polyurethane coating. *Biochem. Eng. J.* **2016**, *113*, 107–113. [CrossRef]
23. Ge, J.; Yang, S.; Fu, F.; Wang, J.; Yang, J.; Cui, L.; Ding, B.; Yu, J.; Sun, G. Amphiphobic fluorinated polyurethane composite microfibrous membranes with robust waterproof and breathable performances. *RSC Adv.* **2013**, *3*, 2248–2255. [CrossRef]
24. Li, P.; Yang, X.; Li, G. Preparation and properties of waterborne cationic fluorinated polyurethane. *J. Polym. Res.* **2012**, *19*, 9786. [CrossRef]
25. Salam, A.; Lucia, L.A.; Jameel, H. Fluorine-based surface decorated cellulose nanocrystals as potential hydrophobic and oleophobic materials. *Cellulose* **2015**, *22*, 397–406. [CrossRef]
26. Wei, C.; Tang, Y.; Zhang, G.; Zhang, Q.; Zhan, X.; Chen, F. Facile fabrication of highly omniphobic and self-cleaning surfaces based on water mediated fluorinated nanosilica aggregation. *RSC Adv.* **2016**, *6*, 74340–74348. [CrossRef]
27. Huang, M.; Liu, Y.; Klier, J.; Schiffman, J.D. High-Performance, UV-Curable Cross-Linked Films via Grafting of Hydroxyethyl Methacrylate Methylene Malonate. *Ind. Eng. Chem. Res.* **2020**, *59*, 4542–4548. [CrossRef]
28. Liu, F.; Liu, A.; Tao, W.; Yang, Y. Preparation of UV curable organic/inorganic hybrid coatings-a review. *Prog. Org. Coat.* **2020**, *145*, 105685. [CrossRef]
29. Su, Y.; Lin, H.; Zhang, S.; Yang, Z.; Yuan, T. One-step synthesis of novel renewable vegetable oil-based acrylate prepolymers and their application in UV-curable coatings. *Polymers* **2020**, *12*, 1165. [CrossRef]

40. Calvez, I.; Szczepanski, C.R.; Landry, V. Preparation and characterization of low gloss UV-curable coatings based on silica surface modification using an acrylate monomer. *Prog. Org. Coat.* **2021**, *158*, 106369. [CrossRef]
41. Magnoni, F.; Rannée, A.; Marasinghe, L.; El-Fouhaili, B.; Allonas, X.; Croutxé-Barghorn, C. Correlation between the scratch resistance of UV-cured PUA-based coatings and the structure and functionality of reactive diluents. *Prog. Org. Coat.* **2018**, *124*, 193–199. [CrossRef]
42. Tan, J.; Liu, W.; Wang, Z. Preparation and performance of waterborne UV-curable polyurethane containing long fluorinated side chains. *J. Appl. Polym. Sci.* **2017**, *134*, 44506. [CrossRef]
43. Fonseca, A.C.; Lopes, I.M.; Coelho, J.F.J.; Serra, A.C. Synthesis of unsaturated polyesters based on renewable monomers: Structure/properties relationship and crosslinking with 2-hydroxyethyl methacrylate. *React. Funct. Polym.* **2015**, *97*, 1–11. [CrossRef]
44. Peng, K.; Zou, T.; Ding, W.; Wang, R.; Guo, J.; Round, J.J.; Tu, W.; Liu, C.; Hu, J. Development of contact-killing non-leaching antimicrobial guanidyl-functionalized polymers via click chemistry. *RSC Adv.* **2017**, *7*, 24903–24913. [CrossRef]
45. *GB 9754-88*; Paints and Varnishes—Measurement of Specular Gloss of Non-Metallic Paint Films at 20 Degree, 60 Degree and 85 Degree. China Standard Press: Beijing, China, 1988.
46. *GB/T6739-1996*; Determination of Film Hardness by Pencil Test. China Standard Press: Beijing, China, 1996.
47. *GB/T1720-1989*; Method of Test for Adhesion of Paint Films. China Standard Press: Beijing, China, 1989.
48. *GB/T 1732-93*; Determination of Impact Resistance of Film. China Standard Press: Beijing, China, 1993.
49. *GB/T 1731-93*; Determination of Flexibility of Films. China Standard Press: Beijing, China, 1993.
50. Tan, H.; Guo, M.; Du, R.; Xie, X.; Li, J.; Zhong, Y.; Fu, Q. The effect of fluorinated side chain attached on hard segment on the phase separation and surface topography of polyurethanes. *Polymer* **2004**, *45*, 1647–1657. [CrossRef]
51. Campos, R.; Guenthner, A.J.; Haddad, T.S.; Mabry, J.M. Fluoroalkyl-functionalized silica particles: Synthesis, characterization, and wetting characteristics. *Langmuir* **2011**, *27*, 10206–10215. [CrossRef]
52. He, X.; Zhang, Y.; He, J.; Liu, F. Synthesis and characterization of cathodic electrodeposition coatings based on octadecyl-modified cationic waterborne polyurethanes. *J. Coat. Technol. Res.* **2020**, *17*, 1255–1268. [CrossRef]
53. Petrović, Z.S.; Zavargo, Z.; Flyn, J.H.; Macknight, W.J. Thermal degradation of segmented polyurethanes. *J. Appl. Polym. Sci.* **1994**, *51*, 1087–1095. [CrossRef]
54. Gradwell, M.; Hourston, D.; Pabunruang, T.; Schafer, F.-U.; Reading, M. High-resolution thermogravimetric analysis of polyurethane/poly (ethyl methacrylate) interpenetrating polymer networks. *J. Appl. Polym. Sci.* **1998**, *70*, 287–295. [CrossRef]
55. Yang, X.; Shen, Y.; Li, P.Z.; Li, G.H. Waterborne cationic fluorinated polyurethane modified by perfluorinated alkyl long chain. *Appl. Eng. Mater.* **2011**, *287*, 2116–2121. [CrossRef]
56. *GB/T 9286-1998*; Paints and Varnishes—Cross Cut Test for Films. China Standard Press: Beijing, China, 1998.

Disclaimer/Publisher's Note: The statements, opinions and data contained in all publications are solely those of the individual author(s) and contributor(s) and not of MDPI and/or the editor(s). MDPI and/or the editor(s) disclaim responsibility for any injury to people or property resulting from any ideas, methods, instructions or products referred to in the content.

Review

An Advanced Review: Polyurethane-Related Dressings for Skin Wound Repair

Wenzi Liang, Na Ni, Yuxin Huang and Changmin Lin *

Department of Histology and Embryology, Shantou University Medical College, Shantou 515041, China; 20wzliang1@stu.edu.cn (W.L.); nani@stu.edu.cn (N.N.); 21yxhuang@stu.edu.cn (Y.H.)
* Correspondence: cocolin@stu.edu.cn

Abstract: The inability of wounds to heal effectively through normal repair has become a burden that seriously affects socio-economic development and human health. The therapy of acute and chronic skin wounds still poses great clinical difficulty due to the lack of suitable functional wound dressings. It has been found that dressings made of polyurethane exhibit excellent and diverse biological properties, but lack the functionality of clinical needs, and most dressings are unable to dynamically adapt to microenvironmental changes during the healing process at different stages of chronic wounds. Therefore, the development of multifunctional polyurethane composite materials has become a hot topic of research. This review describes the changes in physicochemical and biological properties caused by the incorporation of different polymers and fillers into polyurethane dressings and describes their applications in wound repair and regeneration. We listed several polymers, mainly including natural-based polymers (e.g., collagen, chitosan, and hyaluronic acid), synthetic-based polymers (e.g., polyethylene glycol, polyvinyl alcohol, and polyacrylamide), and some other active ingredients (e.g., LL37 peptide, platelet lysate, and exosomes). In addition to an introduction to the design and application of polyurethane-related dressings, we discuss the conversion and use of advanced functional dressings for applications, as well as future directions for development, providing reference for the development and new applications of novel polyurethane dressings.

Keywords: polyurethane; wound dressing; natural polymers; synthetic polymers; composite material

Citation: Liang, W.; Ni, N.; Huang, Y.; Lin, C. An Advanced Review: Polyurethane-Related Dressings for Skin Wound Repair. *Polymers* **2023**, *15*, 4301. https://doi.org/10.3390/polym15214301

Academic Editors: Zhuohong Yang and Chang-An Xu

Received: 21 September 2023
Revised: 25 October 2023
Accepted: 25 October 2023
Published: 1 November 2023

Copyright: © 2023 by the authors. Licensee MDPI, Basel, Switzerland. This article is an open access article distributed under the terms and conditions of the Creative Commons Attribution (CC BY) license (https://creativecommons.org/licenses/by/4.0/).

1. Introduction

The skin is the body's largest organ, and it is not only the first line of physiological defense, but also essential for survival. In addition, the skin has a complex self-regulatory function [1]. In response to harmful stress, such as pathogens, thermal, mechanical, and chemical hazards, the skin responds to regulate local and systemic homeostasis. The structure of the skin consists of three layers: superficial epidermis, deeper dermis, and subcutaneous hypodermis. The epidermis is mainly composed of keratinocytes and undergoes constant renewal where basal epidermal stem cells with high proliferation potential produce new daughter cells or translocation expansion cells [2]. Skin can regulate water and permeate oxygen and carbon dioxide, and its sensory properties can affect thermoregulation and immune function [3]. Wound healing refers to a series of physiological processes in which damaged tissues are repaired through various cells and interstitial tissue after skin injury (Figure 1) [4], and it mainly occurs in skin tissues after traumatic injury, infectious ulcers, or burns. Rapid wound healing and rapid regeneration of damaged skin are essential to restore barrier function. Chronic wounds are also known clinically as hard-to-heal wounds because they are more difficult to heal and take a long time to treat, such as in people with diabetes or those who are chronically bedridden [5]. Chronic wounds occur when the normal healing process stalls, which can seriously affect patient quality of life and place a heavy burden on healthcare systems [6].

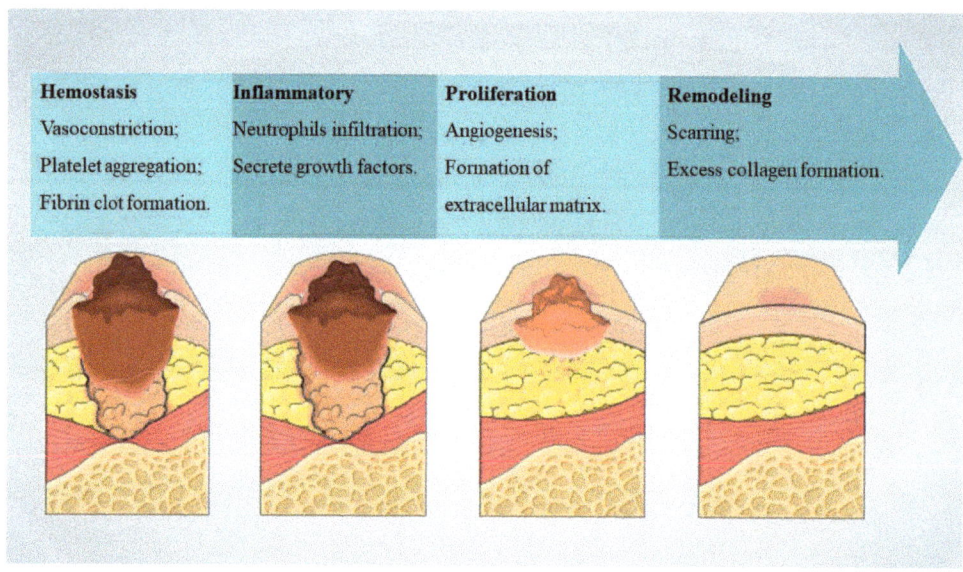

Figure 1. Typical stages of wound healing [4].

Recovery of skin wounds requires optimal temperature, humidity, pH, and oxygen. A moist environment and appropriate pH can maintain the activity of cells and enzymes, which is conducive to wound healing, as well as resistance to infection and protection from harmful external factors. As a result of a moist wound environment, autolysis debridement occurs, pain is reduced, scarring is reduced, collagen deposition occurs, blood vessels form, and migration of keratinocytes is enhanced [7]. To achieve an ideal wound environment, dressings may be selectively applied outside the wound to promote repair or to reduce the risk of infection and relieve or reduce pain. Local treatment requires choosing the right wound dressing based on the wound's characteristics, location, depth, area, level of exudation, presence of infection, stage of healing, and skin type [8]. The ideal dressing should fulfill the following requirements (Figure 2): ensure physical continuity of the wound; have ideal fluid handling capacity; have good antimicrobial activity against a wide range of bacteria and fungi; provide optimal thermoregulation, humidity, and pH; exhibit good cytocompatibility; support growth and proliferation of fibroblasts; protect against rejection (inhibit granulation, fibronectin formation); be non-allergenic, be comfortable to use, permeable to oxygen and carbon dioxide, and cost-effective; do no harm to the ulcer edges [8–11]. A number of different types of dressings, including polyurethane foam, membrane dressings, hydrocolloid dressings, hydrogel dressings, alginate dressings, semi-permeable polyurethane membrane dressings, and genetically engineered flax dressings, are used all over the world [8,12]. Doctors will choose the appropriate dressings according to medical knowledge, treatment experience, clinical characteristics of wounds, and the specific needs of patients.

Polyurethane is an important biomolecular material, which has been the focus of research, and it plays a vital role in the field of artificial organs, medical devices, and medical materials. The reason for choosing polyurethane as the substrate material for dressings is mainly based on the following reasons: (1) Polyurethane materials are composed of soft and hard segments, and the performance of polyurethane materials can be adjusted by changing the type and proportion of soft and hard segments; (2) Polyurethane material has excellent mechanical properties, and it is easy to be processed; (3) Polyurethane materials have excellent biocompatibility and low toxicity. For the synthesis of polyurethane, it is a kind of polymer material containing carbamate group (-NH-COO-), the main synthesis method is by polyether, polyester, or polycarbonate diols and diisocyanate for addition reaction, and then by chain extenders to expand the chain into polymer. The main chain of polyurethane is composed of soft and hard segments. Due to the thermodynamic incompatibility between the soft and hard segments, the performance of polyurethane is related to the chemical structure and proportion of the soft and hard segments, which further affects the performance of polyurethane dressing materials [13–18]. Conventional textile fiber wound dressings usually become infiltrated with wound secretions and newly formed soft tissue. The wound secretions and newly formed soft tissue infiltrate are difficult to remove, often resulting in secondary skin damage. In the early days, modern dressings composed of polyurethane polymers were reported to be more effective, comfortable, convenient, and economical compared to other traditional dressings. The advantages of polyurethane dressings have played an important role in outpatient settings and inpatient care [13–18]. However, drawbacks such as the inability to control leakage, increased cost of care, and poor cost-effectiveness of polyurethane polymer-related dressings have also been reported [16,19,20]. Although polyurethane-related dressings have been commercialized, there are still many functional deficiencies [21,22]. Currently, polyurethane dressings have been further improved to facilitate wound healing in the early stages and to reduce patient pain and discomfort, achieving the goal of minimizing wound healing time and improving cost-effectiveness [21,23]. The most common approaches to the development of innovative and improved polyurethane wound dressings include the synthesis and modification of biocompatible materials to improve biomedical performance, to overcome undesirable biological functions of polyurethane polymers, antimicrobial functions, and to impart mechanical and thermal properties of biomolecules. In addition to delivering unique and versatile functionality, these new polyurethane polymeric materials also perform specific biochemical functions, making them as the ideal wound dressings [24].

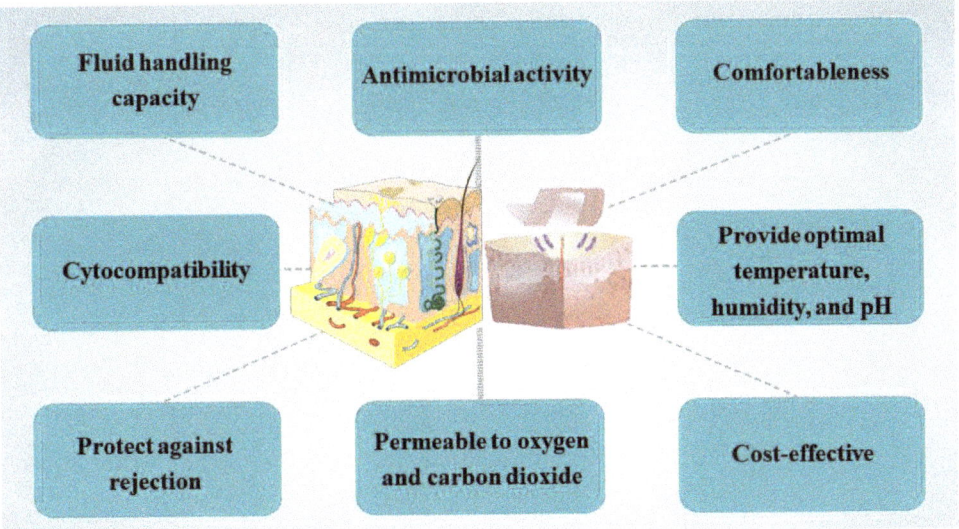

Figure 2. Characteristics possessed by an ideal dressing [8–11].

Up to now, many investigators have published reports on the effects of modified polyurethane polymer dressings on wound healing, but few of them have analyzed research in the field. The purpose of this work is to analyze the research results of polyurethane-related skin wound dressings, and to provide support for further research in related fields by comprehensively sorting out and analyzing the unsolved problems.

2. Methods

2.1. Search Strategies

The literature search was conducted using PubMed and Web of Science. Articles published with "polyurethane", "dressings", and "wound" as titles or abstracts from 2019 January to 2023 October were retrieved. The search equation used for PubMed was ((polyurethane [Title/Abstract]) AND ((dressings [Title/Abstract]) OR (wound [Title/Abstract]))). The search equation used for Web of Science was (TS = (polyurethane) AND (TS = (dressings) OR TS = (wound))).

Study inclusion criteria included (1) studies of polyurethane composite dressing synthesis methods and characterization, (2) studies on biocompatibility and medical performance assessment of dressings, and (3) studies involving natural polymers, synthetic polymers, or other bioactive ingredients.

Study exclusion criteria included (1) studies that did not involve research on wound dressings, such as studies on e-skin composed of polyurethane, (2) reviews and meta-analyses, (3) papers that were not in English, (4) papers that were not in the category of empirical research, (5) papers that had outdated ideas or repetitive arguments.

2.2. Search Results

A total of 125 articles were retrieved from PubMed. A total of 757 articles were retrieved from Web of science. Finally, 45 reports were selected based on the inclusion/exclusion criteria and were included for data extraction, as described in Sections 3.1–3.3.

2.3. Categorization and Display Strategies

Polyurethane composites are widely used as dressings to treat skin wounds. With additives that provide biocompatibility, polyurethane wound dressings can be functionally controlled. Polyurethane is usually combined with a polymer to form a conforming material, and these polymers are generally divided into natural polymers and synthetic polymers [11,25]. A natural polymer is composed of biomolecules derived from nature, such as microorganisms, animals, or plants that can mimic the original cellular environment and extracellular matrix very closely. The main natural polymers used in synthetic dressings include silk protein, pullulan, chitosan, cellulose, alginate, glucan, collagen, elastin, carrageenan, pectin, agarose, hyaluronic acid, fibrin, chitin, and gelatin [26,27]. Synthetic polymers comprising the dressing include polyglycolic acid, polyethylpyrodanone, polylactic acid, polyhydroxyethyl methacrylate, polycaprolactone, polyvinyl alcohol, and polylactic acid-co-glycolic acid [28]. Furthermore, other bioactive ingredients, such as essential oils, dextrans, cells, acellular matrices, propolis, vitamins, growth factors, thyroid hormones, proteins, insulin growth factors, enzymes, and nanoparticles that fight bacteria, are also used in the synthesis of wound complex dressings [27,29].

3. Results

3.1. Natural-Based Polymeric Wound Polyurethane Dressing

Many natural materials, usually including collagen, chitosan, hyaluronic acid, vegetable oil, tannic acid, thymol, lignin, and some animal sources of skin, have been successfully used in the production of polyurethane dressings [26,27].

3.1.1. Collagen

Collagen dressing is highly absorbent, can control wound exudation, and protect the wound. Blending collagen products with polyurethane materials can effectively improve the performance of collagen dressings [30]. Composite hydrogels based on collagen crosslinked with polyurethane and metal-organic frameworks (MOFs) with aluminum as metallic center were synthesized by the microemulsion method. It was found that the entanglement of polyurethane, collagen, and MOFs was made by hydrogen and coordination bonds promoted by the chemical structure of the MOF, leading to a semi-crystalline rough surface with interconnected porosity and aggregates of round shape, enhancing the thermal degradation resistance, mechanical degradation resistance and biocompatibility [31].

3.1.2. Chitosan

Chitosan is a linear polysaccharide with acetyl and amine group branches, and in acidic conditions the amino group is converted to a polycation (type IV amine). It is a biocompatible, non-toxic, and biodegradable biopolymer that can be used in the manufacture of a wide range of medical materials [32]. Studies have confirmed the benefit of adding the chitosan to polyurethane (PU) polymers and the PU/chitosan scaffolds developed by electrostatic spinning, which allows for the formation of a homogeneous structure in the scaffold fibers. The effect of PU/chitosan on the morphology and cellular activity of fibroblasts was assessed and it was found that this scaffold became a more favorable environment for fibroblast survival and growth. This suggests that PU/chitosan dressing can be a potential wound dressing. In addition, the study proposes another beneficial polysaccharide, hyaluronic acid [33], as described in Section 3.1.3.

3.1.3. Hyaluronic Acid (HA)

Hyaluronic acid, a non-sulfated glycosaminoglycan, is a major component of the skin's extracellular matrix and is involved in the processes of angiogenesis, inflammatory response, and tissue regeneration. Due to its excellent biocompatibility, biodegradability and hydrophilicity, HA has been widely used in the synthesis of wound dressings [34]. The preparation of PU/St (starch) and PU/St/HA core-shell nanofibers was accomplished by electrostatic spinning. To evaluate the properties of PU/St nanofibers and PU/St/HA nanofibers in vitro, mouse fibroblasts were used. For the purpose of evaluating cell survival and proliferation, the MTT assay was employed. It was found that PU/St/HA core-shell scaffolds did not significantly alter cell survival and proliferation, and that they were more biocompatible and did not cause cytotoxicity. In vivo studies in rats have shown that core-shell PU/St/HA wound dressings keep the skin moist, do not produce excessive wound exudate, have a higher quality of tissue repair, and confer faster wound healing [35]. At −20 °C, a dihydrazide-modified waterborne biodegradable polyurethane emulsion (PU-ADH) and oxidized hyaluronic acid (OHA) were autonomously crosslinked to form a hybrid hyaluronic acid−PU (HA-PU) cryogel by hydrazone bonding. Through the specific macroporous structure (~220 µm) formed by the polymerized PU-ADH particles and long-chain OHA, the dried cryogel swelled rapidly within minutes and could absorb blood or water up to 16 and 22 times its dry weight. This instantaneous shape recovery capability facilitated rapid hemostasis in minimally invasive procedures. In addition, the cryogel had a greater biocompatibility than gauze, enhanced blood coagulation, and activated endogenous coagulation after about 2 min of use. Using the same composition as HA-PU low-temperature gels, injectable HA-PU hydrogels with good self-healing properties were prepared at room temperature. In vivo evaluations of animals demonstrated that the cryogel was extremely effective in rapid wound healing, reduced immune-inflammation, and promoted angiogenesis and regeneration of hair follicles [36].

3.1.4. Vegetable Oil

Vegetable oils are one of the most important biomass raw materials for synthesizing polymers. The main components of vegetable oils are triglycerides. In addition, there are also some highly reactive active sites in vegetable oil, including double bonds, hydroxyl groups, and ester groups, which provide the possibility for the preparation of various polymers with different structures and functions. Due to the wide source, renewable, non-toxic, and biocompatibility characteristics, vegetable oil-based polymers are widely used in the production of biomedical materials [37]. A novel soybean oil polyol with built-in urethane and quaternary ammonium groups was synthesized by a non-isocyanate route using environmentally friendly and renewable carbonated soybean oil as raw material. Polyurethane wound dressing was prepared by the reaction of isophorone diisocyanate, castor oil, and this polyol. Antimicrobial activity and cytocompatibility of the dressing were good. Depending on the hydration state and the dry state, the dressing can have a tensile strength of 5 MPa and 17 MPa. The equilibrium rate of water absorption was 50%, and the water vapor transmission rate was 390 g per square meter per day. Evaluation of an optimized dressing on a full-layer unsterilized wound showed that wound healing progresses well, with the regenerated skin achieving a tensile strength of about 80% of normal healthy skin at day 21 post-injury [38].

3.1.5. Tannic Acid (TA)

TA is a naturally occurring plant-derived polyphenol. TA can be used as a crosslinking agent by supramolecular or physicochemical methods and is widely used in the production of skin adhesives and wound dressings [39]. Biologically active biomimetic skin hydrogel band-aids could be developed utilizing imidazolidinyl urea-based reinforced polyurethane (PMI) combined with TA (TAP hydrogels). As a result of multiple non-covalent interactions between TA and the PMI hydrogel network, the mechanical properties of the TAP hydrogel were strengthened, and the wound dressing's structural integrity could be maintained under local stresses. For its excellent self-recovery and anti-fatigue properties, TAP30% hydrogel provides excellent comfort. Owing to its good anti-moisture, adhesive, organ hemostasis, excellent anti-inflammatory, antioxidant properties, and antimicrobial activity, when diabetic mice were treated with TAP hydrogel, they were able to recover from skin incisions and defects more quickly. The therapeutic efficacy of TAP hydrogel was further investigated and shown to be effective in a diabetic mouse model infected with *Staphylococcus aureus* [40]. A study has shown a water retaining separable adhesive hydrogel wound dressing composed of TA. The incorporation of TA with abundant catechol moieties provided the hydrogel with improved mechanical properties, good tissue adhesion, and hemostatic ability. Then, a hydrophobic polyurethane-related coating was encapsulated on the surface of the hydrogel to maintain a high water content of the hydrogel for a long time [41].

3.1.6. Thymol

A versatile portable electrostatic spinning device has been created, featuring an adjustable perfusion rate and a high voltage capacity of up to 11 kV. Thymol, a natural antimicrobial compound, was doped into ethanol-soluble polyurethane (EPU) skin-like W&B nanofibrous membranes to give them antimicrobial activity. EPU-like skin-type waterproof and breathable nanofiber membranes with antimicrobial activity were prepared using a customized device. Excellent uniformity of structure was observed in the final nanofibrous membrane, which is composed of fluorinated polyurethane (FPU), EPU, and thymol. The membrane has a tensile stress of 1.83 MPa, and a tensile strain of 453%. The permeability is 3.56 kg m^{-2} d^{-1}, hydrostatic pressure is 17.6 cm H$_2$O, and antimicrobial activity is high [42].

3.1.7. Lignin

In the pathophysiology of wounds, a number of external and internal factors contribute to impaired wound healing. In particular, oxidative stress is an important factor in inhibiting wound healing [43]. There is uncertainty about the biocompatibility of lignin, which is a plant-derived antioxidant, and it remains underdeveloped as a biomaterial limiting its biomedical applications [44]. A lignin nanogel has been developed and explored for its therapeutic effects in skin wounds. Lignin derived from coconut shells shows good antioxidant properties. In a thermosensitive nanogel based on polyethylene glycol (PEG), polypropylene glycol (PPG), and polydimethylsiloxane (PDMS) polyurethane copolymers, lignin was incorporated. No significant cytotoxicity was observed with nanogels containing doped lignin. As the result of lignin nanogel antioxidant properties, oxidative stress-induced apoptosis in LO2 cells was prevented. In a mouse burn wound model, lignin nanogels accelerated wound healing, a result further supported by immunostaining for the cell proliferation marker Ki67. In this regard, lignin nanogels may prove useful as a wound dressing that promotes wound healing through its antioxidant properties [45]. For wound dressings, porous nanocomposite polyurethane foams that contain nanolignin (NL) and coated with natural antimicrobial propolis had also been reported. The foams were soaked in ethanol extract of propolis (EEP) after synthesis. NL and EEP coatings improved the foams' hydrophilicity as measured by contact angle. Furthermore, the EEP coating enabled the dressing to have significant antimicrobial effect and good cytocompatibility. The effectiveness of PU-NL/EEP on wound healing has also been demonstrated in a rat whole skin wound model [46].

In addition, one-step foaming methods have also been proposed for lignin-based polyurethane foams (LPUFs), where fully biobased polyether polyols partially replace petroleum-based feedstocks. LPUF skeletons contain trace amounts of phenolic hydroxyl groups (~4 mmol) that act as a direct reducing and capping agent for silver ions (<0.3 mmol). A lignin composite foam has been developed with improved mechanically and thermally properties [47].

3.1.8. Peppermint Extract

Wound dressings containing herbal extracts with high antimicrobial properties and a nanoscale-controlled release system have been shown to facilitate the healing of ulcerated wounds [48,49]. Herbal extracts of peppermint have been used to treat bacteria and inflammation [50], and a novel mint extract added to polyurethane-based nanofibers has been shown to be useful for diabetic wound healing. In order to optimize the release of the extracts, gelatin nanoparticles (CGN) have been crosslinked with the extracts and ultimately incorporated into nanofibers. Direct incorporation of extracts into a polyurethane matrix also controlled extract release. With an antimicrobial rate of 99.9%, the wound dressing was able to absorb *Staphylococcus aureus* and *Escherichia coli*. The in vivo study found that this extract improved wound healing after using this extract as an active compound. Inflammation is significantly reduced in wounds treated with nanofiber extracts, according to histopathological studies. In addition, skin of treated individuals had characteristics more similar to normal skin, including the epidermis exhibited thinning, the reticular ridges appeared normal, and the appendages grew back [51].

3.1.9. Gelatin

Gelatin is one of the most biodegradable and biocompatible polymers derived from the hydrolysis of collagen. It helps in cell adhesion and speeds up the healing process of wounds, but it is prone to degradation [52]. Blending of natural and synthetic polymers could improve structural stability. Introducing 20% PU to gelatin scaffolds (Gel80−PU20) results in a significant increase in the degradation resistance, yield strength, and elongation of these scaffolds without altering the cell viability. In vivo studies using a mouse excisional wound biopsy grafted with the scaffolds reveals that the Gel80−PU20 scaffold enables greater cell infiltration than clinically established matrices [53]. Personalized medicine is made possible by three-dimensional (3D) printing of soft biomaterials. By developing different forms of 3D-printed biomaterials, artificial organ fabrication can be facilitated and desired properties can be incorporated into biomaterials. In order to develop 3D-printable gelatin methacryloyl (GelMA) polyurethane biodegradable hydrogels and cryogels, GelMA was combined with dialdehyde-functionalized polyurethane (DFPU). A 3D-printed biomaterial with high print resolution, smart functionality, and biocompatibility was presented by the GelMA-PU system, demonstrating a combination of self-healing and 3D-printing capabilities. With GelMA-PU, the ink pool for biomaterial 3D printing has been expanded, allowing applications such as tissue-engineered scaffolds, minimally invasive surgical instruments, and electronic wound dressings [54]. It has also been shown that an absorbable gelatin sponge combined with a polyurethane film could be effectively used for skin reconstruction of bone or tendon exposed wounds [55].

3.1.10. Dextran

Dextran is a polysaccharide with good biocompatibility and degradability and an interfering effect on coagulation and hemostasis, which can be used to compensate for the adverse effects of antimicrobial agents in wound dressings. It has great potential for application in medical materials and tissue engineering [56]. To develop antimicrobial wound dressings, pH-stimulated drug release nanofiber membranes of polyurethane/dextran were developed. Dextran was added to polyurethane to increase hydrophilicity, air permeability, percent adsorption value, and biodegradability. Dextran can also be used as a reinforcing filler in polyurethane matrices. Dextran induces high platelet adhesion and hemostasis, which is essential for promoting the wound healing process. In addition, 20 wt% dextran-loaded membrane (PU/20D) enhanced cell proliferation, attachment, and survival of fibroblasts [57]. It has been shown that polyurethane prepolymers could be made into wound dressings by sol-gel hydrolysis polycondensation reaction and surface modification by dextran. The biological properties of the final dressings were improved, and the dextran anhydride modification resulted in dressings with low hemolysis rates and prolonged clot formation [58].

In summary, the advantages of natural polymers as a source of dressings are that they are widely available, renewable, degradable, non-toxic, and biocompatible. The disadvantages are the complex structure of natural polymers, complicated extraction process, and poor mechanical properties. The chemical structure of natural-based materials is shown in Table 1, and the content of the natural-based polyurethane materials discussed above are shown in Table 2.

Table 1. Chemical structure of natural-based materials.

Material	Structure	Material	Structure
Collagen		Tannic acid	
Chitosan			
Hyaluronic acid		Thymol	
		Lignin	

Table 2. Summary of methods and properties of natural-based polyurethane materials.

Natural Product	Process and Method	Research Models	Characterization
Collagen [31]	Microemulsion.	MTT Assay; antibacterial test; hemolysis test.	Enhances the mechanical properties and biocompatibility.
Chitosan [33]	Electrospinning.	MTT assay; trypan blue exclusion assay; DAPI staining.	Biocompatible.
Hyaluronic acid [35,36]	Coaxial electrospinning technique; crosslinked.	L929 cell viability; rat wound model; rat liver hemostasis model.	Biocompatible; non-toxic; promotes cell adhesion; shape-restoring ability; anti-inflammatory; enhances angiogenesis and regeneration of hair follicles.
Vegetable oil [38]	Polyaddition.	L929 cell viability; rat wound model.	Tensile strength; retention of moisture; cytocompatibility.
Tannic acid [40]	Polyaddition.	Diabetic mouse wound model.	Hemostatic; anti-moisture adhesion; anti-inflammatory; antioxidant.
Thymol [42]	Electrospinning.	Antimicrobial test.	Stretchable; breathable; moisturizing.
Lignin [45,46]	Dialysis; freeze-drying.	Oxidative stress model of LO2 cells; mouse burned skin model.	Antioxidant; promotes cell proliferation; non-cytotoxic; absorbency.
Peppermint extract [51]	Electrospinning.	MTT Assay; antibacterial test; diabetic rat wound model.	Anti-inflammatory; absorbent; promotes functional skin regeneration; antibacterial.
Gelatin [54]	3D printing; dialysis.	hMSCs culture in GelMA-PU cryogel.	High printing resolution; biocompatibility; adhesive, light transmittable; biodegradable.
Dextran [57,58]	Electrospinning.	In vitro degradation studies; vapor transmission rate analysis; blood compatibility evaluation; antibacterial activity.	Good hydrophilicity, water vapor permeability, adsorption rate and biodegradability, and promotes platelet adhesion and hemostasis.

3.2. Synthetic Polymer and Inorganic Modified Polyurethane Dressings

Synthetic polymers are chemically synthesized in the laboratory and are also known as artificial polymers. In order to improve the biological properties of such polymers to reach their potential as wound dressings, various surface and bulk modifications have been applied [28].

3.2.1. Povidone-Iodine (PVP-I)

Povidone-iodine has potent broad-spectrum activity against bacteria, virus, fungus, and protozoa [59]. PVP-I polyurethane dressing (Betafoam) is a new type of polyurethane dressing impregnated with 3% of PVP-I [60]. For the first time, the effect of PVP-I dressings on split-thickness skin graft donor area wounds was validated in a clinical case. The efficacy and safety of PVP-I dressing were compared with that of vanilla oil gauze and cellular water dressing. This was primarily determined by observing the degree of donor site epithelialization. PVP-I dressing provided better wound healing with significantly shorter time to complete epithelialization (approximately 14 days). PVP-I foam dressing allowed for easier wound care, less bleeding and easier removal of dressing adhesion, and better exudate management. It offers significant clinical advantages and is cost-effective [61]. Betafoam has been verified to be effective in wound healing in a rat skin healing model, showing good performance in re-epithelialization, angiogenesis, collagen deposition, and tissue invasion [62].

3.2.2. Polyacrylamide (PAAm)

In tissue engineering, drug delivery, smart materials, and drug delivery systems, PAAm is a polymer used extensively for its excellent mechanical properties, hydrophilicity, and biocompatibility [63]. Waterborne polyurethanes are able to provide many functional groups that make polyurethanes easy to functionalize when interacting with other chemicals [6,64]. After rapid-curing by UV, a mechanically flexible PU-PAAm hydrogel skin dressing with good adhesion was developed. The polyurethane component of the PU-PAAm hydrogel acts as a "bridge", accelerating the interpenetrating polymer network (IPN) formation, which consists of a physically crosslinked polyurethane network surrounded by a chemically crosslinked PAAm network. Due to its unique IPN structure, the hydrogel is exceptionally stretchable and ductile. During application, the hydrogel and skin form hydrogen bonds and electrostatic interactions, which ensured strong adhesion, and the dressing is applied without irritating the skin and causing skin damage. L929 fibroblast experiments were used to validate the biocompatibility of PU-PAAm hydrogel, and rabbit skin wound healing experiments further confirmed the remarkable skin regeneration-stimulating ability of PU-PAAm hydrogel [6]. Using super-tough thermoplastic polyurethane (HTPU) hydrogel and chemically crosslinked PAAm as the first and second network, HTPU/PAAm double-network hydrogels were synthesized by a one-step radical polymerization in a study. The toughness and strength of this polyurethane-related hydrogel were greatly improved, and it has broad application prospects in wound dressing [65].

3.2.3. Polycaprolactone (PCL)

Polycaprolactone is an important polymer with good mechanical properties, miscibility with other polymers, and biodegradability [66]. PCL/Gel scaffolds have shown significant value in skin tissue engineering. However, these scaffolds have poor antimicrobial properties and are unsuitable for water vapor transmission [67,68]. PCL/Gel scaffolds are electrostatically spun on a dense membrane consisting of polyurethane/ethanolic extract of propolis (PU/EEP). As an upper layer, PU/EEP membranes protect the wound from external contamination and dehydration, and the PCL/Gel scaffolds act as a lower layer to promote cell proliferation and adhesion. Antimicrobial assays showed significant antibacterial activity against *Staphylococcus aureus*, *Escherichia coli*, and *Staphylococcus epidermidis*. The PU/EEP-PCL/Gel bilayer dressing had high hydrophilicity, biocompatibility, and biodegradability. In vivo experiments demonstrated that the double-layer wound dressing significantly promoted skin wound healing and collagen deposition in Wistar rats [68]. A two-layer wound dressing has been prepared using an electrostatically spun PCL/CS fiber mat as the inner layer, and polyurethane foam-coated EEP as the top layer. An electrostatically spun mat consisting of uniform nanofibers with enhanced hydrophilicity, swelling rate, and degradation properties is prepared by mixing PCL and CS solutions [69].

3.2.4. Polylactic Acid (PLA)

Polylactic acid (PLA) is biodegradable and biocompatible, and it is a polymer widely used in biomedical materials. However, the brittleness and weak mechanical properties of polylactic acid nanofibers limit their application. PU has excellent elasticity and mechanical properties suitable for specific tissues. When PLA and PU are used together, in addition to improving the mechanical properties of wound dressings, they can also promote biodegradation [70]. When PLA is added to wound polyurethane dressings (PU/PLA, 50/50, w/w), the dressings absorb wound exudates, dry quickly, are comfortable, and have high biocompatibility to support fibroblast growth [71,72]. In a study, novel hollow nanofiber materials were produced by the coaxial electrospinning method from PU/PLA blend nanofibers of different weight ratios (20:80, 40:60, 50:50, 60:40, and 80:20). Moreover, hollow PU/PLA nanofibers were observed to be 2–4 times thinner than solid PU/PLA nanofibers. The production of hollow nanofibers in the range of 235–518 nm was achieved.

It was determined that the biomedical material, which has the highest liquid absorption capacity with a value of 756% and can dry in 10 min, is PU/PLA (50/50, w/w) nanofiber. [72]. In addition, a prospective, comparative, randomized clinical study showed that polylactic acid membrane could improve the prognosis of cracked skin graft donors [73].

3.2.5. Polyethylene Glycol (PEG)

PEG is formed by the stepwise addition polymerization of ethylene oxide with water or glycol. Polyethylene glycol polymer, because of its good water solubility and good compatibility with many organic components, has been widely used. Chen et al. prepared a PEG and triethoxysilane (APTES) modified high absorption polyurethane foam dressing (PUESi) [74]. The research results indicated that PUESi dressings not only had good adhesive resistance and deformation absorption ability, but also could effectively promote wound healing. Pahlevanneshan et al. synthesized polyurethane foam with polyethylene glycol, glycerin, nano-lignin (NL), 1, 6 diisocyanate hexane and water as foaming agents, and soaked it with propolis ethanol extract (EEP) [46]. The results indicated that PU-NL/EEP material had high cell viability and cell adhesion, and in vivo wound healing experiments were conducted using a Wistar rats' full-thickness skin wound model, confirming that PU-NL/EEP material exhibited high wound healing effects. In addition, Vakil's team developed a polyurethane-based polyethylene glycol hydrogel with cell compatible shape memory function, and then physically mixed plant-based phenolic acid onto the hydrogel scaffold so that it could be easily transported to the wound site, thereby increasing the wound healing efficiency and reducing the risk of infection [75].

3.2.6. Polyvinyl Alcohol (PVA)

PVA is a white, stable, and non-toxic water-soluble polymer made from vinyl acetate through polymerization and alcoholysis. It is an extremely safe organic polymer with non-toxic and good biocompatibility, widely used in wound dressings and artificial joints. Hussein's team used dual spinneret electrospinning technology to prepare polyurethane and polyvinyl alcohol gelatin (PVA/Gel) nanofiber scaffolds [76]. By adding cinnamon essential oil (CEO), the inhibitory effect of loaded low-dose nanoceria PU/PVA Gel NFs on Staphylococcus aureus was improved, and the therapeutic effect on diabetes wounds was effectively improved. Carayon's team prepared multiple types of polyurethane-polylactic acid-polyvinyl alcohol composite porous matrices (CPMs), and the research results showed that the average porosity of CPMs was 69–81%, making their pore size more suitable for skin regeneration [77].

3.2.7. Tributylammonium Alginate Surface-Modified Cationic Polyurethane (CPU)

Disintegration of membranes and death of bacteria can be caused by positively charged cationic polymers. Low cytotoxicity and long-lasting antibacterial activity are the main advantages of cationic polymers. A transparent tri-butylammonium alginate CPU skin dressing has been created for use in the treatment of full-thickness wounds. The surface-modified polyurethane of this dressing has improved hydrophilicity and tensile Young's modulus between 1.5–3 MPa, which is close to natural skin. MTS and scratch assays were used to assess cell viability and showed that the dressings were cytocompatible and could promote fibroblast migration. Surface-modified CPU polymers are highly inhibitory to Gram-positive *Staphylococcus aureus* and Gram-negative *Escherichia coli* bacteria. In vivo experiments in rats showed that surface-modified CPU dressings promote the rate of wound healing, shorten the period of persistent inflammation, enhance collagen deposition, and promote blood vessel formation [78].

3.2.8. Cellulose Acetate/Polyurethane Nanofibrous Mats Containing Reduced Graphene Oxide/Silver Nanocomposites and Curcumin

Nanofiber scaffolds can be prepared by electrostatic spinning using polyurethane and cellulose acetate as raw materials. Reduced graphene oxide/silver nanocomposites have strong antimicrobial activity. In order to prevent the aggregation of silver nanoparticles (AgNPs), AgNPs were loaded onto reduced graphene oxide and nanocomposites were prepared using a green and convenient hydrothermal method. The electrostatic spinning method was used to prepare scaffold materials containing reduced graphene oxide/silver nanocomposites, curcumin, or both. The MTT cell proliferation assay showed that the scaffold has good biocompatibility. Evaluation of antimicrobial activity showed that the scaffold is able to inhibit both Gram-negative and Gram-positive bacteria. In vivo experiments and histopathological studies showed that scaffolds containing graphene oxide/silver nanocomposites and curcumin can promote the rate of skin wound healing, suggesting that nanomaterials have a good biomedical potential for wound healing [79].

3.2.9. Nanosized Copper-Based Metal-Organic Framework

Nano-Cu-BTC (copper (II)-benzene-1, 3, 5-tricarboxylate) is doped into polyurethane foams (PUF), through a polyaddition reaction of castor oil and chitosan with toluene 2, 4-diisocyanate, to improve the functionality of the dressing by modifying the PUF surface. The physical and thermal properties of Nano-Cu-BTC-PUF (PUF@Cu-BTC) were compared with those of control PUF, including swelling rate, phase transition, thermal gravity loss, and cytocompatibility, and they were evaluated for inhibitory activity against methicillin-resistant *Staphylococcus aureus*, Pseudomonas aeruginosa, and Klebsiella pneumoniae. The antimicrobial activity of PUF@Cu-BTC against the tested bacteria is significant and selective, and its cytotoxicity toward mouse embryonic fibroblasts is low. According to Cu (II) ion release assay, PUF@Cu-BTC is stable for 24 h in phosphate-buffered saline. Because PUF@Cu-BTC displays selective bactericidal activity and low cytotoxicity, it has potential for use as a skin wound dressing [80].

3.2.10. Silver

Silver has strong antimicrobial activity and is commonly used in wound care. However, silver also has the potential to have toxic effects on skin cells, which can affect wound healing. There is some evidence that short-term use of dressings containing nanosilver is feasible in infected wounds, but the use of silver-containing dressings on clean wounds and closed surgical incisions is not appropriate. Ideal silver preparations are silver nanoparticles (AgNPs) and silver-coated polyurethane dressings for negative pressure wound therapy [81]. A new method of incorporating AgNPs onto the surface of polyurethane nanofibers has been proposed. Before the electrospinning process, $AgNO_3$ and tannic acid were added to PU solution to make the AgNPs uniformly distributed on the surface of PU nanoparticles [82]. Thiol-terminated polyurethane prepolymers with two different molecular weight PEGs and terminated propargyl polyurethane crosslinkers have been prepared and polymerization reactions carried out with and without the addition of silver salts. A radical-mediated step-growth polymerization reaction and resultant thioether linkages created during polymerization lead to high conversion of the starting macromonomer and the formation of a hydrophilic network. Even in the hydrated state, the materials offer desirable dimensional strength and flexibility, as evidenced by its high tensile strength, good extensibility, and minimal permanent set. The reduction in silver salt during network formation both from reaction with free radicals and residual DMF solvent available in the reaction medium led to the formation of AgNPs. This dressing showed little toxicity to fibroblasts, high bactericidal and fungicidal activity, and good biocompatibility. No significant reduction in cell migration was observed with AgNPs dressings [83]. In addition to antibacterial properties, studies have also shown that the addition of AgNPs can improve mechanical properties (tensile strength and elongation at break) [71,84].

In summary, the advantages of synthetic polymers as a source of dressings are their low cost, defined structure, tunable properties, good mechanical properties, high chemical stability, and good antibacterial ability. The disadvantages are the complexity of synthetic polymer synthesis, single performance, and environmental unfriendliness. The monomer chemical structure of synthetic materials is shown in Table 3, and the contents of synthetic polymer and inorganic modified polyurethane materials discussed above are shown in Table 4.

Table 3. Monomer chemical structure of synthetic materials.

Acrylamide	(structure)	Ethylene glycol	(structure)
Caprolactone	(structure)	Vinyl alcohol	(structure)
Lactic acid	(structure)	Cellulose acetate	(structure)

Table 4. Summary of methods and properties of synthetic polymer and inorganic modified polyurethane materials.

Synthetic Polymer and Inorganic Modified Polyurethane	Process and Method	Research Models	Characterization
Povidone-iodine [61,62]	Maceration.	Rat full-thickness skin defect model; prospective randomized case studies.	Promotes re-epithelialization, angiogenesis, collagen deposition, tissue invasion; absorbent.
Polyacrylamide [6]	One-pot method.	L929 fibroblast cytocompatibility assay; rabbit full-thickness skin defect model.	Superior stretch and ductility; adhesion; water absorption; moisture retention; antimicrobial; breathability.
Polycaprolactone [55]	Electrospinning.	L929 fibroblast cytocompatibility assay; rat wound model.	Hydrophilic; biodegradable; promotes collagen deposition; antimicrobial.
Polylactic Acid [71,72]	Polyaddition, electrospinning.	L929 fibroblast cytocompatibility assay.	Water absorption; biocompatibility.
Polyethylene glycol [74,75]	Self-foaming reactions.	Nondiabetic and diabetes mellitus rat wound models.	Absorbency and antiadhesion properties.
Polyvinyl alcohol [77]	SC/PL technique.	MTT assay; microbiology tests; cytotoxicity assay.	Antimicrobial; cytocompatible.
Tributylammonium alginate surface-modified cationic polyurethane [78]	Supramolecular ionic interactions.	Human dermal fibroblast model; infected and non-infected wounds in a rat full-thickness skin defect model.	Promotes fibroblast migration; hydrophilic; anti-inflammatory; promotes collagen deposition, angiogenesis; antibacterial.
Cellulose acetate/polyurethane nanofibrous mats containing reduced graphene oxide/silver nanocomposites and curcumin [79]	Improved Hummer method; hydrothermal method; electrospinning.	MTT assay using MEF cells; antibacterial test; C57 mouse wound model.	Moisturization; antimicrobial; promotes regeneration of the epidermal layer.

Table 4. Cont.

Synthetic Polymer and Inorganic Modified Polyurethane	Process and Method	Research Models	Characterization
Nanosized copper-based metal-organic framework [80]	Crosslinking.	Antibacterial test; cytotoxicity assay mouse embryonic fibroblasts.	Selective antimicrobial capacity; cytocompatibility.
Silver [83]	Blending and light curing.	L929 fibroblast cytocompatibility assay and scratch assay; antibacterial test.	Antimicrobial; permeable to oxygen and carbon dioxide; tensile strength.

3.3. Polyurethane Dressings Loaded with Other Bioactive Ingredients

Although polyurethane-related dressings are physiologically, mechanically, and economically superior to other dressing materials, they have poor healing capabilities and are considered passive wound dressings. Therefore, bioactive additives such as growth factors, biomolecules, or cells have been applied to polyurethane foam dressings to improve their healing qualities, and they are particularly suitable for the treatment of complex wounds (e.g., infected wounds, burn wounds, and diabetic wounds) that cannot be healed with conventional dressings [85].

3.3.1. Multipotent Adult Progenitor Cells (MAPCs)

MAPCs are non-hematopoietic adherent cells derived from bone marrow, and preclinical evaluations of MAPCs have shown significant therapeutic benefits in improving tissue regeneration [86]. Advanced dressings for the delivery of MAPCs have been greatly developed in recent years [86,87], and a polyurethane-related dressing for the delivery of MAPCs is described below. A free radical-rich layer has been produced, by plasma immersion ion implantation (PIII) on medical polyurethane dressings that can attach biomolecules rapidly and covalently. The reactivity of polyurethane treated with PIII was used to immobilize the extracellular matrix protein tropoelastin, which could still maintain a functional conformation after sterilization with medical-grade ethylene oxide. MAPC adhesion and proliferation were promoted by tropoelastin-functionalized patches treated with PIII while preserving their cellular phenotype. In a topically applied MAPC patch, cells transfer to wounds on the skin, and untransferred MAPCs fill in the patch surface for subsequent cell transfer. Using such a wound patch, MAPCs and cytokines can be continuously delivered, enabling its use as a large-area dressing [88].

3.3.2. Platelet Lysate

Chronic skin lesions are difficult to heal due to reduced levels and activity of endogenous growth factors. The platelet lysate, obtained by repeated freeze–thawing of platelet-enriched blood samples, is an easily attainable source of a wide range of growth factors and bioactive mediators involved in tissue repair [89]. A bilayer fibrin/polyether polyurethane scaffold loaded with platelet lysate is made by a combination of electrostatic spinning and spray phase-inversion. Enzyme-linked immunosorbent assays and fibroblast proliferation have been used to detect release and bioactivity of growth factors released from platelet cleavage scaffolds. Bilayer fibrin/polyether polyurethane scaffolds loaded with platelet lysate sustain the release of biologically active platelet-derived growth factors in vitro. An in vivo experiment revealed that the scaffold helped diabetic mice heal wounds more quickly. Histological results showed that platelet lysate and growth factor-loaded scaffold promoted collagen deposition and re-epithelialization in wounds of diabetic mice [90].

3.3.3. Exosomes

Elevated oxidative stress, infection, reduced angiogenesis, and subsequent hypoxia are key factors in the non-healing of chronic diabetic wounds. The management and successful treatment of diabetic wounds remains a major therapeutic challenge, and the development of biological dressings with the ability to deliver oxygen, induce angiogenesis, and protect against oxidative stress and infection is important for the treatment of diabetic wounds [91]. Exosomes are cell-derived vesicles that carry large amounts of growth factors and tiny RNAs that maintain cellular homeostasis and regulate intercellular communication, including wound healing and angiogenesis [92]. OxOBand wound dressing is loaded with oxygen-releasing antioxidant exosomes, and it was developed specifically for promoting wound healing and skin regeneration in diabetic wounds. OxOBand is comprised of antioxidant polyurethane, a highly porous cryomaterial capable of sustained oxygen release, supplemented with adipose-derived stem cell (ADSC) exosomes. When applying ADSC exosomes and oxygen-releasing antioxidant scaffolds to a wound, fibroblasts and keratinocytes are able to attach, survive, migrate, and proliferate. In vivo results showed that OxOBand increases wound healing rate, re-epithelialization, and granulation tissue formation in diabetic rats. OxOBand treats diabetic wounds by promoting collagen remodeling, angiogenesis, and reducing oxidative stress [93,94].

3.3.4. Adipose Stem Cell (ADSC)-Seeded Cryogel/Hydrogel Biomaterials

Adipose-derived stem cells have emerged as a promising tool for skin wound healing, but their therapeutic potential is largely dependent on the cell delivery system [95]. Hydrogels and cryogel biomaterials with antimicrobial properties are prepared from glycol chitosan and a novel biodegradable Schiff base cross-linking agent, difunctional polyurethane (DF-PU). Such a cryogel has a water absorption of ~2730 ± 400%, abundant macropores, 86.5 ± 1.6% formed by ice crystals, and a cell proliferation rate of ~240%, and hydrogels exhibit considerable antimicrobial activity and biodegradability. An adipose stem cell-seeded cryogel/hydrogel dressing was applied to the wounds of diabetic rats, and then acupuncture in Chinese medicine was performed to promote wound healing. The wound healing rate was as high as 90.34 ± 2.3%, with the wounds forming granulation tissues with sufficient micro-vessels and completing re-epithelialization within 8 days. By activating C5a and C3a, increasing the expression of cytokines TGF-β1 and SDF-1, and down-regulating proinflammatory cytokines IL-1β and TNF-α, the combination of acupuncture and stem cell-seeded cryogel/hydrogel biomaterials led to synergistic immunomodulation of the wound [96].

3.3.5. L-Arginine (L-Arg)

L-Arginine is recognized as a conditionally essential amino acid for tissue growth in mature and juvenile mammals and has been used as a scavenger of reactive oxygen species in various species. In one study, polyurethane was used as a base polymer and blended with L-arginine to obtain desirable dressing properties such as better cell viability, cell attachment and proliferation, and enhanced antioxidant capacity of the dressing by blocking reactive oxygen species production [97]. A novel tissue adhesive (G-DLPU), constructed from L-Arg-based degradable polyurethane (DLPU) and GelMA, was prepared for wound care using the pro-angiogenic properties of L-Arg. After systematic characterization, G-DLPUs were found to have excellent shape-adaptive adhesion. In addition, the release of L-Arg during degradation and the production of NO were confirmed to contribute to wound healing. Biocompatibility was verified in in vivo experiments, and testing the hemostatic effect on damaged organs in a rat liver hemorrhage model showed that G-DLPUs reduced hepatic hemorrhage, with no significant inflammatory cells seen near the wound. Its therapeutic role in wound treatment was demonstrated in a mouse model of total skin defects, which showed that the hydrogel adhesive significantly improved the thickness of the neodermis and enhanced vascularization [98].

3.3.6. LL37 Peptide

Antimicrobial peptides (AMPs) have therapeutic potential for treating bacteria and promoting skin regeneration. AMP LL37 peptide, an endogenous peptide in human skin, belongs to the antimicrobial family, and has antimicrobial, angiogenic, and immunomodulatory properties. LL37 peptide interacts with surface receptors, such as the epidermal growth factor receptor, on keratin-forming and endothelial cells. EGFRs and surface receptors such as formyl peptide receptor-like-1 (FPRL1) of keratinocytes and endothelial cells, respectively, mediate the migration of these cells and promote wound healing [99]. These studies evaluated the antimicrobial and pro-regenerative effects of LL37 peptides immobilized on a polyurethane-based wound dressing (PU-adhesive-LL37 dressing). The PU-adhesive-LL37 dressing killed Gram-negative and Gram-positive bacteria in human serum after 16 antimicrobial test cycles without inducing bacterial resistance. Importantly, re-epithelialization and wound healing were enhanced and wound macrophage infiltration was reduced in mice, with type 2 diabetes, treated with this new dressing compared to polyurethane-treated wounds of animals. Treatment of wounds of diabetic mice for 6 days with PU-adhesive-LL37 dressing resulted in a decrease in pro-inflammatory factor expression. In addition, the new dressing did not induce an acute inflammatory response compared with the control group. In summary, PU-adhesiative-LL37NP dressing may prevent bacterial infection, promote tissue contact for wound healing, and induce anti-inflammatory and re-epithelialization processes in diabetic wounds [100].

3.3.7. Plasma Rich in Growth Factor (PRGF)

Growth factors such as PRGF serve as a rich source of active proteins that accelerate tissue regeneration. Animal experiments showed that PRGF-associated scaffolds contributed to skin wound healing and accelerated the formation of epidermal layers and skin appendages in rats [101]. In multilayered scaffolds created using PRGF from platelet-rich plasma, the outer layer is composed of polyurethane-cellulose acetate (PU-CA) fibers, while the inner layer is composed of PRGF-containing gelatin fibers. This approach, to prepare electrospun, biologically active scaffolds containing PRGF to induce cell proliferation and migration in vitro, is novel. Fluorescent images of fibroblast activity monitoring, using enhanced green fluorescent protein-labeled fibroblasts, showed that the migrating cell number on PRGF scaffolds was increased on day six. Real-time polymerase chain reaction analysis also revealed approximately 3-fold, 2-fold, and 2-fold increases in SGPL1, *DDR2*, and VEGF, respectively, on PRGF-containing scaffolds compared with cells migrating on PU-CA [102].

3.3.8. Tri-Cell-Laden (Fibroblasts, Keratinocytes, and Endothelial Progenitor Cells) Hydrogels

Inadequate supply of donor skin limits the potential for treating severe wounds, and ex vivo engineered cell regeneration methods have been introduced as a viable alternative that promises to replace autologous skin grafting as the standard of care. The prevascularized mucosal cell sheet containing cultured keratinocytes, plasma fibrin, fibroblasts, and endothelial progenitor cells showed in vivo efficacy and tissue plasticity in cutaneous wounds by promoting accelerated healing [103]. A promising therapeutic strategy for treating inhomogeneous wounds is to fabricate customizable tissue-engineered skin. A planar/curved bioprintable hydrogel has been created that holds promise for the production of tissue-engineered skin. The dressing was evaluated in a rat irregular and chronic wound model. Gelatin and polyurethane are the main components of the hydrogel. There is excellent 3D printing ability and structural stability with polymer loaded with the three cell types. Treatment of circular wounds in normal and diabetic rats with planar-printed triple-cell-loaded hydrogels showed complete re-epithelialization and healing of the wound, and there was an abundance of new vessels and collagen after 4 weeks. Large, irregular skin wounds in rats treated with curvilinear-bioprinted, triple cell-loaded hydrogels showed wound repair was achieved after four weeks [104].

3.3.9. Membranes Containing Mesoglycan and Lactoferrin

Thermoplastic polyurethane fiber membranes have been prepared using a uniaxial electrostatic spinning process. Fibers were then separately charged with two pharmacological agents, mesoglycan (MSG) and lactoferrin (LF), by supercritical CO_2 impregnation. MSG and LF are uniformly distributed in a microscale structure. Angular contact analysis confirmed the fulfillment of MSG-loaded hydrophobic and LF-loaded hydrophilic membranes. The impregnation kinetics indicated that the maximum loadings of MSG and LT were $0.18 \pm 0.20\%$ and $0.07 \pm 0.05\%$, respectively. Franz diffusion cells were used to simulate human skin contact in in vitro experiments. After 28 h, MSG release plateaued, whereas LF release plateaued after 15 h. Representing as human keratinocytes and fibroblasts, HaCaT and BJ cell lines have been evaluated for their compatibility with electrospun membranes in vitro [105].

In summary, the advantages of bioactive ingredients used as a source of dressings are significant therapeutic effects, good biocompatibility, and good antioxidant properties. The disadvantages are the high cost of bioactive ingredients, the scarcity of raw materials, the harsh storage conditions, and the difficulty of preservation. The contents of polyurethane dressings loaded with bioactive ingredients discussed above are shown in Table 5.

Table 5. Summary of methods and properties of polyurethane dressings loaded with bioactive ingredients.

Bioactive Ingredients	Process and Method	Research Models	Characterization
Multipotent adult progenitor cells [72]	Plasma immersion ion implantation; covalent attachment.	Human skin repair model.	Moisturizing; anti-hydrolytic; anti-inflammatory; modulates immune response; promote dermal and vascular regeneration; recruits other stem cells.
Platelet lysate [73]	A combination of electrospinning and spray.	Cell proliferation of mouse fibroblasts; diabetic mouse wound model.	Promotes capillary and collagen deposition; re-epithelialization; anti-inflammatory.
Exosomes [76,77]	Embedding.	Diabetic rat wound model; HaCaT, SH-SY5Y and NIH3T3 cell viability.	Enhances collagen deposition; increase neovascularization; reduces oxidative stress; promotes development of mature epithelial structures and hair follicle regeneration.
Adipose stem cell-seeded cryogel/hydrogel biomaterials [78]	Chemical synthesis.	Diabetic rat wound model; antibacterial testing.	Biodegradability; down-regulation of pro-inflammatory cytokines; angiogenesis; re-epithelialization.
L-Arginine [79]	Dialysis; freeze-drying.	Murine full-thickness skin defect model.	Shape-adaptive adhesion; biocompatibility; hemostasis; vascular regeneration; anti-inflammatory.
LL37 peptide [81]	Gum.	Antibacterial testing; cytotoxicity of human dermal fibroblasts; type II diabetic mouse wound model.	Antibacterial; anti-inflammatory; induces epithelialization.
Plasma rich in growth factor [83]	Electrospinning.	Human foreskin fibroblast cell viability.	Induction of fibroblast proliferation and migration.

Table 5. Cont.

Bioactive Ingredients	Process and Method	Research Models	Characterization
Tri-cell-laden (fibroblasts, keratinocytes, endothelial progenitor cells) [84]	3D Planar-/Curvilinear- Bioprinting.	Rat fibroblast and keratinocyte viability; circular wound models in normal and diabetic rats.	Promotes vascularization, collagen regeneration; re-epithelialization.
Mesoglycan and lactoferrin [85]	Uniaxial electrospinning; supercritical impregnation.	Human immortalized keratinocytes and human immortalized fibroblast viability.	Biocompatibility; moisture control capability.

4. Discussion

4.1. Fabrication Techniques

Polyurethane polymers used as wound dressings are receiving more and more attention from scholars [8,106]. Polyurethane polymers have a relatively clear basic chemical structure that can be easily altered to add specific functional groups, and the materials are rarely associated with disease transmission and immunogenicity problems. Polyurethane polymers with adjustable soft and hard segments and modifiable chain extensions are widely used as materials for biological applications. The ratio or composition of hard and soft segments can be manipulated to alter the physicochemical properties of polyurethane [36,107,108]. As the clinical requirements for the functionality of the dressings increase, researchers in related fields have tried to modify the polyurethane with different ingredients or polymers in order to enhance multiple biological functions and physical properties [25]. By performing key biological functions, natural polymers sustain life and allow organisms to adapt to their environment. The worlds of synthetic and natural polymers are almost separate because synthetic polymers lack some specific biological functions; biochemical reactions caused by synthetic polymers can sometimes be uncontrolled and unwanted. Biologically active synthetic polymers with antimicrobial activity, among others, have been developed due to recent advances in synthetic polymerization techniques, such as antimicrobial activity, among others [109]. However, synthetic materials are less biodegradable and biocompatible, and the materials usually need to be combined with natural or other synthetic polymers to achieve the desired healing effect.

The preparation methods of dressings are various. In this review, electrospinning, molding, blending, composite, foaming, fiber bonding, hybridization, perfusion, solvent-free ring opening polymerization, bionic strategy, microemulsion method, one-step foaming, sol-gel, melt blowing, photopolymerization, and solvent-free phase separation are mainly involved. Due to the specific functional requirements of dressings used for wound healing, they are usually required to have specific performance such as antibacterial, biocompatible, adhesive, hydrophilic, antibacterial, mechanical properties, etc. Therefore, in order to meet the above performance requirements, the preparation methods of polyurethane dressings are usually not single, and most involve two or more manufacturing methods. The wound dressing is mainly based on polyurethane resin, which is blended or copolymerized with organic/inorganic materials to prepare composite materials with specific functions. In addition, reducing or loading methods can also be used to combine antibacterial, anti-inflammatory, and pro-healing therapeutic factors with polyurethane materials to prepare wound dressings suitable for special requirements.

Biomass materials have the advantages of wide sources, abundant raw materials, non-toxicity, and good biocompatibility, which will show huge development space in the field of wound dressings. The disadvantage is that its material properties are relatively single, so combining natural polymers and synthetic polymers can better meet the requirements of biomedical research, facilitate the utilization of their respective advantages, and achieve synergistic enhancement effects.

4.2. Biocompatibility Evaluations

Good biocompatibility is a prerequisite for the safe application of wound dressings in the clinic, so any dressing must undergo an adequate biocompatibility evaluation before it is applied in the clinic. Biocompatibility refers to the ability of a biomaterial to have an acceptable host response during the wound healing process, with the ability to be non-toxic, non-sensitizing, non-mutagenic, and non-carcinogenic [110]. Biomaterial biocompatibility studies to date can be categorized into animal, cellular, and molecular levels. Animal level evaluation can truly and comprehensively reflect the overall condition of the material after acting on the organism, but the cycle is long, expensive, complicated, not easy to control individual differences, and the results of animal experiments are not necessarily well applied to human beings. Cellular experiments have the advantages of simplicity, speed, sensitivity and economy, and they are easy to standardize and have good repeatability, but they cannot well simulate the complex physiological environment in the body. Changes at the cellular level and even at the overall level are caused by changes at the molecular level of the organism. Therefore, an in-depth study of the effects of biomaterials at the molecular level can reveal the mechanism of the interaction between the material and the organism. However, most of the biocompatibility investigations of polyurethane dressings mentioned above have been accomplished through cellular and animal experiments, and fewer molecular experiments have been carried out. Molecular experiments can be used to carry out basic research on biocompatibility, clarify the interaction relationship, elucidate the mechanism of dressing action, guide the research, development and application of new wound dressings, and lay the foundation for reducing the number of experimental animals and establishing new standards and methods for the safety evaluation of biological dressings.

4.3. Healing Evaluations

Skin wound repair is accomplished through a series of complex and highly coordinated processes [4], and is particularly susceptible to impairment by infection, inflammation, and oxidative stress, which prolong wound healing during the recovery process. In the face of the destruction of various factors, the rational design of intelligent and multifunctional wound dressings is imminent [111]. Elimination of bacterial infections is essential for better wound recovery. The inappropriate use and misuse of antibiotics in recent years has led to increased difficulty in treating wound infections, as bacteria reduce the penetration of antibiotics by forming biofilms or forcing antibiotics out of the body to reduce the concentration of antibiotics within the bacteria, weakening the antibiotic antimicrobial effect and creating resistance to the antibiotics [112,113]. Therefore, to overcome the hardship of antimicrobial resistance, the development of novel non-antibiotic strategies for difficult-to-treat drug-resistant bacterial infections has become a focus of much research, along with research on polyurethane-related antimicrobial materials.

The polyurethane-related dressings discussed in this article focus on some of the basic and necessary features of dressings, including antimicrobial properties, adhesion and hemostasis, anti-inflammation and anti-oxidation, substance delivery, and self-healing. However, less attention has been paid to the function of the skin after healing, especially the formation of skin appendages (such as hair follicles, sweat glands, and sebaceous glands), and scar, which are important structures affecting the function of the skin and have only been explored in a few articles [93,114,115]. Therefore, the development of polyurethane dressings that promote functional repair (protection, thermoregulation, modification) of the skin is the next priority.

Functional exploration of polyurethane dressings has primarily used rat, mouse, and rabbit models, with wound repair models from pigs being rarely used [116]. The main disadvantages of using rabbit or rodent models are the differences in skin physiology, healing patterns, and skin-attached hairs. In the field of skin healing, the pig model is considered to be a useful analog of human skin because it has many anatomical and physiological similarities and to human wound healing, and it is a better model for studying skin regeneration. Functional judgment of dressings in pigs would make the conclusions more clinically applicable [117].

Although polyurethane-related dressings have been observed to promote re-epithelialization, collagen deposition, and nerve repair in animal models, research has mainly focused on observation and analysis of experimental phenomena and lacks in-depth exploration of basic principles. Most of the studies did not assess whether the dressings could dynamically adapt to microenvironmental changes and healing at different stages displayed by chronic wounds, so as to realize the precise intervention of the dressings in order to accelerate the inflammation–proliferation–remodeling phase, and to realize the rapid and high-quality repair of chronic wounds. Overall, research on reforming clinical polyurethane dressings is still in the exploratory stage, but the preliminary findings display its potential clinical therapeutic value.

5. Conclusions and Future Perspectives

Nowadays, a variety of new polyurethane functional dressings have emerged as effective medical material candidates. As the clinical needs of skin wounds continue, the function of wound dressings needs to change from a single physical barrier or capability to the current multifunctional composite with a trend towards further intelligence [118]. Therefore, this paper presents a review of functional polyurethane dressings covered in the existing studies, which are mostly composed of composite materials. The composites include synthetic polymers, natural polymers, and other active ingredients, among others. In conclusion, the addition of polymers or active ingredients to polyurethanes can improve the functionality of the dressing, with natural polymers excelling in increasing the degradability and cytocompatibility properties of the dressing, synthetic polymers in antimicrobial and moisturizing properties, and other bioactive ingredients in promoting wound healing effects. In order to meet the actual application, they need to be compounded to utilize their advantages and realize cost savings.

Although these improved polyurethane-related dressings have been shown in studies to be multifunctional and intelligent, there are not many commercially available polyurethane dressings, and much progress has yet to be made in clinical applications. In the future research, with the development of composite material synthesis technology, the deepening of cognition, the innovation of treatment means, and the update of treatment guidelines, including electrospinning, 3D printing, scalp transplantation therapy, stem cell treatment, and genetic therapy, the biological materials for skin wound dressings have made great progress. Researchers can now address the significant need for new strategies for the treatment of chronic wounds, with the goal of breaking through the application bottleneck, and provide new design ideas and theoretical bases for treatment.

Author Contributions: Conceptualization, W.L. and Y.H.; supervision, C.L. and N.N.; funding acquisition, C.L.; writing, W.L. All authors have read and agreed to the published version of the manuscript.

Funding: This work is supported by the National Natural Science Foundation of China (No. 007-41369023).

Acknowledgments: We thank Stanley Lin for polishing the English and making corrections.

Conflicts of Interest: The authors declared no potential conflict of interest with respect to the research, authorship, and/or publication of this article.

References

1. Bocheva, G.; Slominski, R.M.; Slominski, A.T. Neuroendocrine Aspects of Skin Aging. *Int. J. Mol. Sci.* **2019**, *20*, 2798. [CrossRef]
2. Pleguezuelos-Beltran, P.; Galvez-Martin, P.; Nieto-Garcia, D.; Marchal, J.A.; Lopez-Ruiz, E. Advances in spray products for skin regeneration. *Bioact. Mater.* **2022**, *16*, 187–203. [CrossRef]
3. Walker, M. Human skin through the ages. *Int. J. Pharm.* **2022**, *622*, 121850. [CrossRef]
4. Aitcheson, S.M.; Frentiu, F.D.; Hurn, S.E.; Edwards, K.; Murray, R.Z. Skin Wound Healing: Normal Macrophage Function and Macrophage Dysfunction in Diabetic Wounds. *Molecules* **2021**, *26*, 4917. [CrossRef] [PubMed]

5. An, Y.; Lin, S.; Tan, X.; Zhu, S.; Nie, F.; Zhen, Y.; Gu, L.; Zhang, C.; Wang, B.; Wei, W.; et al. Exosomes from adipose-derived stem cells and application to skin wound healing. *Cell Prolif.* **2021**, *54*, e12993. [CrossRef] [PubMed]
6. Hou, Y.; Jiang, N.; Sun, D.; Wang, Y.; Chen, X.; Zhu, S.; Zhang, L. A fast UV-curable PU-PAAm hydrogel with mechanical flexibility and self-adhesion for wound healing. *RSC Adv.* **2020**, *10*, 4907–4915. [CrossRef]
7. Nuutila, K.; Eriksson, E. Moist Wound Healing with Commonly Available Dressings. *Adv. Wound Care* **2021**, *10*, 685–698. [CrossRef] [PubMed]
8. Skorkowska-Telichowska, K.; Czemplik, M.; Kulma, A.; Szopa, J. The local treatment and available dressings designed for chronic wounds. *J. Am. Acad. Dermatol.* **2013**, *68*, e117–e126. [CrossRef] [PubMed]
9. Jones, R.E.; Foster, D.S.; Longaker, M.T. Management of Chronic Wounds-2018. *JAMA* **2018**, *320*, 1481–1482. [CrossRef]
10. Dornseifer, U.; Lonic, D.; Gerstung, T.I.; Herter, F.; Fichter, A.M.; Holm, C.; Schuster, T.; Ninkovic, M. The ideal split-thickness skin graft donor-site dressing: A clinical comparative trial of a modified polyurethane dressing and aquacel. *Plast. Reconstr. Surg.* **2011**, *128*, 918–924. [CrossRef]
11. Peng, W.; Li, D.; Dai, K.; Wang, Y.; Song, P.; Li, H.; Tang, P.; Zhang, Z.; Li, Z.; Zhou, Y.; et al. Recent progress of collagen, chitosan, alginate and other hydrogels in skin repair and wound dressing applications. *Int. J. Biol. Macromol.* **2022**, *208*, 400–408. [CrossRef]
12. Zeng, Q.; Qi, X.; Shi, G.; Zhang, M.; Haick, H. Wound Dressing: From Nanomaterials to Diagnostic Dressings and Healing Evaluations. *ACS Nano* **2022**, *16*, 1708–1733. [CrossRef] [PubMed]
13. Nikoletti, S.L.G.; Gandossi, S.; Coombs, G.; Wilson, R. A prospective, randomized: A prospective, randomized, controlled trial comparing transparent polyurethane and hydrocolloid dressings for central venous catheters. *Am. J. Infect. Control* **1999**, *27*, 488–496. [CrossRef] [PubMed]
14. Vaingankar, N.V.S.P.; Eagling, V.; King, C.; Elender, F. Comparison of hydrocellular foam and calcium alginate in the healing and comfort of split thickness skin graft donor sites. *J. Wound Care* **2001**, *11*, 21–29. [CrossRef]
15. Akita, S.; Akino, K.; Imaizumi, T.; Tanaka, K.; Anraku, K.; Yano, H.; Hirano, A. A polyurethane dressing is beneficial for split-thickness skin-graft donor wound healing. *Burns* **2006**, *32*, 447–451. [CrossRef] [PubMed]
16. Kneilling, M.; Breuninger, H.; Schippert, W.; Hafner, H.M.; Moehrle, M. A modified, improved, easy and fast technique for split-thickness skin grafting. *Br. J. Dermatol.* **2011**, *165*, 581–584. [CrossRef]
17. Yuki, Y.; Takenouchi, T.; Takatsuka, S.; Fujikawa, H.; Abe, R. Investigating the use of tie-over dressing after skin grafting. *J. Dermatol.* **2017**, *44*, 1317–1319. [CrossRef] [PubMed]
18. Satake, T.; Muto, M.; Nagashima, Y.; Haga, S.; Homma, Y.; Nakasone, R.; Kadokura, M.; Kou, S.; Fujimoto, H.; Maegawa, J. Polyurethane Foam Wound Dressing Technique for Areola Skin Graft Stabilization and Nipple Protection After Nipple-Areola Reconstruction. *Aesthetic Plast. Surg.* **2018**, *42*, 442–446. [CrossRef]
19. Guillén-Solà, M.S.M.A.; Tomàs-Vidal, A.M.; GAUPP-Expert Panel. A multi center, randomized, clinical trial comparing adhesive polyurethane foam dressing and adhesive hydrocolloid dressing in patients. *BMC Fam. Pract.* **2013**, *19*, 1–10.
20. Kaiser, D.H.J.; Mayer, D.; French, L.E.; Läuchli, S. Alginate dressing and polyurethane film versus paraffin gauze in the treatment of split thickness skin graft donor sites. *Adv. Ski. Wound Care* **2013**, *18*, 22–29. [CrossRef]
21. Daar, D.A.; Wirth, G.A.; Evans, G.R.; Carmean, M.; Gordon, I.L. The Bagautdinov dressing method: Negative pressure wound therapy in a patient with an allergy to acrylate adhesive. *Int. Wound J.* **2017**, *14*, 198–202. [CrossRef]
22. Greenwood, J.E. The evolution of acute burn care—Retiring the split skin graft. *Ann. R. Coll. Surg. Engl.* **2017**, *99*, 432–438. [CrossRef]
23. Rezapour-Lactoee, A.; Yeganeh, H.; Ostad, S.N.; Gharibi, R.; Mazaheri, Z.; Ai, J. Thermoresponsive polyurethane/siloxane membrane for wound dressing and cell sheet transplantation: In-vitro and in-vivo studies. *Mater. Sci. Eng. C Mater. Biol. Appl.* **2016**, *69*, 804–814. [CrossRef]
24. Park, J.U.; Jung, H.D.; Song, E.H.; Choi, T.H.; Kim, H.E.; Song, J.; Kim, S. The accelerating effect of chitosan-silica hybrid dressing materials on the early phase of wound healing. *J. Biomed. Mater. Res. B Appl. Biomater.* **2017**, *105*, 1828–1839. [CrossRef]
25. Morales-Gonzalez, M.; Diaz, L.E.; Dominguez-Paz, C.; Valero, M.F. Insights into the Design of Polyurethane Dressings Suitable for the Stages of Skin Wound-Healing: A Systematic Review. *Polymers* **2022**, *14*, 2990. [CrossRef] [PubMed]
26. Hussain, Z.; Thu, H.E.; Shuid, A.N.; Katas, H.; Hussain, F. Recent Advances in Polymer-based Wound Dressings for the Treatment of Diabetic Foot Ulcer: An Overview of State-of-the-art. *Curr. Drug Targets* **2018**, *19*, 527–550. [CrossRef]
27. Yaşayan, G.; Alarçin, E.; Bal-Öztürk, A.; Avci-Adali, M. Natural polymers for wound dressing applications. *Stud. Nat. Prod. Chem.* **2022**, *74*, 367–441.
28. Alven, S.; Peter, S.; Mbese, Z.; Aderibigbe, B.A. Polymer-Based Wound Dressing Materials Loaded with Bioactive Agents: Potential Materials for the Treatment of Diabetic Wounds. *Polymers* **2022**, *14*, 724. [CrossRef] [PubMed]
29. Boateng, J.C.O. Advanced therapeutic dressings for effective wound healing—A review. *J. Pharm. Sci.* **2015**, *8*, 9–21. [CrossRef] [PubMed]
30. Qian, B.; Li, J.; Guo, K.; Guo, N.; Zhong, A.; Yang, J.; Wang, J.; Xiao, P.; Sun, J.; Xiong, L. Antioxidant biocompatible composite collagen dressing for diabetic wound healing in rat model. *Regen. Biomater.* **2021**, *8*, rbab003. [CrossRef]
31. Cabrera-Munguia, D.A.; Claudio-Rizo, J.A.; Becerra-Rodríguez, J.J.; Flores-Guia, T.E.; Rico, J.L.; Vásquez-García, S.R. Enhanced biocompatibility and bactericidal properties of hydrogels based on collagen–polyurethane–aluminium MOFs for biomedical applications. *Bull. Mater. Sci.* **2023**, *46*, 1–10. [CrossRef]

32. Andreica, B.I.; Anisiei, A.; Rosca, I.; Sandu, A.I.; Pasca, A.S.; Tartau, L.M.; Marin, L. Quaternized chitosan/chitosan nanofibrous mats: An approach toward bioactive materials for tissue engineering and regenerative medicine. *Carbohydr. Polym.* **2023**, *302*, 120431. [CrossRef]
33. Hashemi, S.S.R.S.; Mahmoudi, R.; Ghanbari, A.; Zibara, K.; Barmak, M.J. Polyurethane/chitosan/hyaluronic acid scaffolds: Providing an optimum environment for fibroblast growth. *J. Wound Care* **2020**, *29*, 586–596. [CrossRef] [PubMed]
34. Graca, M.F.P.; Miguel, S.P.; Cabral, C.S.D.; Correia, I.J. Hyaluronic acid-Based wound dressings: A review. *Carbohydr. Polym.* **2020**, *241*, 116364. [CrossRef] [PubMed]
35. Movahedi, M.; Asefnejad, A.; Rafienia, M.; Khorasani, M.T. Potential of novel electrospun core-shell structured polyurethane/starch (hyaluronic acid) nanofibers for skin tissue engineering: In vitro and in vivo evaluation. *Int. J. Biol. Macromol.* **2020**, *146*, 627–637. [CrossRef]
36. Wang, M.; Hu, J.; Ou, Y.; He, X.; Wang, Y.; Zou, C.; Jiang, Y.; Luo, F.; Lu, D.; Li, Z.; et al. Shape-Recoverable Hyaluronic Acid–Waterborne Polyurethane Hybrid Cryogel Accelerates Hemostasis and Wound Healing. *ACS Appl. Mater. Interfaces* **2022**, *14*, 17093–17108. [CrossRef]
37. Miao, S.; Wang, P.; Su, Z.; Zhang, S. Vegetable-oil-based polymers as future polymeric biomaterials. *Acta Biomater.* **2014**, *10*, 1692–1704. [CrossRef]
38. Gholami, H.; Yeganeh, H. Vegetable oil-based polyurethanes as antimicrobial wound dressings: In vitro and in vivo evaluation. *Biomed. Mater.* **2020**, *15*, 045001. [CrossRef]
39. Chen, C.; Yang, H.; Yang, X.; Ma, Q. Tannic acid: A crosslinker leading to versatile functional polymeric networks: A review. *RSC Adv.* **2022**, *12*, 7689–7711. [CrossRef]
40. Yang, Y.; Zhao, X.; Yu, J.; Chen, X.; Wang, R.; Zhang, M.; Zhang, Q.; Zhang, Y.; Wang, S.; Cheng, Y. Bioactive skin-mimicking hydrogel band-aids for diabetic wound healing and infectious skin incision treatment. *Bioact. Mater.* **2021**, *6*, 3962–3975. [CrossRef]
41. Zhang, Z.; Zhang, Y.; Liu, Y.; Zheng, P.; Gao, T.; Luo, B.; Liu, X.; Ma, F.; Wang, J.; Pei, R. Water-retaining and separable adhesive hydrogel dressing for wound healing without secondary damage. *Sci. China Mater.* **2023**, *66*, 3337–3346. [CrossRef]
42. Yue, Y.; Gong, X.; Jiao, W.; Li, Y.; Yin, X.; Si, Y.; Yu, J.; Ding, B. In-situ electrospinning of thymol-loaded polyurethane fibrous membranes for waterproof, breathable, and antibacterial wound dressing application. *J. Colloid Interface Sci.* **2021**, *592*, 310–318. [CrossRef]
43. Zhou, X.; Ruan, Q.; Ye, Z.; Chu, Z.; Xi, M.; Li, M.; Hu, W.; Guo, X.; Yao, P.; Xie, W. Resveratrol accelerates wound healing by attenuating oxidative stress-induced impairment of cell proliferation and migration. *Burns* **2021**, *47*, 133–139. [CrossRef] [PubMed]
44. Wang, J.; Tian, L.; Luo, B.; Ramakrishna, S.; Kai, D.; Loh, X.J.; Yang, I.H.; Deen, G.R.; Mo, X. Engineering PCL/lignin nanofibers as an antioxidant scaffold for the growth of neuron and Schwann cell. *Colloids Surf. B Biointerfaces* **2018**, *169*, 356–365. [CrossRef] [PubMed]
45. Xu, J.; Xu, J.J.; Lin, Q.; Jiang, L.; Zhang, D.; Li, Z.; Ma, B.; Zhang, C.; Kai, D.; et al. Lignin-Incorporated Nanogel Serving As an Antioxidant Biomaterial for Wound Healing. *ACS Appl. Bio Mater.* **2021**, *4*, 3–13. [CrossRef]
46. Pahlevanneshan, Z.; Deypour, M.; Kefayat, A.; Rafienia, M.; Sajkiewicz, P.; Esmaeely Neisiany, R.; Enayati, M.S. Polyurethane-Nanolignin Composite Foam Coated with Propolis as a Platform for Wound Dressing: Synthesis and Characterization. *Polymers* **2021**, *13*, 3191. [CrossRef] [PubMed]
47. Li, S.; Zhang, Y.; Ma, X.; Qiu, S.; Chen, J.; Lu, G.; Jia, Z.; Zhu, J.; Yang, Q.; Chen, J.; et al. Antimicrobial Lignin-Based Polyurethane/Ag Composite Foams for Improving Wound Healing. *Biomacromolecules* **2022**, *23*, 1622–1632. [CrossRef] [PubMed]
48. Abazari, M.; Akbari, T.; Hasani, M.; Sharifikolouei, E.; Raoufi, M.; Foroumadi, A.; Sharifzadeh, M.; Firoozpour, L.; Khoobi, M. Polysaccharide-based hydrogels containing herbal extracts for wound healing applications. *Carbohydr. Polym.* **2022**, *294*, 119808. [CrossRef]
49. He, L.; Liu, Y.; Lau, J.; Fan, W.; Li, Q.; Zhang, C.; Huang, P.; Chen, X. Recent progress in nanoscale metal-organic frameworks for drug release and cancer therapy. *Nanomedicine* **2019**, *14*, 1343–1365. [CrossRef] [PubMed]
50. Mahendran, G.; Rahman, L.U. Ethnomedicinal, phytochemical and pharmacological updates on Peppermint (*Mentha* × *piperita* L.)—A review. *Phytother. Res.* **2020**, *34*, 2088–2139. [CrossRef]
51. Almasian, A.; Najafi, F.; Eftekhari, M.; Shams Ardekani, M.R.; Sharifzadeh, M.; Khanavi, M. Preparation of Polyurethane/Pluronic F127 Nanofibers Containing Peppermint Extract Loaded Gelatin Nanoparticles for Diabetic Wounds Healing: Characterization, In Vitro, and In Vivo Studies. *Evid. Based Complement. Altern. Med.* **2021**, *20*, 6646702. [CrossRef]
52. Lin, L.; Regenstein, J.M.; Lv, S.; Lu, J.; Jiang, S. An overview of gelatin derived from aquatic animals: Properties and modification. *Trends Food Sci. Technol.* **2017**, *68*, 102–112. [CrossRef]
53. Sheikholeslam, M.; Wright, M.E.E.; Cheng, N.; Oh, H.H.; Wang, Y.; Datu, A.K.; Santerre, J.P.; Amini-Nik, S.; Jeschke, M.G. Electrospun Polyurethane-Gelatin Composite: A New Tissue-Engineered Scaffold for Application in Skin Regeneration and Repair of Complex Wounds. *ACS Biomater. Sci. Eng.* **2020**, *6*, 505–516. [CrossRef] [PubMed]
54. Cheng, Q.P.; Hsu, S.H. A self-healing hydrogel and injectable cryogel of gelatin methacryloyl-polyurethane double network for 3D printing. *Acta Biomater.* **2023**, *164*, 124–138. [CrossRef] [PubMed]
55. Yu, P.; Hong, N.; Chen, M.; Zou, X. Novel application of absorbable gelatine sponge combined with polyurethane film for dermal reconstruction of wounds with bone or tendon exposure. *Int. Wound J.* **2023**, *20*, 18–27. [CrossRef] [PubMed]

56. Qu, J.; Liang, Y.; Shi, M.; Guo, B.; Gao, Y.; Yin, Z. Biocompatible conductive hydrogels based on dextran and aniline trimer as electro-responsive drug delivery system for localized drug release. *Int. J. Biol. Macromol.* **2019**, *140*, 255–264. [CrossRef]
57. Sagitha, P.; Reshmi, C.R.; Sundaran, S.P.; Binoy, A.; Mishra, N.; Sujith, A. In-vitro evaluation on drug release kinetics and antibacterial activity of dextran modified polyurethane fibrous membrane. *Int. J. Biol. Macromol.* **2019**, *126*, 717–730.
58. Gharibi, R.; Kazemi, S.; Yeganeh, H.; Tafakori, V. Utilizing dextran to improve hemocompatibility of antimicrobial wound dressings with embedded quaternary ammonium salts. *Int. J. Biol. Macromol.* **2019**, *131*, 1044–1056. [CrossRef] [PubMed]
59. Shoji, M.I.K.; Sriwilaijaroen, N.; Mayumi, H.; Morikane, S.; Takahashi, E.; Kido, H.; Suzuki, Y.; Takeda, K.; Kuzuhara, T. Anti-influenza Activity of Povidone-Iodine-Integrated Materials. *Biol. Pharm. Bull.* **2023**, *46*, 1231–1239. [CrossRef] [PubMed]
60. Jung, J.A.H.S.; Jeong, S.H.; Dhong, E.S.; Park, K.G.; Kim, W.K. In Vitro Evaluation of Betafoam, a New Polyurethane Foam Dressing. *Adv. Skin. Wound Care* **2017**, *30*, 262–271. [CrossRef]
61. Pak, C.S.; Park, D.H.; Oh, T.S.; Lee, W.J.; Jun, Y.J.; Lee, K.A.; Oh, K.S.; Kwak, K.H.; Rhie, J.W. Comparison of the efficacy and safety of povidone-iodine foam dressing (Betafoam), hydrocellular foam dressing (Allevyn), and petrolatum gauze for split-thickness skin graft donor site dressing. *Int. Wound J.* **2019**, *16*, 379–386. [CrossRef]
62. Lee, J.W.; Song, K.Y. Evaluation of a polyurethane foam dressing impregnated with 3% povidone-iodine (Betafoam) in a rat wound model. *Ann. Surg. Treat. Res.* **2018**, *94*, 1–7. [CrossRef] [PubMed]
63. Wei, Y.; Chen, L.; Jiang, Y. Self-healing polyacrylamide (PAAm) gels at room temperature based on complementary guanine and cytosine base pairs. *Soft Matter* **2022**, *18*, 7394–7401. [CrossRef]
64. Kang, S.Y.; Ji, Z.; Tseng, L.F.; Turner, S.A.; Villanueva, D.A.; Johnson, R.; Albano, A.; Langer, R. Design and Synthesis of Waterborne Polyurethanes. *Adv. Mater.* **2018**, *30*, e1706237. [CrossRef] [PubMed]
65. Wu, F.; Chen, L.; Wang, Y.; Fei, B. Tough and stretchy double-network hydrogels based on in situ interpenetration of polyacrylamide and physically cross-linked polyurethane. *J. Mater. Sci.* **2019**, *54*, 12131–12144. [CrossRef]
66. Labet, M.; Thielemans, W. Synthesis of polycaprolactone: A review. *Chem. Soc. Rev.* **2009**, *38*, 3484–3504. [CrossRef]
67. Djagny, V.B.; Wang, Z.; Xu, S. Gelatin: A valuable protein for food and pharmaceutical industries: Review. *Crit. Rev. Food Sci. Nutr.* **2001**, *41*, 481–492. [CrossRef] [PubMed]
68. Eskandarinia, A.; Kefayat, A.; Agheb, M.; Rafienia, M.; Amini Baghbadorani, M.; Navid, S.; Ebrahimpour, K.; Khodabakhshi, D.; Ghahremani, F. A Novel Bilayer Wound Dressing Composed of a Dense Polyurethane/Propolis Membrane and a Biodegradable Polycaprolactone/Gelatin Nanofibrous Scaffold. *Sci. Rep.* **2020**, *10*, 3063. [CrossRef]
69. Shie Karizmeh, M.; Poursamar, S.A.; Kefayat, A.; Farahbakhsh, Z.; Rafienia, M. An in vitro and in vivo study of PCL/chitosan electrospun mat on polyurethane/propolis foam as a bilayer wound dressing. *Mater. Sci. Eng. C Mater. Biol. Appl.* **2022**, *135*, 112667. [CrossRef]
70. Samatya Yılmaz, S.; Yazıcı Özçelik, E.; Uzuner, H.; Kolayli, F.; Karadenizli, A.; Aytac, A. The PLA-Ag NPs/PU bicomponent nanofiber production for wound dressing applications: Investigation of core/shell displacement effect on antibacterial, cytotoxicity, mechanical, and surface properties. *Int. J. Polym. Mater. Polym. Biomater.* **2023**, *9*, 1–18. [CrossRef]
71. Samatya Yilmaz, S.; Aytac, A. Fabrication and characterization as antibacterial effective wound dressing of hollow polylactic acid/polyurethane/silver nanoparticle nanofiber. *J. Polym. Res.* **2022**, *29*, 8–17. [CrossRef]
72. Samatya Yilmaz, S.; Aytac, A. The highly absorbent polyurethane polylactic acid blend electrospun tissue scaffold for dermal wound dressing. *Polym. Bull.* **2022**, *13*, 9–21. [CrossRef]
73. Moellhoff, N.; Lettner, M.; Frank, K.; Giunta, R.E.; Ehrl, D. Polylactic Acid Membrane Improves Outcome of Split-Thickness Skin Graft Donor Sites: A Prospective, Comparative, Randomized Study. *Plast. Reconstr. Surg.* **2022**, *150*, 1104–1113. [CrossRef]
74. Chen, C.F.; Chen, S.H.; Chen, R.F.; Liu, K.F.; Kuo, Y.R.; Wang, C.K.; Lee, T.M.; Wang, Y.H. A Multifunctional Polyethylene Glycol/Triethoxysilane-Modified Polyurethane Foam Dressing with High Absorbency and Antiadhesion Properties Promotes Diabetic Wound Healing. *Int. J. Mol. Sci.* **2023**, *15*, 12506. [CrossRef] [PubMed]
75. Vakil, A.U.; Ramezani, M.; Monroe, M.B.B. Antimicrobial Shape Memory Polymer Hydrogels for Chronic Wound Dressings. *ACS Appl. Bio Mater.* **2022**, *5*, 5199–5209. [CrossRef]
76. Hussein, M.A.M.; Gunduz, O.; Sahin, A.; Grinholc, M.; El-Sherbiny, I.M.; Megahed, M. Dual Spinneret Electrospun Polyurethane/PVA-Gelatin Nanofibrous Scaffolds Containing Cinnamon Essential Oil and Nanoceria for Chronic Diabetic Wound Healing: Preparation, Physicochemical Characterization and In-Vitro Evaluation. *Molecules* **2022**, *27*, 2146. [CrossRef]
77. Carayon, I.; Szarlej, P.; Gnatowski, P.; Pilat, E.; Sienkiewicz, M.; Glinka, M.; Karczewski, J.; Kucinska-Lipka, J. Polyurethane based hybrid ciprofloxacin-releasing wound dressings designed for skin engineering purpose. *Adv. Med. Sci.* **2022**, *67*, 269–282. [CrossRef] [PubMed]
78. Hosseini Salekdeh, S.S.; Daemi, H.; Zare-Gachi, M.; Rajabi, S.; Bazgir, F.; Aghdami, N.; Nourbakhsh, M.S.; Baharvand, H. Assessment of the Efficacy of Tributylammonium Alginate Surface-Modified Polyurethane as an Antibacterial Elastomeric Wound Dressing for both Noninfected and Infected Full-Thickness Wounds. *ACS Appl. Mater. Interfaces* **2020**, *12*, 3393–3406. [CrossRef]
79. Esmaeili, E.; Eslami-Arshaghi, T.; Hosseinzadeh, S.; Elahirad, E.; Jamalpoor, Z.; Hatamie, S.; Soleimani, M. The biomedical potential of cellulose acetate/polyurethane nanofibrous mats containing reduced graphene oxide/silver nanocomposites and curcumin: Antimicrobial performance and cutaneous wound healing. *Int. J. Biol. Macromol.* **2020**, *152*, 418–427. [CrossRef]
80. Lee, D.N.; Gwon, K.; Nam, Y.; Lee, S.J.; Tran, N.M.; Yoo, H. Polyurethane Foam Incorporated with Nanosized Copper-Based Metal-Organic Framework: Its Antibacterial Properties and Biocompatibility. *Int. J. Mol. Sci.* **2021**, *22*, 13622. [CrossRef]

81. Khansa, I.; Schoenbrunner, A.R.; Kraft, C.T.; Janis, J.E. Silver in Wound Care-Friend or Foe? A Comprehensive Review. *Plast. Reconstr. Surg. Glob. Open* **2019**, *7*, e2390. [CrossRef]
82. Pant, B.; Park, M.; Park, S.J. One-Step Synthesis of Silver Nanoparticles Embedded Polyurethane Nano-Fiber/Net Structured Membrane as an Effective Antibacterial Medium. *Polymers* **2019**, *11*, 1185. [CrossRef]
83. Rabiee, T.; Yeganeh, H.; Gharibi, R. Antimicrobial wound dressings with high mechanical conformability prepared through thiol-yne click photopolymerization reaction. *Biomed. Mater.* **2019**, *14*, 045007. [CrossRef] [PubMed]
84. Rahmani, S.; Ghaemi, F.; Khaleghi, M.; Haghi, F. Synthesis of novel silver nanocomposite hydrogels based on polyurethane/poly (ethylene glycol) via aqueous extract of oak fruit and their antibacterial and mechanical properties. *Polym. Compos.* **2021**, *42*, 6719–6735. [CrossRef]
85. Song, E.H.; Jeong, S.H.; Park, J.U.; Kim, S.; Kim, H.E.; Song, J. Polyurethane-silica hybrid foams from a one-step foaming reaction, coupled with a sol-gel process, for enhanced wound healing. *Mater. Sci. Eng. C Mater. Biol. Appl.* **2017**, *79*, 866–874. [CrossRef]
86. Kirby, G.T.; Mills, S.J.; Vandenpoel, L.; Pinxteren, J.; Ting, A.; Short, R.D.; Cowin, A.J.; Michelmore, A.; Smith, L.E. Development of Advanced Dressings for the Delivery of Progenitor Cells. *ACS Appl. Mater. Interfaces* **2017**, *9*, 3445–3454. [CrossRef] [PubMed]
87. Mills, S.J.; Kirby, G.T.; Hofma, B.R.; Smith, L.E.; Statham, P.; Vaes, B.; Ting, A.E.; Short, R.; Cowin, A.J. Delivery of multipotent adult progenitor cells via a functionalized plasma polymerized surface accelerates healing of murine diabetic wounds. *Front. Bioeng. Biotechnol.* **2023**, *11*, 1213021. [CrossRef] [PubMed]
88. Yeo, G.C.; Kosobrodova, E.; Kondyurin, A.; McKenzie, D.R.; Bilek, M.M.; Weiss, A.S. Plasma-Activated Substrate with a Tropoelastin Anchor for the Maintenance and Delivery of Multipotent Adult Progenitor Cells. *Macromol. Biosci.* **2019**, *19*, e1800233. [CrossRef]
89. Losi, P.; Briganti, E.; Sanguinetti, E.; Burchielli, S.; Al Kayal, T.; Soldani, G. Healing effect of a fibrin-based scaffold loaded with platelet lysate in full-thickness skin wounds. *J. Bioact. Compat. Polym.* **2015**, *30*, 222–237. [CrossRef]
90. Losi, P.; Al Kayal, T.; Buscemi, M.; Foffa, I.; Cavallo, A.; Soldani, G. Bilayered Fibrin-Based Electrospun-Sprayed Scaffold Loaded with Platelet Lysate Enhances Wound Healing in a Diabetic Mouse Model. *Nanomaterials* **2020**, *10*, 2128. [CrossRef]
91. Holl, J.; Kowalewski, C.; Zimek, Z.; Fiedor, P.; Kaminski, A.; Oldak, T.; Moniuszko, M.; Eljaszewicz, A. Chronic Diabetic Wounds and Their Treatment with Skin Substitutes. *Cells* **2021**, *10*, 655. [CrossRef] [PubMed]
92. Kalluri, R.; LeBleu, V.S. The biology, function, and biomedical applications of exosomes. *Science* **2020**, *367*, 1–8. [CrossRef]
93. Shiekh, P.A.; Singh, A.; Kumar, A. Exosome laden oxygen releasing antioxidant and antibacterial cryogel wound dressing OxOBand alleviate diabetic and infectious wound healing. *Biomaterials* **2020**, *249*, 120020. [CrossRef] [PubMed]
94. Shiekh, P.A.; Singh, A.; Kumar, A. Data supporting exosome laden oxygen releasing antioxidant and antibacterial cryogel wound dressing OxOBand alleviate diabetic and infectious wound healing. *Data Brief* **2020**, *31*, 105671. [CrossRef] [PubMed]
95. Jiang, Y.L.; Wang, Z.L.; Fan, Z.X.; Wu, M.J.; Zhang, Y.; Ding, W.; Huang, Y.Z.; Xie, H.Q. Human adipose-derived stem cell-loaded small intestinal submucosa as a bioactive wound dressing for the treatment of diabetic wounds in rats. *Biomater. Adv.* **2022**, *136*, 212793. [CrossRef]
96. Chen, T.Y.; Wen, T.K.; Dai, N.T.; Hsu, S.H. Cryogel/hydrogel biomaterials and acupuncture combined to promote diabetic skin wound healing through immunomodulation. *Biomaterials* **2021**, *269*, 120608. [CrossRef]
97. Allur Subramaniyan, S.; Sheet, S.; Balasubramaniam, S.; Berwin Singh, S.V.; Rampa, D.R.; Shanmugam, S.; Kang, D.R.; Choe, H.S.; Shim, K.S. Fabrication of nanofiber coated with l-arginine via electrospinning technique: A novel nanomatrix to counter oxidative stress under crosstalk of co-cultured fibroblasts and satellite cells. *Cell Commun. Adhes.* **2018**, *24*, 19–32. [CrossRef] [PubMed]
98. Zou, F.; Wang, Y.; Zheng, Y.; Xie, Y.; Zhang, H.; Chen, J.; Hussain, M.I.; Meng, H.; Peng, J. A novel bioactive polyurethane with controlled degradation and L-Arg release used as strong adhesive tissue patch for hemostasis and promoting wound healing. *Bioact. Mater.* **2022**, *17*, 471–487. [CrossRef]
99. de Breij, A.R.M.; Cordfunke, R.A.; Malanovic, N.; de Boer, L.; Koning, R.I.; Ravensbergen, E.; Franken, M.; van der Heijde, T.; Boekema, B.K.; Kwakman, P.H.S.; et al. The antimicrobial peptide SAAP-148 combats drug-resistant bacteria and biofilms. *Sci. Transl. Med.* **2022**, *1*, 1–9. [CrossRef]
100. Rai, A.; Ferrao, R.; Marta, D.; Vilaca, A.; Lino, M.; Rondao, T.; Ji, P.; Paiva, A.; Ferreira, L. Antimicrobial Peptide-Tether Dressing Able to Enhance Wound Healing by Tissue Contact. *ACS Appl. Mater. Interfaces* **2022**, *14*, 24213–24228. [CrossRef]
101. Biazar, E.; Heidari Keshel, S.; Rezaei Tavirani, M.; Kamalvand, M. Healing effect of acellular fish skin with plasma rich in growth factor on full-thickness skin defects. *Int. Wound J.* **2022**, *19*, 2154–2162. [CrossRef] [PubMed]
102. Shams, F.; Moravvej, H.; Hosseinzadeh, S.; Kazemi, B.; Rajabibazl, M.; Rahimpour, A. Evaluation of in vitro fibroblast migration by electrospun triple-layered PU-CA/gelatin.PRGF/PU-CA scaffold using an AAVS1 targeted EGFP reporter cell line. *Bioimpacts* **2022**, *12*, 219–231. [CrossRef] [PubMed]
103. Lee, J.; Shin, D.; Roh, J.L. Promotion of skin wound healing using prevascularized oral mucosal cell sheet. *Head Neck* **2019**, *41*, 774–779. [CrossRef] [PubMed]
104. Wu, S.D.; Dai, N.T.; Liao, C.Y.; Kang, L.Y.; Tseng, Y.W.; Hsu, S.H. Planar-/Curvilinear-Bioprinted Tri-Cell-Laden Hydrogel for Healing Irregular Chronic Wounds. *Adv. Healthc. Mater.* **2022**, *11*, e2201021. [CrossRef]
105. Mottola, S.; Viscusi, G.; Iannone, G.; Belvedere, R.; Petrella, A.; De Marco, I.; Gorrasi, G. Supercritical Impregnation of Mesoglycan and Lactoferrin on Polyurethane Electrospun Fibers for Wound Healing Applications. *Int. J. Mol. Sci.* **2023**, *24*, 9269. [CrossRef]
106. Webster, J.; Gillies, D.; O'Riordan, E.; Sherriff, K.L.; Rickard, C.M. Gauze and tape and transparent polyurethane dressings for central venous catheters. *Cochrane Database Syst. Rev.* **2011**, CD003827.

107. Carre, C.; Ecochard, Y.; Caillol, S.; Averous, L. From the Synthesis of Biobased Cyclic Carbonate to Polyhydroxyurethanes: A Promising Route towards Renewable Non-Isocyanate Polyurethanes. *ChemSusChem* **2019**, *12*, 3410–3430. [CrossRef]
108. Hu, J.J.; Liu, C.C.; Lin, C.H.; Tuan-Mu, H.Y. Synthesis, Characterization, and Electrospinning of a Functionalizable, Polycaprolactone-Based Polyurethane for Soft Tissue Engineering. *Polymers* **2021**, *13*, 1527. [CrossRef]
109. Jung, K.; Corrigan, N.; Wong, E.H.H.; Boyer, C. Bioactive Synthetic Polymers. *Adv. Mater.* **2022**, *34*, e2105063. [CrossRef]
110. Lu, X.; Bao, X.; Huang, Y.; Qu, Y.; Lu, H.; Lu, Z. Mechanisms of cytotoxicity of nickel ions based on gene expression profiles. *Biomaterials* **2009**, *30*, 141–148. [CrossRef]
111. Zhou, L.; Min, T.; Bian, X.; Dong, Y.; Zhang, P.; Wen, Y. Rational Design of Intelligent and Multifunctional Dressing to Promote Acute/Chronic Wound Healing. *ACS Appl. Bio Mater.* **2022**, *35*, 14–18. [CrossRef]
112. Shao, H.; Zhou, J.; Lin, X.; Zhou, Y.; Xue, Y.; Hong, W.; Lin, X.; Jia, X.; Fan, Y. Bio-inspired peptide-conjugated liposomes for enhanced planktonic bacteria killing and biofilm eradication. *Biomaterials* **2023**, *30*, 122183. [CrossRef]
113. Guo, Z.; He, J.X.; Mahadevegowda, S.H.; Kho, S.H.; Chan-Park, M.B.; Liu, X.W. Multifunctional Glyco-Nanosheets to Eradicate Drug-Resistant Bacteria on Wounds. *Adv. Healthc. Mater.* **2020**, *9*, e2000265. [CrossRef]
114. Beanes, S.R.; Dang, C.; Soo, C.; Ting, K. Skin repair and scar formation: The central role of TGF-beta. *Expert. Rev. Mol. Med.* **2003**, *5*, 1–22. [CrossRef]
115. Tapking, C.; Popp, D.; Branski, L.K. Pig Model to Test Tissue-Engineered Skin. *Methods Mol. Biol.* **2019**, *1993*, 239–249. [PubMed]
116. Yang, J.; Huang, Y.; Dai, J.; Shi, X.; Zheng, Y. A sandwich structure composite wound dressing with firmly anchored silver nanoparticles for severe burn wound healing in a porcine model. *Regen. Biomater.* **2021**, *8*, rbab037. [CrossRef] [PubMed]
117. Hamilton, D.W.; Walker, J.T.; Tinney, D.; Grynyshyn, M.; El-Warrak, A.; Truscott, E.; Flynn, L.E. The pig as a model system for investigating the recruitment and contribution of myofibroblasts in skin healing. *Wound Repair. Regen.* **2022**, *30*, 45–63. [CrossRef] [PubMed]
118. Liang, Y.; He, J.; Guo, B. Functional Hydrogels as Wound Dressing to Enhance Wound Healing. *ACS Nano* **2021**, *15*, 12687–12722. [CrossRef] [PubMed]

Disclaimer/Publisher's Note: The statements, opinions and data contained in all publications are solely those of the individual author(s) and contributor(s) and not of MDPI and/or the editor(s). MDPI and/or the editor(s) disclaim responsibility for any injury to people or property resulting from any ideas, methods, instructions or products referred to in the content.

MDPI AG
Grosspeteranlage 5
4052 Basel
Switzerland
Tel.: +41 61 683 77 34

Polymers Editorial Office
E-mail: polymers@mdpi.com
www.mdpi.com/journal/polymers

Disclaimer/Publisher's Note: The statements, opinions and data contained in all publications are solely those of the individual author(s) and contributor(s) and not of MDPI and/or the editor(s). MDPI and/or the editor(s) disclaim responsibility for any injury to people or property resulting from any ideas, methods, instructions or products referred to in the content.

www.ingramcontent.com/pod-product-compliance
Lightning Source LLC
LaVergne TN
LVHW070723100526
838202LV00013B/1160